Fitting the Human

The aim of this book is to provide "human engineering" for workplaces, tools, machinery, computers, shift work, lighting, sound, climate, work demands, offices, vehicles, healthcare, and the home—and everything else that we can produce—to suit the human body and mind. Now being published in its eighth edition, *Fitting the Human* focuses on the primary ergonomic task of accommodating the human with the appropriate selection of equipment and tools, work requirements and procedures, physical and social conditions at work, and working hours and shift arrangements.

This book provides the ergonomic information needed for the user-friendly design of tasks, equipment, and workplaces. It follows the successful format of previous editions, with updated information and practical guidelines that augment the previous information. It offers refreshed information on ergonomic design for the home and workplace, contemporary ways of working, healthcare and medicine, and artificial intelligence and autonomy. This text also recognizes that cultural differences in living and working vary around the world, so additional insights are offered into ergonomics in global cultures and regions. This title will help the reader understand how to plan and design an overall system and its details to fit the human.

Published under the mantra of "solid information, easy to read, easy to understand, easy to apply," *Fitting the Human* is written for students and professionals in ergonomics, human factors, product and work design, safety, architecture, management, and all fields of engineering.

Fitting the Human

Introduction to Ergonomics / Human Factors Engineering

Eighth Edition

Katrin Kroemer Elbert

CRC Press
Taylor & Francis Group
Boca Raton London New York

CRC Press is an imprint of the
Taylor & Francis Group, an **informa** business

Eighth edition published 2025
by CRC Press
2385 NW Executive Center Drive, Suite 320, Boca Raton FL 33431

and by CRC Press
4 Park Square, Milton Park, Abingdon, Oxon, OX14 4RN

CRC Press is an imprint of Taylor & Francis Group, LLC

First edition published by Etienne Grandjean as *Physiologische Artbeitsgestaltung* (in German) in 1963
Seventh edition published by CRC Press 2017

Library of Congress Cataloging-in-Publication Data
Names: Kroemer, K. H. E., 1933- author. | Kroemer-Elbert, K. E. (Katrin E.), author.
Title: Fitting the human : introduction to ergonomics/human factors engineering /
Katrin E. Kroemer-Elbert.
Other titles: Physiologische Arbeitsgestaltung. English
Description: Eighth edition. | Boca Raton, FL : CRC Pres, 2025. |
Revised edition of: Fitting the human : introduction to ergonomics /
Karl H.E. Kroemer. Boca Raton : CRC Press/Taylor & Francis Group, 2017. |
Includes bibliographical references and index.
Identifiers: LCCN 2024040684 (print) | LCCN 2024040685 (ebook) |
ISBN 9781032619583 (hbk) | ISBN 9781032674551 (pbk) | ISBN 9781032674575 (ebk)
Subjects: LCSH: Work–Physiological aspects. | Human mechanics. |
Human engineering. | Industrial hygiene.
Classification: LCC QP309 .G7313 2025 (print) | LCC QP309 (ebook) |
DDC 612/.042–dc23/eng/20241205
LC record available at https://lccn.loc.gov/2024040684
LC ebook record available at https://lccn.loc.gov/2024040685

ISBN: 978-1-032-61958-3 (hbk)
ISBN: 978-1-032-67455-1 (pbk)
ISBN: 978-1-032-67457-5 (ebk)

DOI: 10.1201/9781032674575

Typeset in Times
by codeMantra

Contents

SECTION II
The human mind

SECTION IV
Organizing and managing work

Foreword

The eighth edition of Fitting the Human

The seventh edition of *Fitting the Human* (2017) was a CRC bestseller: it provides ergonomic information needed for the user-friendly design of things and tasks. This eighth edition maintains the successful format of previous editions, with updated information and practical guidelines that augment the previous content. In particular, this edition offers refreshed information and references on body sizes, design for home, office, and workplace (in particular, design for accessibility, continuous improvements, and addressing safety and risks), contemporary ways of working (physical work, remote/virtual work, and within organizations), healthcare and medicine (with a focus on human-centered design and the needs of patients and healthcare providers), artificial intelligence (including ergonomic considerations for autonomous vehicles and other applications of artificial intelligence/machine learning), and general principles of design. This text also recognizes that ways of living and working do differ around the world, so additional insights are offered into ergonomics within various cultural settings.

Ergonomics

Fitting the Human focuses on the primary ergonomic task: accommodating the human through appropriate selection of

- equipment and tools,
- work requirements and procedures,
- physical and social conditions at work, and
- working hours and shift arrangements.

Matching individuals with job requirements through personnel selection and training falls mainly within the domain of industrial psychology, which overlaps significantly with Ergonomics.

This book endorses the fundamental and most successful approach: *plan and design the overall system and its details to suit the human.*

Ease of use

This new edition follows the motto of previous versions: "Solid information, easy to read, easy to understand, easy to apply." Small markers (*) in the text indicate where references or additional explanations are helpful but don't break the flow of reading. These appear in a separate *Notes* section at the end of each chapter, which the reader may skip or consult.

Each chapter has *Steps to Fit the Human* after the chapter Summary, which provides a step-by-step process for making assessments based on the information presented in that chapter. There are also *References and Further Reading,* which allows the readers of this book to find additional information to guide them on their own journey into the rich scientific literature of ergonomics and human factors. Previous editions of this book also contain some of the older references.

Fitting the Human since 1963

Etienne Grandjean, who died in 1991, authored the first printing in 1963 as *Physiologische Arbeitsgestaltung.* He edited his text in 1969 and 1979. Ott Verlag published all three German editions. In 1980, Taylor & Francis released the third edition, translated into English, titled *Fitting the Task to the Man*, and in 1988, it printed a revised version. The fifth edition, co-authored by Karl HE Kroemer, appeared in 1997 as *Fitting the Task to the Human.* Dr. Kroemer completely revised the next edition to reflect that demands on humans continually change—not only in task content but also in utensils and equipment, and in the physical and social environments as well. Therefore, the titles of the sixth and seventh editions were simplified to *Fitting the Human.* Now the authorship has been passed on again to me: an engineer trained in mechanics and biomedicine, with decades of hands-on work experience applying engineering design principles and a lifetime of learning from Dr. Kroemer, who is my father. I bring additional insight into practical ergonomics and its applications, in particular in healthcare, medicine, and industry, while maintaining the important fundamentals that support all scientific and engineering designs.

Human engineering

The aim remains the same: to provide "human engineering" for workplaces, tools, machinery, computers, shift work, lighting, sound, climate, work demands, offices, vehicles, healthcare, and the home—and everything else that we can produce—to suit the human body and mind. In this new edition, I strive to continue Etienne Grandjean's and Karl Kroemer's direct, lucid, down-to-earth style of providing essential knowledge about design that fits the human body and mind. I hope this book will appeal to students and professionals in engineering, design, architecture, safety, and management, and to everybody else interested in making work safe, efficient, satisfying, and more enjoyable.

Do you want more information?

This book is an introduction to ergonomic issues that are—and continue to be—discussed in the international literature. If you want to stay with me as author/co-author, you may consider using these books that approach related topics: *Engineering Physiology*, 5th edition (2020); *Ergonomics: How to Design for Ease and Efficiency*, 3rd edition (2018), and Kroemer's *Extra-Ordinary Ergonomics* (2006) and *Office Ergonomics*, 2nd edition (2017).

Thank you!

To everyone striving to make the world a better place through ergonomic design, to everyone enhancing our work and personal lives through human factors principles, to everyone reading this book – I thank you.

Special thanks are due to my father, Karl HE Kroemer, for trusting me with this next edition, to my husband, Richard Elbert, for supporting and inspiring me in all things, and to our children Kristianna and Justin. May our lives be fitting.

Katrin Kroemer Elbert, BS, MS, PhD

About the author

Katrin Kroemer Elbert, Ph.D., has worked in research and development as a mechanical engineer for over three decades in the medical device industry, focusing on implants and medical devices for surgical, therapeutic, and diagnostic applications with Stryker-Howmedica orthopedics and Johnson & Johnson's Ethicon. She holds a bachelor's degree in engineering science and mechanics from Virginia Polytechnic Institute and State University (Virginia Tech) and master's and doctoral degrees in mechanical engineering from Cornell University, all with an additional concentration in biomedical engineering. She has a certification in engineering management from the Stevens Institute of Technology and has completed other project management and leadership trainings and programs. She brings wide-ranging theoretical and practical experience in biomedical engineering and the design and development of useful medical devices and systems. In addition, she collaborates across supply chain, regulatory, quality, and other functions to support devices that meet the needs of users and patients across the globe. She holds intellectual property and has numerous publications in the surgical and orthopedic literature. She has previously collaborated with her father, Karl H. Eberhard Kroemer (Dr.-Ing. and Professor Emeritus at Virginia Tech), on other ergonomics textbooks.

This edition is an update to the previous edition, and thus, credit is given to its previous author: Karl Heinrich Eberhard Kroemer, Dr.-Ing. studied mechanical engineering at the

Technical University Hanover, Germany, earning Vor-Diplom, Diplom-Ingenieur, and Doktoringenieur (Dr.-Ing) degrees. He first worked as a research engineer at the Max Planck Institute for Work Physiology, Germany (1960–1966). Then, he brought his family to the United States to be a research industrial engineer (1966–1973) in the Human Engineering Division of the Aerospace Medical Research Laboratory at the Wright-Patterson Air Force Base, Ohio, USA: note that Wright-Patterson is generally recognized as the birthplace of human factors engineering. He returned to Germany as the director of and professor in Ergonomics and Occupational Safety and Accident Research (1973–1976). Finally, he moved to the United States permanently to become a professor of ergonomics and industrial engineering at Wayne State University, Michigan (1976–1981) and then a professor in Virginia Tech's Department of Industrial and Systems Engineering as well as the director of the Industrial Ergonomics Laboratory (1981–1998, emeritus). He has been elected as a fellow of the Human Factors and Ergonomics Society and of the Ergonomics Society. He has served on the committee on Human Factors of the National Research Council, National Academy of Sciences, and has worked as a United Nations-International Labor Organization expert on ergonomics in Bucharest, Romania, and Mumbai, India. He holds intellectual property and has hundreds of publications as well as numerous awards for research and contributions to the science of ergonomics and human factors.

The first four editions of this text were written by Etienne Grandjean. Professor Grandjean (1914–1991) was one of the leading figures in ergonomics in Europe. He was a Swiss physician who studied physiology and then worked and taught for many decades at the Swiss Federal Institute of Technology: a pioneering researcher who could write easily and simply about topics of great complexity.

As scientists and engineers, we all stand on the shoulders of giants.

The first page

Humans are similar, and individuals are unique. This dichotomy influences the study of ergonomics (also known as human factors engineering).

- Overall, we are alike. Our bodies follow the same design. We all have similar bones and muscles, similar circulatory and metabolic functions, and a similar brain layout.
  Therefore, we can formulate ergonomic principles and guidelines that apply to all people. Recognizing similarities among people is one hallmark of ergonomics.
- However, individuals differ from each other; for example, in size and appearance, strengths, skills, abilities, interests, preferences, and expectations.

Thus, work tasks and work systems must accommodate people of diverse sizes, capabilities, and wants. Recognizing differences among people is the other hallmark of ergonomics.

Ergonomics, literally meaning the science of work, is the application of scientific principles, data, and methods drawn from a variety of disciplines to the development of engineered systems in which people play significant roles. This discipline relies on the knowledge of human characteristics relevant to the system and on the understanding that the system exists to serve humans. Human-centered design acknowledges human variability and human needs. Thus, ergonomics seeks to optimize human well-being and overall system performance. (Visit the International Ergonomics & Human Factors Association at https://iea.cc for more details.)

All the same?

(courtesy of Kristianna Elbert 2024)

The ergonomic goal is to make work safe, efficient, satisfying, and even enjoyable.

DOI: 10.1201/9781032674575-1

SECTION I

The human body

The first section of this book discusses our human bodies. We humans have remarkably similar physiques: typically, one head, two arms, and two legs attached to a central trunk. We use similar sets of muscles and bones to move our bodies. Respiration, circulation, nervous control, and other basic functions are fundamentally the same for each of us. However, as groups of people living in unique climates, consuming specific diets, and engaging in distinct activities over long periods of time, our bodies have adapted to the particular conditions in which we live and work. Thus, within the general similarity of humankind, there are differences in the bodies between groups of people. Furthermore, individuals vary in size and, importantly, in capabilities and ambitions.

We recognize and measure these differences so that we can design tasks, gear, equipment, and procedures to fit their specific needs. Therefore, the following chapters provide relevant information for ergonomic design.

- Chapter 1, Body Sizes
- Chapter 2, Body Mobility
- Chapter 3, Muscular Work
- Chapter 4, Body Strength

DOI: 10.1201/9781032674575-2

CHAPTER ONE

Body sizes

Body sizes differ

We all experience changes in body size throughout our lives: rapid growth during childhood, followed by a period of relatively constant dimensions during adulthood for about 20 to 40 years or more, and then final variations that come with senescence (aging). As a general rule, men grow to be taller as adults than women do. Body sizes vary across different populations around the globe and are affected by both genetics and environmental factors such as nutrition. At any given age, some people are smaller or larger than their peers, and the proportions of the body also differ widely among individuals. We need to understand body sizes in order to fit the tasks to the human.

1.1 Our Earth's populations

Height and weight

The most common way to describe the bodies of populations and individuals is by standing height (stature) and body weight. Table 1.1 contains recent (2021) estimates of adult stature for six different regions of the Earth as well as the calculated body mass index (BMI), which is a measure that combines height and weight into a single value.

Such estimates provide an overall impression of the differences in adults' body sizes, but designers often need more precise anthropometric information. Variances, even though small compared to our similarities, must be quantified precisely so that we can design appropriate clothing tariffs, widths of eyeglass frames, sizes of protective footwear, heights of workbenches, and dimensions of cockpits—all tailored to fit various user groups.

DOI: 10.1201/9781032674575-3

Table 1.1 Average stature and BMI in several regions of the earth

Region (country)	Females Height (cm)	Females BMI (kg/m²)	Males Height (cm)	Males BMI (kg/m²)
Africa				
North (Morocco)	157.8	26.7	170.4	25.6
East (Ethiopia)	155.7	21.1	166.2	20.1
West (Senegal)	162.5	24.1	173.1	21.9
Southeast (Namibia)	158.8	25.5	167.0	23.1
America, South				
Guatemala	149.4	27.5	163.4	25.8
Peru	152.9	27.3	165.2	26.4
Argentina	159.2	27.8	174.6	28.0
America, North				
Canada	163.9	26.8	178.1	27.5
United States of America	163.5	29.1	177.1	29.0
Asia				
Northeast (Japan)	158.3	21.8	170.8	23.7
Northeast (China)	159.7	23.6	171.8	24.3
Northwest (Turkey)	160.5	28.9	174.2	27.3
Northwest (Yemen)	154.0	24.2	159.9	23.7
Southeast (India)	152.6	22.0	164.9	21.8
Southeast (Indonesia)	152.8	23.9	163.6	22.6
Australia	165.9	26.9	179.2	27.7
Europe				
North (Iceland)	165.9	25.6	180.5	27.2
North (Russia)	165.3	26.7	176.5	26.0
Central (Netherlands)	168.7	25.3	182.5	26.1
Central (Germany)	165.9	26.0	179.9	27.5
Southeast (Portugal)	163.0	25.3	172.9	26.3
Southeast (Spain)	163.4	24.9	176.6	27.2
Southcentral (Italy)	164.6	24.9	177.8	26.6

Source: Adapted from data for 1996 cohort per Roser M, Appel C, Ritchie H (2021)—"Human Height" Published online at OurWorldInData.org. Retrieved on December 30, 2023, from https://ourworldindata.org/human-height#how-does-human-height-vary-across-the-world".

1.2 Measuring the human body

Anthropometrics around the globe

Anthropometrics (defined as the measurement of the human individual) such as body height and weight can be estimated, but this may not be sufficiently detailed for the intended use of the descriptive data. To gather the data, we could simply ask individuals about their body size, and if they know, they can inform us. However, with the current concerns about obesity in many countries, some respondents are likely to slightly exaggerate their height and underreport their weight. Therefore, actual measurements are needed to have confidence in the data.

Measurements have been taken over many years and in many regions of the Earth. These show variations in stature, influenced by genetics, gender, nutrition, and health. Currently, on average, the tallest men are found in northern Europe and the shortest women in Central America, with a greater than 30 cm difference in height, as shown in Table 1.1.

Table 1.1 also lists the BMI (which is the weight in kilograms, kg, divided by the square of height in meters, m) of individuals measured around the globe. BMI is a convenient way to combine weight and height into a single value, but it is not a perfect predictor of health or fitness. For example, since muscle tissue is denser than fat, very athletic (i.e., muscular) adults may have a high BMI. Pregnant women may also have a (temporarily) elevated BMI. BMI also varies across age groups, with adults in their fourth and fifth decades of life tending to have a higher mean BMI compared to adults in their 20s or 70s, reflecting that adults often gain weight in their middle and later decades. Yet altogether, these data reflect the magnitudes of differences in average body sizes and weights among groups of people.

How to measure

The designer needs specific data: describing the size and shape of the head for fitting helmets, for example, or understanding reach distances for devising a proper workspace. As a rule, such information cannot be deduced from stature or BMI (or weight) but must be specifically measured. Selecting a representative sample of people from the group of interest and performing the measurements on them is often a formidable task. The traditional technique has been to take physical measurements with handheld devices such as tapes, anthropometers, and gauges, and then to record and analyze the data. Newer techniques may include automated scanning and storing of the three-dimensional digitized surface of the human body*; yet current information still relies predominantly on the results of the hands-on measurement approach. Figure 1.1 depicts 36 of the most commonly taken measurements, and Table 1.2 describes these measurements and their use* in anthropometry and design.

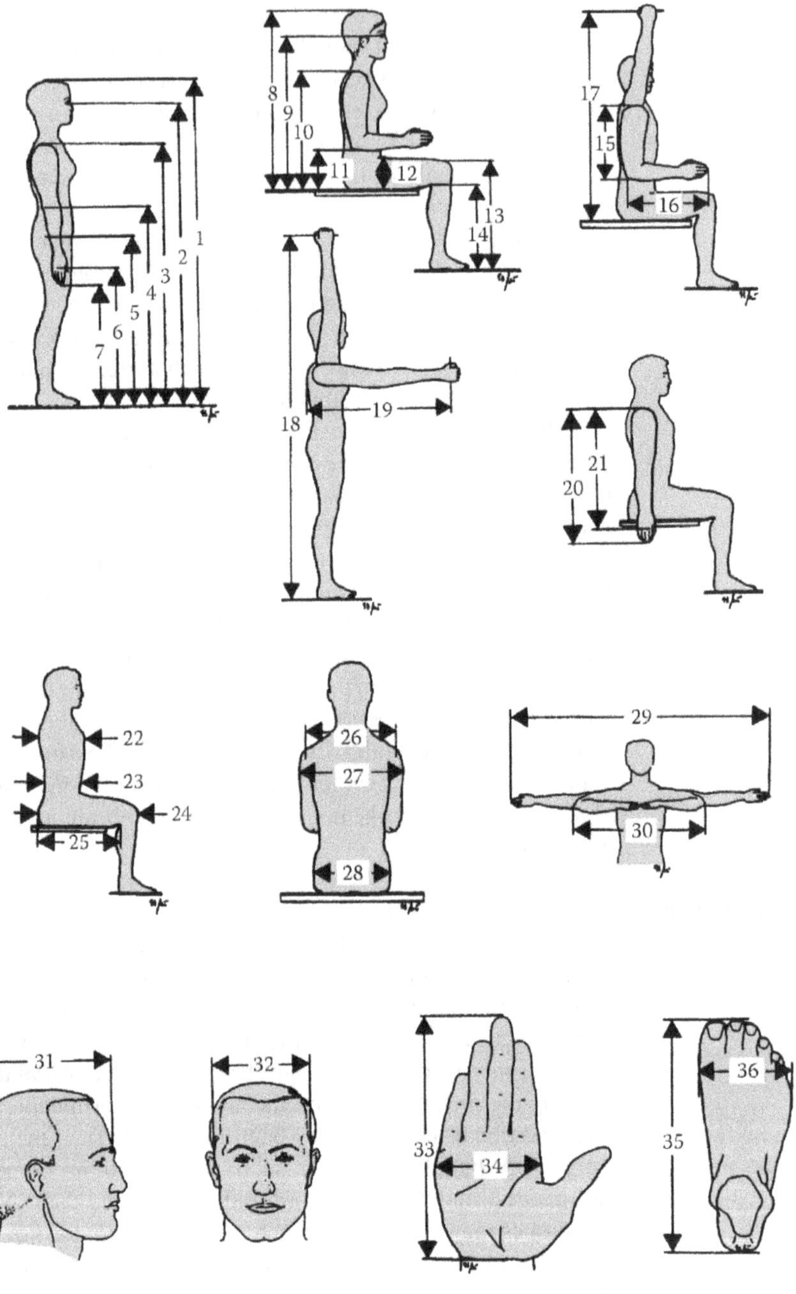

FIGURE 1.1 Illustrations of commonly measured body dimensions.

Table 1.2 Descriptions of common body measures and their applications

Dimensions	Applications
1. Stature The vertical distance from the floor to the top of the head, when standing	A main measure for comparing population samples. Reference for the minimal height of overhead obstructions. Add height for more clearance, hat, shoes, and stride.
2. Eye height, standing The vertical distance from the floor to the outer corner of the right eye, when standing	Origin of the visual field of a standing person. Reference for the location of visual obstructions and of targets such as displays; consider slump and motion.
3. Shoulder height (acromion), standing The vertical distance from the floor to the tip (acromion) of the right shoulder, when standing	Starting point for arm length measurements; near the center of rotation of the upper arm. Reference point for hand reaches; consider slump and motion.
4. Elbow height, standing The vertical distance from the floor to the lowest point of the right elbow, when standing, with the elbow flexed at 90°	Reference for height and distance of the work area of the hand and the location of controls and fixtures; consider slump and motion.
5. Hip height (trochanter), standing The vertical distance from the floor to the trochanter landmark on the upper side of the right thigh, when standing	Traditional anthropometric measure, indicator of the leg length and the height of the hip joint. Used for comparing population samples.
6. Knuckle height, standing The vertical distance from the floor to the knuckle (metacarpal bone) of the middle finger of the right hand, when standing	Reference for low locations of controls, handles, and handrails; consider slump and motion of the standing person.
7. Fingertip height, standing The vertical distance from the floor to the tip of the extended index finger of the right hand, when standing	Reference for the lowest location of controls, handles, and handrails; consider slump and motion of the standing person.
8. Sitting height The vertical distance from the sitting surface to the top of the head, when sitting	Reference for the minimal height of overhead obstructions. Add height for more clearance, hat, and trunk motion of the seated person.
9. Sitting eye height The vertical distance from the sitting surface to the outer corner of the right eye, when sitting	Origin of the visual field of a seated person. Reference point for the location of visual obstructions and of targets such as displays; consider slump and motion.
10. Sitting shoulder height (acromion) The vertical distance from the sitting surface to the tip (acromion) of the shoulder, when sitting	Starting point for arm length measurements; near the center of rotation of the upper arm. Reference for hand reaches; consider slump and motion.

(Continued)

Table 1.2 (Continued) Descriptions of common body measures and their applications

Dimensions	Applications
11. Sitting elbow height The vertical distance from the sitting surface to the lowest point of the right elbow, when sitting, with the elbow flexed at 90°	Reference for the height of an armrest, of the work area of the hand and of keyboard and controls; consider slump and motion of the seated person.
12. Sitting thigh height (clearance) The vertical distance from the sitting surface to the highest point on the top of the horizontal right thigh, with the knee flexed at 90°	Reference for the minimal clearance needed between the seat pan and the underside of a structure, such as a table or desk; add clearance for clothing and motions.
13. Sitting knee height The vertical distance from the floor to the top of the right kneecap, when sitting, with the knees flexed at 90°	Traditional anthropometric measure for lower leg length. Reference for the minimal clearance needed below the underside of a structure, such as a table or desk; add height for shoe.
14. Sitting popliteal height The vertical distance from the floor to the underside of the thigh directly behind the right knee, when sitting with the knees flexed at 90°	Reference for the height of a seat; add height for shoe.
15. Shoulder–elbow length The vertical distance from the underside of the right elbow to the right acromion, with the elbow flexed at 90° and the upper arm hanging vertically	Traditional anthropometric measure for comparing population samples.
16. Elbow–fingertip length The distance from the back of the right elbow to the tip of the extended middle finger, with the elbow flexed at 90°	Traditional anthropometric measure. Reference for fingertip reach when moving the forearm in the elbow.
17. Overhead grip reach, sitting The vertical distance from the sitting surface to the center of a cylindrical rod firmly held in the palm of the right hand	Reference for the height of overhead controls operated by a seated person. Consider ease of motion, reach, and finger/hand/arm strength.
18. Overhead grip reach, standing The vertical distance from the standing surface to the center of a cylindrical rod firmly held in the palm of the right hand	Reference for the height of overhead controls operated by a standing person. Add shoe height. Consider ease of motion, reach, and finger/hand/arm strength.
19. The horizontal distance from the back of the right shoulder blade to the center of a cylindrical rod firmly held in the palm of the right hand	Reference for forward reach distance. Consider ease of motion, reach, and finger/hand/arm strength.

(Continued)

Table 1.2 (Continued) Descriptions of common body measures and their applications

Dimensions	Applications
20. Arm length, vertical The vertical distance from the tip of the right middle finger to the right acromion, with the arm hanging vertically	A traditional measure for comparing population samples. Reference for the location of controls very low on the side of the operator. Consider ease of motion, reach, and finger/hand/arm strength.
21. Downward grip reach The vertical distance from the right acromion to the center of a cylindrical rod firmly held in the palm of the right hand, with the arm hanging vertically	Reference for the location of controls low on the side of the operator. Consider ease of motion, reach, and finger/hand/arm strength.
22. Chest depth The horizontal distance from the back to the right nipple	A traditional measure for comparing population samples. Reference for the clearance between the seat backrest and the location of obstructions in front of the trunk.
23. Abdominal depth, sitting The horizontal distance from the back to the most protruding point on the abdomen	A traditional measure for comparing population samples. Reference for the clearance between the seat backrest and the location of obstructions in front of the trunk.
24. Buttock–knee depth, sitting The horizontal distance from the back of the buttocks to the most protruding point on the right knee, when sitting with the knees flexed at 90°	Reference for the clearance between the seat backrest and the location of obstructions in front of the knees.
25. Buttock–popliteal depth, sitting The horizontal distance from the back of the buttocks to the back of the right knee just below the thigh, when sitting with the knees flexed at 90°	Reference for the depth of a seat.
26. Shoulder breadth (biacromial) The distance between the right and left acromions	A traditional measure for comparing population samples. Indicator of the distance between the centers of rotation of the two upper arms.
27. Shoulder breadth (bideltoid) The maximal horizontal breadth across the shoulders between the lateral margins of the right and left deltoid muscles	Reference for the lateral clearance required at shoulder level. Add space for ease of motion and tool use.
28. Hip breadth, sitting The maximal horizontal breadth across the hips or the thighs, whatever is greater, when sitting	Reference for seat width. Add space for clothing and ease of motion.

(Continued)

Table 1.2 (Continued) Descriptions of common body measures and their applications

Dimensions	Applications
29. Span The distance between the tips of the middle fingers of the horizontally outstretched arms and hands	A traditional measure for comparing population samples. Reference for sideway reach.
30. Elbow span The distance between the tips of the elbows of the horizontally outstretched upper arms when the elbows are flexed so that the fingertips of the hands meet in front of the trunk	Reference for the lateral space needed at upper body level for ease of motion and tool use.
31. Head length The distance from the glabella (between the browridges) to the most rearward protrusion (the occiput) on the back, in the middle of the skull	A traditional measure for comparing population samples. Reference for headgear size.
32. Head breadth The maximal horizontal breadth of the head above the attachment of the ears	A traditional measure for comparing population samples. Reference for headgear size.
33. Hand length The length of the right hand between the crease of the wrist and the tip of the middle finger, with the hand flat	A traditional measure for comparing population samples. Reference for hand tool and gear size. Consider manipulations, gloves, tool use.
34. Hand breadth The breadth of the right hand across the knuckles of the four fingers	A traditional measure for comparing population samples. Reference for hand tool and gear size and for the opening through which a hand may fit. Consider manipulations, gloves, and tool use.
35. Foot length The maximal length of the right foot, when standing	A traditional measure for comparing population samples. Reference for shoe and pedal size.
36. Foot breadth The maximal breadth of the right foot, at right angle to the long axis of the foot, when standing	A traditional measure for comparing population samples. Reference for shoe size and spacing of pedals.
37. Weight (kg) Nude body weight taken to the nearest tenth of a kilogram	A traditional measure for comparing population samples. Reference for body size, clothing, strength, health, etc. Add weight for clothing and equipment worn on the body.

Source: Adapted from Kroemer (2006).

Russian and Chinese adults

For common body measurements, specific data may be available in the literature. Table 1.3 is an example of the results of such detailed measures taken on Russian students between 1984 and 1986. Table 1.4 gives anthropometric data measured on young Chinese adults in Taiwan between 1996 and 2000. Different populations (in these cases, with geographical, sex, and age differences) have different measurements, but the relevance of these differences depends on the intended use of the data.

More compilations

There are compilations of data* describing the body sizes of populations in Russia, France, Japan, the United Kingdom, the United States, and a few other regions. Unfortunately, similarly detailed information remains missing on most populations, probably because it is so laborious and expensive to obtain. Thus, when we need information on other groups of people, we may have to estimate the data or, better yet, conduct some measurements ourselves.

1.3 No "average person"

There is one serious problem with average data, such as shown in Tables 1.1, 1.3, and 1.4. We may not design a glove or a workplace for the "average person" because that is just a statistical artifact. If we tried to design for that phantom, the resulting glove or workplace would be too small for half the people and too big for the others. Instead, it is necessary to consider specific users: for example, we can design for very large and tall people so that they can safely fit into an airplane seat, or we may need to pay particular attention to shorter individuals to ensure they can reach items stored on a high shelf. If doorways were fixed to suit people of average height, many individuals passing through them would strike their heads against the lintels.

The "normal" distribution

To avoid headaches, we need some advice about the statistics* that describe distributions of anthropometric measurements and then, to compile them into tables. Figure 1.2 depicts a typical collection of anthropometric measurements: for example, the stature of a group of men. Statisticians call this symmetric and bell-shaped curve a normal (or Gaussian) distribution: the data are clustered in the middle, and the farther away from the center, the fewer data points exist.

Table 1.3 Body size measures taken between 1984 and 1986 on Russian and Uzbek students 18–22 years of age

Dimension (in mm)	Females				Males			
	5th percentile	Mean	95th percentile	SD	5th percentile	Mean	95th percentile	SD
1. Stature								
Ethnic Russians (Moscow)	1542	1637	1731	57	1664	1757	1849	56
Uzbeks (Tashkent)	1498	1578	1658	49	1615	1700	1786	52
2. Eye height, standing								
Ethnic Russians (Moscow)	1433	1526	1618	57	1547	1637	1728	55
Uzbeks (Tashkent)	1387	1463	1538	46	1496	1581	1665	51
3. Shoulder height (acromion), standing								
Ethnic Russians (Moscow)	1245	1334	1422	54	1351	1440	1529	54
Uzbeks (Tashkent)	1217	1284	1371	47	1313	1391	1469	48
4. Elbow height, standing								
Ethnic Russians (Moscow)	941	1010	1080	42	1004	1083	1162	48
Uzbeks (Tashkent)	909	970	1031	37	985	1042	1099	35
5. Hip height (trochanter)	nda	nda	nda	nda	nda	nda	nda	nda
6. Knuckle height, standing								
Ethnic Russians (Moscow)	676	731	786	34	710	773	836	39
Uzbeks (Tashkent)	632	687	742	33	676	734	792	35
7. Fingertip height, standing								
Ethnic Russians (Moscow)	582	635	687	32	508	668	729	37

(*Continued*)

Table 1.3 (Continued) Body size measures taken between 1984 and 1986 on Russian and Uzbek students 18–22 years of age

Dimension (in mm)	Females				Males			
	5th percentile	Mean	95th percentile	SD	5th percentile	Mean	95th percentile	SD
Uzbeks (Tashkent)	546	599	652	32	579	635	691	34
8. Sitting height								
Ethnic Russians (Moscow)	806	859	911	32	860	912	964	32
Uzbeks (Tashkent)	793	839	885	28	858	905	952	29
9. Sitting eye height								
Ethnic Russians (Moscow)	694	742	790	29	737	790	844	33
Uzbeks (Tashkent)	676	723	771	29	737	784	830	28
10. Sitting shoulder height (acromion)	nda	nda	nda	nda	nda	nda	nda	nda
11. Sitting elbow height								
Ethnic Russians (Moscow)	196	236	275	24	202	243	284	25
Uzbeks (Tashkent)	191	229	267	23	186	229	272	26
12. Sitting thigh height (clearance)								
Ethnic Russians (Moscow)	126	148	172	14	122	151	179	18
Uzbeks (Tashkent)	114	142	170	17	120	143	165	14
13. Sitting knee height								
Ethnic Russians (Moscow)	487	527	567	24	520	562	603	25
Uzbeks (Tashkent)	446	487	528	25	494	531	569	23
14. Sitting popliteal height								
Ethnic Russians (Moscow)	386	423	461	23	429	468	508	24
Uzbeks (Tashkent)	366	398	430	20	400	430	460	18
15. Shoulder–elbow length	nda	nda	nda	nda	nda	nda	nda	nda

(Continued)

Table 1.3 (Continued) Body size measures taken between 1984 and 1986 on Russian and Uzbek students 18–22 years of age

	Females				Males			
Dimension (in mm)	5th percentile	Mean	95th percentile	SD	5th percentile	Mean	95th percentile	SD
16. Elbow–fingertip length	nda	nda	nda	nda	nda	nda	nda	nda
17. Overhead grip reach, sitting								
Ethnic Russians (Moscow)	1094	1169	1244	46	1199	1276	1354	47
Uzbeks (Tashkent)	1085	1152	1219	41	1193	1256	1319	38
18. Overhead grip reach, standing	nda	nda	nda	nda	nda	nda	nda	nda
19. Forward grip reach								
Ethnic Russians (Moscow)	641	702	763	37	697	759	821	38
Uzbeks (Tashkent)	609	673	737	39	686	745	803	36
20. Arm length, vertical	nda	nda	nda	nda	nda	nda	nda	nda
21. Downward grip reach	nda	nda	nda	nda	nda	nda	nda	nda
22. Chest depth								
Ethnic Russians (Moscow)	209	242	256	21	207	245	312	20
Uzbeks (Tashkent)	200	233	265	20	211	244	276	20
23. Abdominal depth, sitting	nda	nda	nda	nda	nda	nda	nda	nda
24. Buttock–knee depth, sitting								
Ethnic Russians (Moscow)	536	584	631	29	561	610	660	30
Uzbeks (Tashkent)	515	564	612	30	541	595	648	33
25. Buttock–popliteal depth, sitting								
Ethnic Russians (Moscow)	446	496	540	29	476	517	557	25

(Continued)

Table 1.3 (Continued) Body size measures taken between 1984 and 1986 on Russian and Uzbek students 18–22 years of age

	Females				Males			
Dimension (in mm)	5th percentile	Mean	95th percentile	SD	5th percentile	Mean	95th percentile	SD
Uzbeks (Tashkent)	423	472	520	29	459	504	550	28
26. Shoulder breadth (biacromial)								
Ethnic Russians (Moscow)	334	360	386	16	369	397	425	17
Uzbeks (Tashkent)	320	347	373	16	349	377	404	17
27. Shoulder breadth (bideltoid)								
Ethnic Russians (Moscow)	377	412	446	21	416	458	492	23
Uzbeks (Tashkent)	352	381	410	17	409	438	466	17
28. Hip breadth, sitting								
Ethnic Russians (Moscow)	334	372	411	23	323	362	410	23
Uzbeks (Tashkent)	329	364	399	21	316	349	381	20
29. Span								
Ethnic Russians (Moscow)	1516	1640	1763	75	1671	1782	1893	68
Uzbeks (Tashkent)	1461	1579	1698	72	1640	1747	1855	66
30. Elbow span								
Ethnic Russians (Moscow)	803	870	933	38	874	935	995	37
Uzbeks (Tashkent)	781	837	894	34	842	909	976	41
31. Head length	nda	nda	nda	nda	nda	nda	nda	nda
32. Head breadth	nda	nda	nda	nda	nda	nda	nda	nda

(Continued)

Table 1.3 (Continued) Body size measures taken between 1984 and 1986 on Russian and Uzbek students 18–22 years of age

	Females				Males			
Dimension (in mm)	5th percentile	Mean	95th percentile	SD	5th percentile	Mean	95th percentile	SD
33. Hand length								
Ethnic Russians (Moscow)	155	168	182	8	174	188	202	9
Uzbeks (Tashkent)	nda	nda	nda	nda	175	188	201	8
34. Hand breadth								
Ethnic Russians (Moscow)	71	76	82	3	80	87	95	5
Uzbeks (Tashkent)	73	79	87	4	82	89	96	4
35. Foot length								
Ethnic Russians (Moscow)	222	239	256	11	247	266	286	12
Uzbeks (Tashkent)	220	237	254	10	242	260	279	11
36. Foot breadth								
Ethnic Russians (Moscow)	82	88	95	4	87	97	107	6
Uzbeks (Tashkent)	81	90	98	5	85	96	107	7
37. Weight (kg)								
Ethnic Russians (Moscow)	49	60	73	7	57	71	85	9
Uzbeks (Tashkent)	45	56	68	7	53	65	76	7

Source: Adapted from Strokina and Pakhomova (1999).

Note: The measured sample consisted of 166 male and 207 female Russians from Moscow and 150 male and 164 female Uzbeks from Tashkent.

SD, standard deviation; nda, no data available.

Table 1.4 Body size measures taken between 1996 and 2000 on nearly 1200 Chinese adults in Taiwan, 25–34 years of age

Dimension (in mm)	Females 5th percentile	Females Mean	Females 95th percentile	Females SD	Males 5th percentile	Males Mean	Males 95th percentile	Males SD
1. Stature	1485	1572	1659	53	1608	1705	1801	59
2. Eye height, standing	nda	nda	nda	nda	nda	nda	nda	nda
3. Shoulder height (acromion), standing	1204	1285	1367	50	1309	1396	1484	53
4. Elbow height, standing	915	978	1040	38	993	1059	1126	40
5. Hip height (trochanter)	735	802	869	41	780	860	939	48
6. Knuckle height, standing	653	708	762	33	705	757	809	32
7. Fingertip height, standing	566	618	670	32	610	659	708	30
8. Sitting height	794	846	898	32	861	910	959	30
9. Sitting eye height	681	732	783	31	742	791	839	29
10. Sitting shoulder height (acromion)	516	561	605	27	560	602	645	26
11. Sitting elbow height	211	252	294	25	226	264	303	24
12. Sitting thigh height (clearance)	nda	nda	nda	nda	nda	nda	nda	nda
13. Sitting knee height	431	471	510	24	474	521	569	29
14. Sitting popliteal height	350	379	408	18	380	411	442	19
15. Shoulder–elbow length	280	309	339	18	308	338	369	19
16. Elbow–fingertip length	339	384	429	27	382	427	472	27
17. Overhead grip reach, sitting	1033	1105	1177	44	1128	1208	1289	49
18. Overhead grip reach, standing	1721	1831	1942	67	1872	2002	2133	79
19. Forward grip reach	597	651	705	33	650	710	770	36
20. Arm length, vertical	618	669	720	31	684	738	793	33

(Continued)

Table 1.4 (Continued) Body size measures taken between 1996 and 2000 on nearly 1200 Chinese adults in Taiwan, 25–34 years of age

	Females				Males			
Dimension (in mm)	5th percentile	Mean	95th percentile	SD	5th percentile	Mean	95th percentile	SD
21. Downward grip reach	nda	nda	nda	nda	nda	nda	nda	nda
22. Chest depth	182	213	244	19	187	217	248	19
23. Abdominal depth, sitting	nda	nda	nda	nda	nda	nda	nda	nda
24. Buttock–knee depth, sitting	487	530	572	26	507	558	608	31
25. Buttock–popliteal depth, sitting	nda	nda	nda	nda	nda	nda	nda	nda
26. Shoulder breadth (biacromial)	282	324	366	25	323	369	415	28
27. Shoulder breadth (bideltoid)	367	406	445	24	422	460	499	23
28. Hip breadth, sitting	316	353	390	23	315	360	404	27
29. Span	1469	1571	1672	62	1625	1738	1852	69
30. Elbow span	737	801	866	39	820	894	968	45
31. Head length	176	187	198	6	185	197	209	7
32. Head breadth	146	161	175	9	154	167	181	8
33. Hand length	154	167	181	8	168	183	199	10
34. Hand breadth	68	75	82	4	77	86	94	5
35. Foot length	nda	nda	nda	nda	nda	nda	nda	nda
36. Foot breadth	nda	nda	nda	nda	nda	nda	nda	nda
37. Weight (kg)	40	52	64	7	53	67	81	9

Source: Adapted from Wang, Wang, and Lin (2002).

Note: The measured sample consisted of nearly 1200 male and female Chinese adults from Taiwan.

SD, standard deviation; nda, no data available.

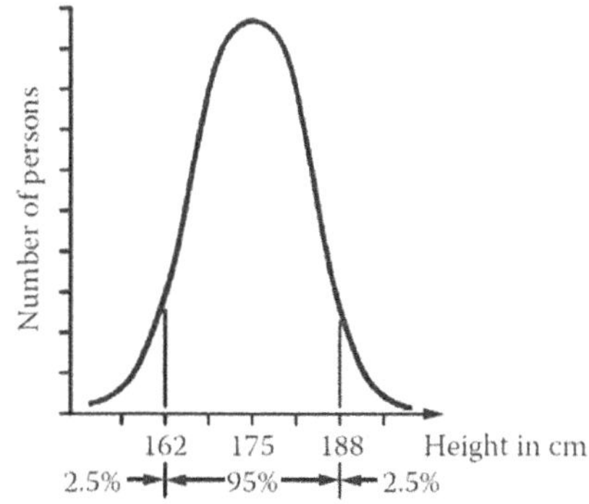

FIGURE 1.2 Typical distribution of anthropometric data.

Mean and average Such a normal distribution is easy to describe by two statistical values: one descriptor is the well-known mean (also often called average), and the other descriptor is the standard deviation, a measure of the peakedness or flatness of the distribution (in effect, the width of the bell shape about the mean). (Note that these descriptors do not apply to non-normal distributions, which require more complex statistical treatments. There is more about this in Chapter 4, which concerns muscle strength data.)

Percentiles If we know the mean (*m*) and the standard deviation (*SD*), we can calculate the numerical value of any point, *percentile* (*p*), in a normal distribution. The fifth percentile is often of design interest: 5% of all data are smaller and 95% are larger. The mean (average) is the same as the fiftieth percentile: half the data lie below and the other half above this value. The fifth percentile, *p5* for short, is 1.65 standard deviations below the mean; conversely, the 95th percentile, *p95*, is 1.65 *SD* above *m*. Table 1.5 contains multiplication factors *k* needed to calculate various percentage points in a normal distribution.

Hand sizes For designing gloves, hand tools, or hand-operated controls, we need information about hand sizes. Figure 1.3 shows relevant measurements, and Table 1.6 contains some available data. The table shows that for many groups of people, hand data are missing, so we must either measure or estimate the missing data.

Further advice Collections of anthropometric data provide a wealth of information to the skilled statistician, but others, including engineers

Table 1.5 Values of factor *k* for computing percentiles (*p*) from mean (*m*) and standard deviation (SD) in a normal distribution

k	Percentile (p) located below the mean (m) $p = m - k \times SD$	Percentile (p) located above the mean (m) $p = m + k \times SD$
4.25	0.001	99.999
2.33	1	99
2.06	2	98
1.96	2.5	97.5
1.88	3	97
1.65	5	95
1.28	10	90
1.04	15	85
1.00	16.5	83.5
0.84	20	80
0.67	25	75
0	50	50

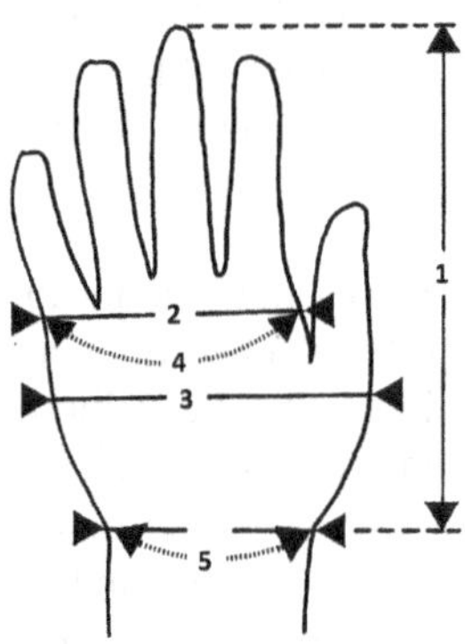

FIGURE 1.3 Common measurements of hand and wrist.

or designers, can use the data as well if we follow proper statistical methodology. As an example, we can calculate percentiles to use as cut-off values to determine adjustment ranges of uniform or composite populations, using statistical formulae and factor k as in Table 1.5. Another application of data collections is to estimate data on groups of people, such as to fill in the vacant cells in Table 1.6. In doing so, it is beneficial to employ the advice given in books written for ergonomists and human factors engineers, such as those listed at the end of the chapter as References and Further Reading.

Table 1.6 Hand and wrist sizes (in mm)

		Females		Males	
Hand measures (data in mm)	Population	Mean	SD	Mean	SD
1. Hand length	British	175	9	180	10
	Chinese, Taiwan	167	8	183	10
	French	173	nda	190	nda
	Germans	174	9	189	9
	Germans, soldiers	176	nda	191	nda
	Japanese	nda	nda	nda	nda
	Russians	168	8	188	9
	US soldiers	180	10	193	10
	US Vietnamese	165	9	177	12
2. Breadth at knuckles	British	75	4	85	5
	Chinese, Taiwan	75	4	86	5
	French	76	nda	86	nda
	Germans	78	4	88	5
	Japanese	90	5	nda	nda
	Russians	76	3	87	5
	US soldiers	78	4	88	4
	US Vietnamese	71	4	79	7
3. Maximal breadth	British	92	5	105	5
	Chinese	nda	nda	nda	nda
	French	nda	nda	nda	nda
	Germans	94	6	107	6
	Japanese	nda	nda	nda	nda
	Russians	nda	nda	nda	nda
	US soldiers	nda	nda	nda	nda
	US Vietnamese	87	6	100	6
4. Circumference at knuckles	British	nda	nda	nda	nda
	Chinese	nda	nda	nda	nda
	French	nda	nda	nda	nda
	Germans	nda	nda	nda	nda
	Japanese	nda	nda	nda	nda
	Russians	nda	nda	nda	nda
	US soldiers	187	9	212	10
	US Vietnamese	nda	nda	nda	nda

(Continued)

Table 1.6 (Continued) Hand and wrist sizes (in mm)

Hand measures (data in mm)	Population	Females Mean	Females SD	Males Mean	Males SD
5. Wrist circumference	British	nda	nda	nda	nda
	Chinese	nda	nda	nda	nda
	French	nda	nda	nda	nda
	Germans	nda	nda	nda	nda
	Japanese	nda	nda	nda	nda
	Russians	nda	nda	nda	nda
	US soldiers	154	8	176	9
	US Vietnamese	137	18	163	15

Source: Adapted from Gordon, Blackwell Bradtmiller et al. (2014), Imrhan, Nguyen, and Nguyen (1993), Kroemer (2006), and Leyk, Kuechmeister, and Jürgens (2006).
SD, standard deviation; nda, no data available.

1.4 Designing to fit the body

People differ

Nearly every adult on Earth can fit into an airplane or use common hand tools if these products are sized well. Yet, individuals within the species vary from one another; we need differently sized shoes to fit our individual feet comfortably. Even among seemingly similar groups, body sizes or segment measurements can differ significantly. For example, in the USA, agricultural workers are on average 2.5 cm shorter in height than other workers. Female American agricultural and manufacturing workers have larger waist circumferences than those in other occupations. Firefighters, police officers, and guards are taller and also heavier (males by 7 kg, females by over 10 kg) than individuals in all other occupations*.

Design principles

Before we begin to design a glove, helmet or other object that must fit its user exactly to be functional, we must decide which range of relevant body sizes we want to accommodate. We have a choice among five approaches:

1. *Custom-fit each individual*: This is a laborious and expensive solution, justifiable in exceptional cases.
2. *Have several fixed sizes*: This can be a reasonable solution, but all sizes must be available, and "between-sized" users may not be accommodated well.
3. *Make it adjustable*: This is usually the approach that provides the best fit for all people, but the adjustment features must be easy to use.

4. *Design for the "extreme" body sizes*: This is the appropriate solution when we must ensure that everybody
    - Can operate a gadget, such as locating an emergency stop button within the shortest reach; or
    - Can fit through an opening, such as making a door or escape hatch wide enough for even the largest person; or
    - Cannot pass through an opening behind which danger lurks, such as firmly attaching railings or safety guards at machinery.
5. *Select those persons whose bodies fit the existing design*: This is a "last and worst resort" if we have failed to achieve, and thus must violate, the fundamental principle of good design: all intended users should be able to employ our design effectively and efficiently.

Fitting a range

Solutions 2 and 3 are the most common. For this, we need to determine the range of body sizes that we intend to accommodate with our design: we must set the lower and upper endpoints of the fit range. Often, we aim to fit all persons larger than the smallest 5% and smaller than the largest 5%; in other words, we are accommodating the central 90% of a group. In doing so, we knowingly exclude 10%, half of whom are very small and the others very big.

Selecting design limits

The design endpoints (i.e., the minimum and maximum of the range to be fitted) depend on the design purpose and therefore must be selected carefully. We may decide to divide our designs into several sizes (Solution 2), each appropriate for a subgroup of all users. This is a routine approach for ready-made clothing: clothes are tailored in size clusters. Within each range, adjustment features (Solution 3) can provide further fitting: shoes are fitted with laces, and the seats of office chairs can be raised and lowered.

Statics and dynamics

The measures shown in Figures 1.1 and 1.2 and in Tables 1.1, 1.3, 1.4, and 1.6 are dimensions taken on the body while standing or sitting still. In reality, hardly anybody is stiffly static at work: we usually move about. To design for movement, the engineer needs to modify the static data to reflect dynamic motion. The recommendations in Table 1.7 can help accomplish this task.

Table 1.7 Guidelines for the conversion of standard measuring postures to functional stances and motions

To address	Consider doing this
Slumped standing or sitting	Deduct 5%–10% from appropriate height measurements
Relaxed trunk	Add 5%–10% to trunk circumferences and depths
Wearing shoes	Add approximately 25 mm to standing and sitting heights; add more for higher heeled shoes
Wearing light clothing	Add about 5% to appropriate dimensions
Wearing heavy clothing	Add 15% or more to appropriate dimensions. (Note that mobility may be strongly reduced by heavy clothing.)
Extended reaches	Add 10% or more for strong motions of the trunk
Use of hand tools	Center the handle at about 40% of hand length, measured from the wrist
Comfortable seat height	Add or subtract up to 10% from the standard seat height

Source: Adapted from Kroemer (2006).

Summary

Fitting equipment and tasks to individuals of various body sizes requires both anthropometric data and proper procedures. Data on many populations are available but should be used with care to avoid inadvertently drawing false conclusions. Depending on the variability in the data and the criticality of the application, we can estimate missing information or, better yet, measure the intended population following standard procedures. Be careful not to design for the "average person," as that person does not exist. Design procedures often involve selecting percentile values that serve as lower and upper limits of accommodation ranges.

Steps to fit the human

Step 1: Determine which body dimensions are important for your design. Consider both the task and the operator.

Step 2: Decide on the range(s) to be fitted and the cut-off point(s). Consider whether you need to custom-fit, can offer only fixed sizes, are able to add adjustability, should include body size extremes, or must limit your design to only selected operators.

Step 3: Design, then test. Modify and iterate as necessary.

Notes

The text contains markers, *, to indicate specific references and comments, which follow.

1.2 Measuring the human body

Techniques: Bradtmiller (2015), Gordon, Churchill, Clauser et al. (1989), Gordon, Blackwell, Bradtmiller et al. (2014), Paquette, Gordon, and Bradtmiller (2009), Roebuck (1995), Koepke, Zwahlen, Wells et al. (2017).

Measurements and their use: ISO Standards 7250, 15535, 20685; Kroemer (2006, 2020).

Compilations of data: for example, Bradtmilller (2015), Kroemer (2020), online (internet) resources such as OurWorldInData.org (https://ourworldindata.org/human-height).

1.3 No "average person"

Statistics: Gordon, Blackwell, Bradtmiller et al. (2014), Marras and Karwowski (2006), Paquette, Gordon, and Bradtmiller (2009), Pheasant and Haslegrave (2006) (Appendix), Roebuck (1995).

1.4 Designing to fit the body

Differences among occupational groups: Hsiao, Long, and Snyder (2002).

References and further reading

The literature contains recent anthropometric information, published usually in journal articles or as research reports. Handbooks, encyclopedias and standards are often a few years behind.

Bradtmiller B (2015). Anthropometry in human systems integration. Chapter 8 in Boehm-Davis DA, Durso FT, Lee JD(eds), *APA Handbook of Human Systems Integration*. Washington, DC, American Psychological Association. https://doi.org/10.1037/14528-000

Gordon CC, Blackwell CL, Bradtmiller B, Parham JL, Barrientos P, Paquette SP, Corner BD, Carson JM, Venezia JC, Rockwell BM, Mucher M, Kristensen S (2014). 2012 Anthropometric Survey of U.S. Army Personnel: Methods and Summary Statistics. Technical Report NATICK/TR-15/007. U.S. Army Natick Soldier Research, Development and Engineering Center, Natick, MA. https://apps.dtic.mil/sti/citations/ADA611869

Hsiao H, Long D, Snyder K (2002). Anthropometric differences among occupational groups. *Ergonomics*, 45:136–152. https://doi.org/10.1080/00140130110115372

Imrhan SN, Nguyen MT, Nguyen NN (1993). Hand anthropometry of Americans of Vietnamese origin. *International Journal of Industrial Ergonomics*, 12:281–287. https://doi.org/10.1016/0169-8141(93)90098-X

ISO 15535 (2023). *General Requirements for Establishing Anthropometric Databases*. Geneva, CH, International Organization for Standardization.

ISO 20685-1 (2018). *3-D Scanning Methodologies for Internationally Compatible Anthropometric Databases – Part 1: Evaluation Protocol for Body Dimensions Extracted from 3-D Body Scans*. Geneva, CH, International Organization for Standardization.

ISO 7250-1 (2017). *Basic Human Body Measurements for Technological Design -- Part 1: Body Measurement Definitions and Landmarks*. Geneva, CH, International Organization for Standardization.

Koepke N, Zwahlen M, Wells JC, Bender N, Henneberg M, Rühli FJ, Staub K (2017). Comparison of 3D laser-based photonic scans and manual anthropometric measurements of body size and shape in a validation study of 123 young Swiss men. *Peer Journal*, 5:e2980. https://doi.org/10.7717/peerj.2980

Kroemer KHE (2006). *"Extra-Ordinary" Ergonomics: How to Accommodate Small and Big Persons, the Disabled and Elderly, Expectant Mothers and Children*. Boca Raton, FL, CRC Press. https://doi.org/10.1201/9780203025246

Kroemer KHE (2008). Anthropometry and biomechanics: anthromechanics. Chapter 2 in Kumar S (ed), *Biomechanics in Ergonomics*, 2nd ed. Boca Raton, FL, CRC Press. https://doi.org/10.1201/9780849379093

Kroemer KHE, Kroemer HJ, Kroemer-Elbert KE (2020). *Engineering Physiology: Bases of Human Factors Engineering/Ergonomics*, 5th ed. Heidelberg, Springer. https://doi.org/10.1007/978-3-030-40627-1

Leyk D, Küchmeister G, Jürgens HW (2006). Combined physiological and anthropometrical databases as ergonomic tools. *Journal of Physiological Anthropology*, 25(6):363–369. https://doi.org/10.2114/jpa2.25.363

Marras WS, Karwowski W (2006). *The Occupational Ergonomics Handbook: Fundamentals and Assessment Tools for Occupational Ergonomics*, 2nd ed. Boca Raton, FL, CRC Press. https://doi.org/10.1201/9781420003635

Paquette S, Gordon C, Bradtmiller B (2009). Anthropometric Survey (ANSUR) II Pilot Study: Methods and Summary Statistics. Technical Report Natick/TR-09/014. Anthrotch, US Army Natick Soldier Research, Development and Engineering Center, Natick, MA. https://apps.dtic.mil/sti/pdfs/ADA498172.pdf

Pheasant S, Haslegrave CM (2006). *Anthropometry, Ergonomics and the Design of Work*, 3rd ed. London, UK, Taylor & Francis. https://doi.org/10.1201/9781315375212

Roebuck JA (1995). *Anthropometric Methods – Designing to Fit the Human Body.* Santa Monica, CA, Human Factors and Ergonomics Society.

Roser M, Appel C, Ritchie H (revised 2024). Human Height: The average height of a population can inform us about the nutrition and living conditions of populations in the past for which we have little other data. https://ourworldindata.org/human-height#how-does-human-height-vary-across-the-world (retrieved May 3 2024)

Strokina AN, Pakhomova BA (1999). *Anthropo-ergonomic Atlas.* Moscow, Russia, Moscow State University Publishing House.

Wang MJJ, Wang EMY, Lin YC (2002). *Anthropometric Data Book of the Chinese People in Taiwan*. Hsinchu, Taiwan, The Ergonomics Society of Taiwan.

CHAPTER TWO

Body mobility

Tasks in motion

Ancient experiences in agriculture, forestry, fishing, and other traditional work such as masonry and carpentry have taught people how to perform the embedded tasks effectively. However, new tasks, tools, and workplaces in modern industry and commerce, as well as in transportation on land and in water, air, and space, must be designed purposefully to suit the human body and mind. One foundation for human-centered designs is the insight that our bodies function best in motion, not in a maintained static stance. We need to understand how the body moves in order to fit the task to the human.

2.1 Basics of work and motion

Made for motion

We are continually changing our body's positioning while we walk or sit, even when we sleep. If injury or disease imposes a fixed body position, circulatory and metabolic functions become impaired; for example, people who must lie in place tend to develop bedsores as well as physical degeneration and aches. Holding still is tiresome, almost impossible for even just an hour; we find ourselves shifting our posture or fidgeting. It is clear that the human body wants to move and that it functions best in motion. Therefore, we should design our equipment and tasks to facilitate movement.

No static templates

Speaking practically, it is convenient to measure the human body while it is held still in a defined upright static posture, whether standing or sitting, as shown in Figure 1.1 in the first chapter of this book. Unfortunately, such artificial measuring stances, as shown in Figure 2.1, have become misleading templates for

DOI: 10.1201/9781032674575-4

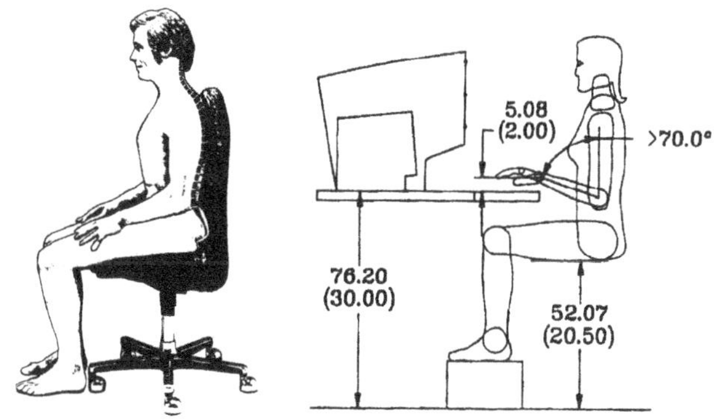

FIGURE 2.1 Unrealistic depictions, circa 1980, of "orthopedically good" sitting of a computer operator at work.

workstation layout; these patterns not only created the false image of a static operator but also gave the wrong impression that being "stiff upright" is desirable or healthy. People like to move, not to stay still; if left alone, they will sit in any position they prefer, as illustrated in Figure 2.2. Designing for movement* is not difficult, as we will see later in this chapter.

Excessive motions

Of course, overdoing motions can lead to trouble. Highly repetitive movement requirements* have been documented since the early 1700s to be associated with hand, wrist, and arm problems. Excessive demands on the limited motion capabilities of the spinal column can easily result in overloading, especially if torso twisting combines with bending; low back pain* has seemed to accompany humankind since the earliest times. Thus, more detail on hand, wrist, and spine anatomy is provided in the next sections.

2.2 Body joints and appendages

Mobility of legs and arms

Legs provide us with powerful mobility: these long body members rotate about their articulations with the trunk at the hip joints, which allow for wide-ranging angular freedom*. Simpler angular motions occur in the knee joints. Changes in foot angle at the ankles are small but important for balance and subtle actions. Our arms provide us with long reaches, and our shoulder and elbow joints offer extensive mobility. Thumbs and fingers are capable of performing complex and finely controlled motions through the articulations of the wrists and digits.

FIGURE 2.2 People sit any way they like. Adapted from Kroemer and Kroemer (2017).

Rotations in body joints

Hip and shoulder joints have a pan-like bone structure, within which the knob-shaped proximal bones of the thigh and upper arm can rotate. The technical analogy is that of a ball joint that can move about three axes of rotation (it has three degrees of freedom for mobility): the upper leg and arm can rotate (1) fore–aft, (2) left–right, and they can (3) twist. In particular, the hip has a more stable ball-and-socket joint configuration. The knee and elbow are simpler joints, having mostly only one axis of rotation, so the lower leg and forearm primarily swing forth and back in their hinge-type joints. The ankle articulation is another ball joint type but with more limited movement in the three axes. The articulations of the spine are limited but allow for some pivoting.

Hand mobility

2.2.1 Our hands The wrist provides the hand with wide-ranging mobility across three axes: bending up and down, left and right, and twisting. (To be precise, the ability to twist is actually in the forearm.) The 27 bones of the hand provide a solid structure; Figure 2.3 identifies them and names their articulations.

Whereas the main body of the hand is only slightly deformable among its eight carpal bones, the attached five digits (one thumb and four fingers) provide great dexterity.

Hand digits

As Figure 2.3 shows, the five digits attach to the main body of the hand by their metacarpal bones at the carpal–metacarpal (CMC) joints, which act as hinges but in different planes. This allows the palm of the hand to snugly cup cylindrical objects and enables the fingers and opposing thumb to touch. The thumb is particularly mobile in its CMC joint, but it has only two final segments (phalanges), whereas the four fingers each have three phalanges. The knuckles, formally known as metacarpal–phalangeal joints, provide two axes of mobility to the fingers, but

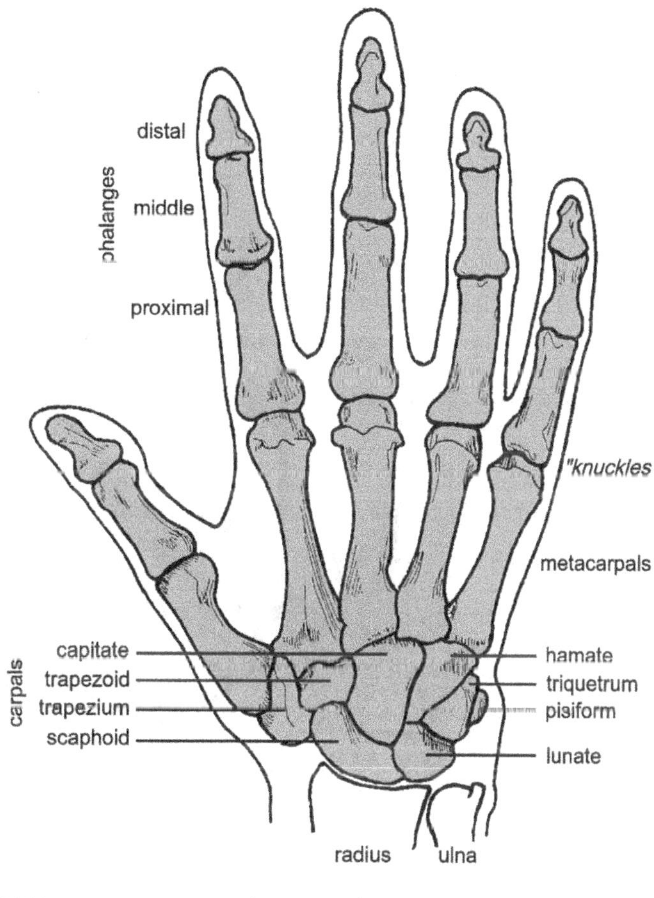

FIGURE 2.3 Bones and joints of the right hand, as seen in a top view.

their interphalangeal (IP) joints (proximal interphalangeal, PIP, and distal interphalangeal, DIP) are simple one-axis hinges.

Movers of the hand

Some of the muscles that bend (flex) and straighten (extend) the hand are located within the hand itself. These intrinsic muscles control most of the finer details of manipulation. However, other muscles are located in the forearm (away from the hand, therefore called extrinsic), from where they produce most of the powerful hand flexion and extension activities. Their contractile forces are transmitted to the hand by pulling on tendons that cross the wrist. Figures 2.4 and 2.5 show the extensor tendons at the back of the hand and the flexor tendons on the palmar side. Sheaths encapsulate the tendons, keeping them in place while providing lubrication to allow them to glide.

Carpal tunnel

Because only limited space is available for the passage of the tendons at the wrist, bending this joint can create increased pressure between a tendon and its sheath. This usually does not

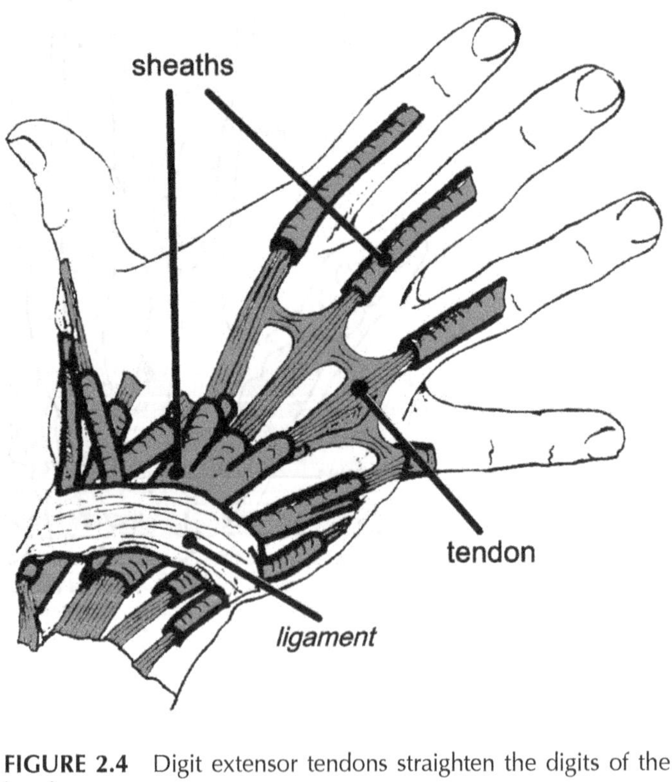

FIGURE 2.4 Digit extensor tendons straighten the digits of the hand.

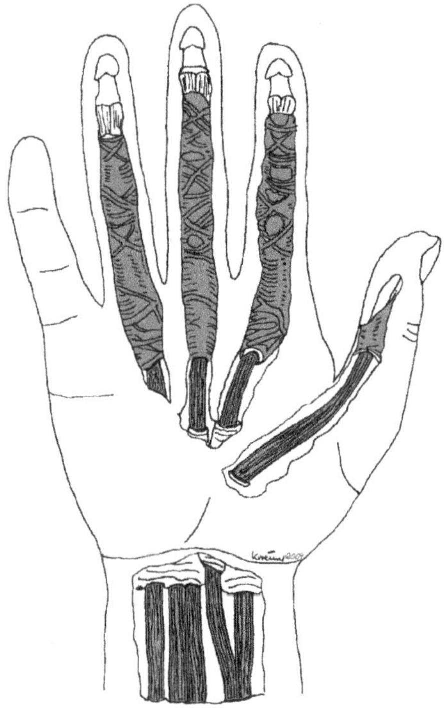

FIGURE 2.5 Digit flexor tendons bend the digits of the hand.

affect the extensor tendons at the back of the hand but can be problematic for the flexors on the palm side (see Figure 2.6). A particularly critical region is the canal-like opening at the base of the hand formed by the carpal bones, covered and made into a "tunnel" by the strong transverse ligament.

Carpal tunnel syndrome

The nine tendons of the extrinsic muscles that flex the hand's digits must pass through this tight tunnel, along with the median nerve and blood vessels. The median nerve is a cord about the diameter of a pencil containing thousands of nerve fibers that supply sensation to the thumb, index and middle fingers, and part of the ring finger. If swelling occurs in this tightly packed carpal tunnel (such as from inflammation that often results from highly repetitive digit actions, for example, typing on keyboards), the pressure inside the tunnel increases. The resulting compression can make tendon movements difficult and painful and can also affect nervous feedback and control through the median nerve; this is the *carpal tunnel syndrome** often associated with temporary disability and possibly even permanent injury.

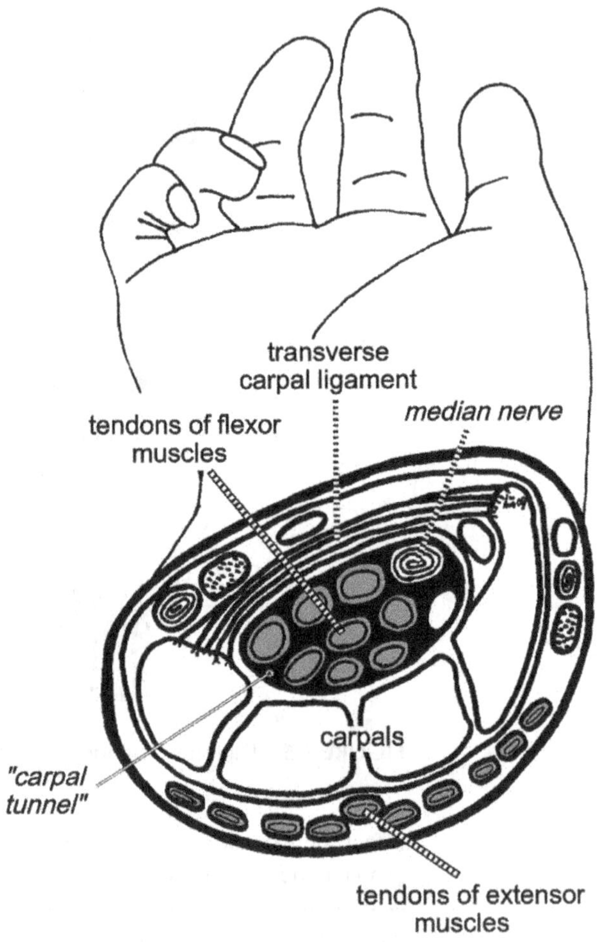

FIGURE 2.6 Cross section of the right hand, distal from the wrist. Carpal bones and the transverse carpal ligament form the carpal tunnel through which pass the digit flexor tendons and the median nerve.

Tendon sheath

The sheaths of tendons are complex configurations of connective tissue related to their location and function. In addition to providing lubrication to facilitate the gliding of the enclosed tendon, sheaths also supply attachments to the bones. They act as pulleys at which the tendons pull to flex or extend the sections of the digits against each other, as sketched (much simplified) for a finger in Figure 2.7. Overall, the complex design of the hand, with its multiple functions, provides wide-ranging, forceful mobility with fine manipulative control.

2.2.2 Our spine The spinal column supports the trunk and head. It is essentially a stack of 24 bones, each called a vertebra, on top of the fused sacrum and coccyx (tailbone)*. The spine is the only solid structure in the trunk of the human body that keeps the rib cage from falling into the pelvis. It supports the entire mass of the upper body. Weights carried in the hands or on the shoulders, or in the belly (abdominal fat) increase the load on the spinal column, affecting compression, bending, and twisting. As seen from the front or back of a person, the healthy spinal column is essentially straight; but viewed from the side, it bends in a series of curves, as sketched in Figure 2.8. At the neck and in the low back, the spine has a forward bend called lordosis; at chest height, the curvature points backward, called kyphosis.

Spine mobility

The joints and motion capabilities of the spinal column are quite different from those of the hands. The main body of each spinal vertebra sits on top of an intervertebral disc (also

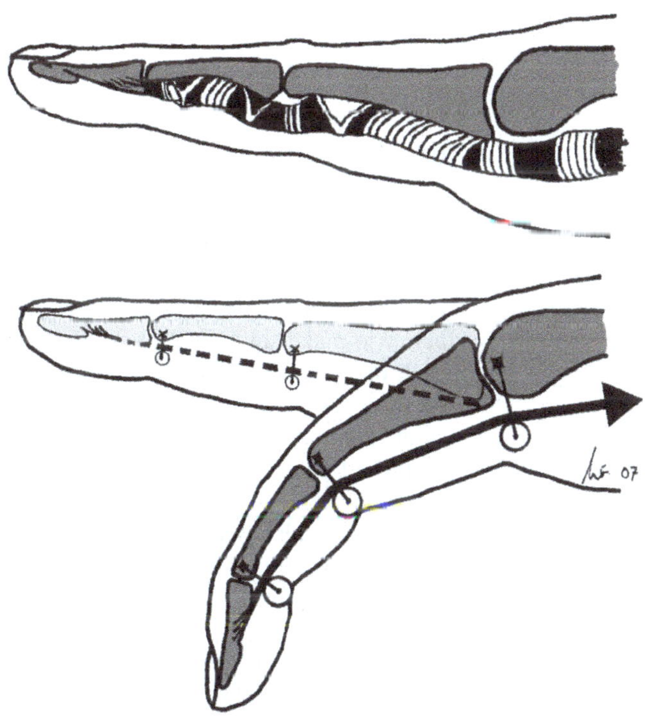

FIGURE 2.7 Ring-like and cruciform ligament attachments keep the flexor tendon in place at the bones and allow it to bend the finger by pulling.

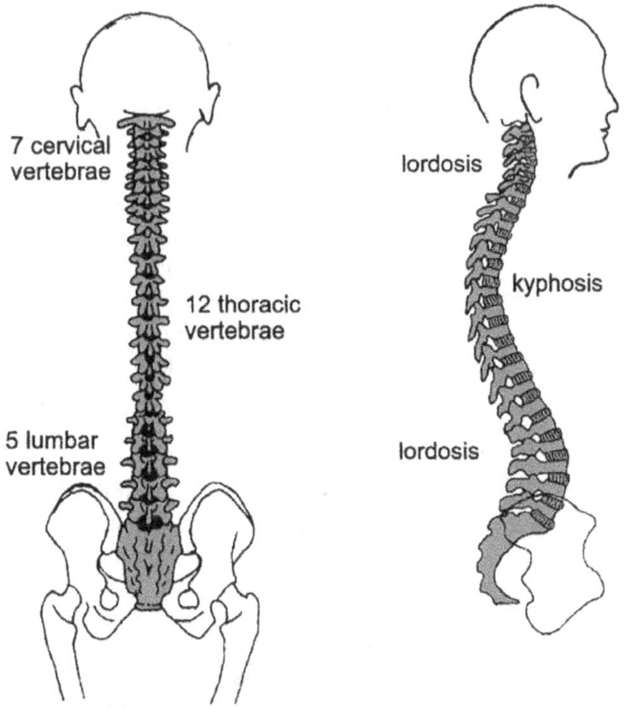

FIGURE 2.8 Vertebrae and curvature of the spinal column.

spelled disk), a tough cushion enclosed by a strong fibrous ring. The disc, filled with a viscous fluid, serves as an elastic pad that absorbs shocks and allows the vertebrae above and below to make small changes in their angles of tilt relative to each other. However, these small angular displacements add up throughout the stack of vertebrae and discs, allowing the spinal column to provide considerable twisting and bending in both the fore–aft and left–right directions. Figure 2.9 shows, schematically, a disc between the main bodies of two adjacent vertebrae. It also depicts the spinous processes of the vertebrae and the spinal cord with its emanating extensions, often called nerve roots.

Spinal vertebra

Figure 2.10 is a perspective sketch of a vertebra: behind the main vertebral body (VB) is an opening, the vertebral foramen, which protects the spinal cord that runs vertically within the foramen. The vertebra's bony structure has three protrusions: the spinous process (SP) points backward, and the two transverse processes point left and right. As Figure 2.11 shows,

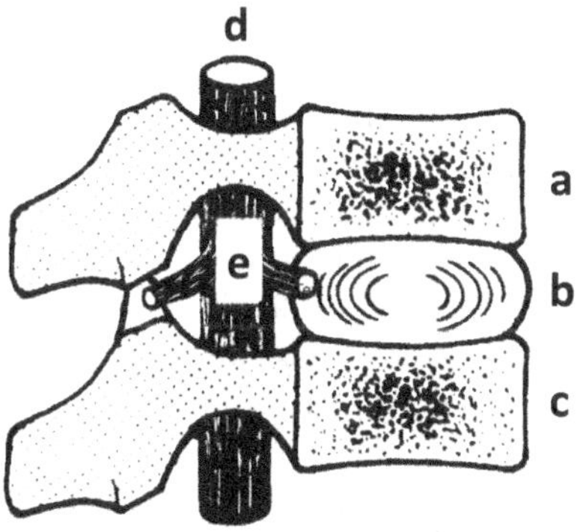

FIGURE 2.9 View, from the right, of the main components of two vertebrae (a and c) with a disc (b) between them; at the rear is the spinal cord (d) with its nerve extensions (e).

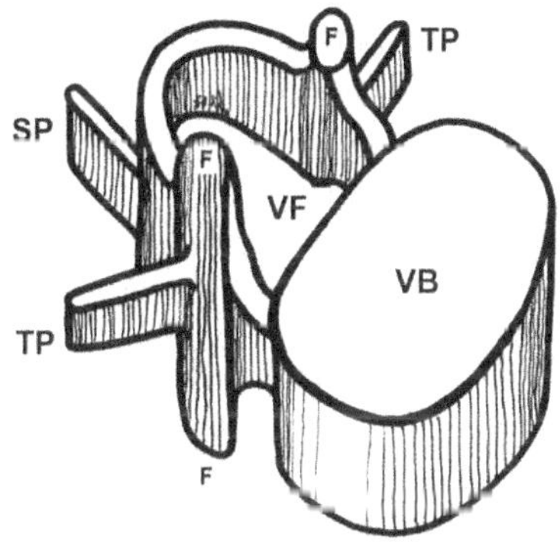

FIGURE 2.10 Basic structure of a vertebra, in an oblique view, with spinous process (SP), two transverse processes (TP), facets (F), vertebral foramen (VF), and vertebral body (VB) shown.

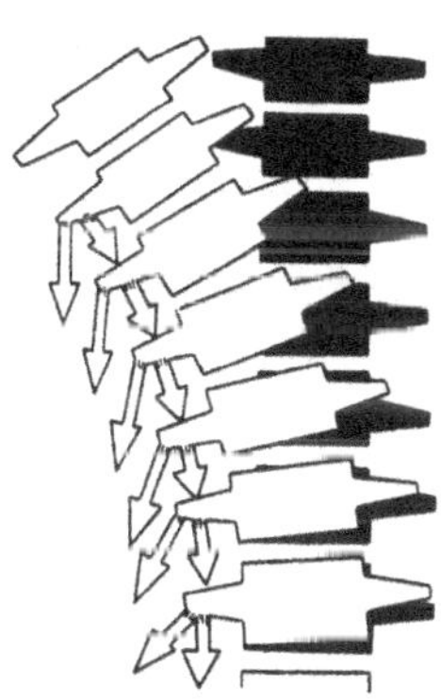

FIGURE 2.11 Forces applied on vertebral processes make the vertebrae tilt. Adapted from Kapandji (1982, more recent edition in 2007).

these protrusions act as lever arms for ligaments and muscles attached to them, which stabilize or bend the spinal column through their coordinated pulls.

Figure 2.12 depicts the five lumbar vertebrae (L1 through L5) atop the sacrum, as seen from the right. Together with Figures 2.9 and 2.10, this schematic illustrates that each vertebra has six bearing surfaces with the vertebrae above and below: the two interfaces at the bottom and top of the main body (VB in Figure 2.10) have cushioning discs, but there are no cushions at the facets (F in Figure 2.10), which provide four bony articulations at the two superior and two inferior surfaces. In this way, the spinal joints provide complex yet limited mobility.

Limited trunk flexibility

Though only small displacements are possible between adjacent bones of the spine due to the deformation of the intermediate discs, the head and neck can twist and bend (fore–aft and side-to-side) within fairly narrow limits, utilizing the upper section of the spinal column. Larger movements of

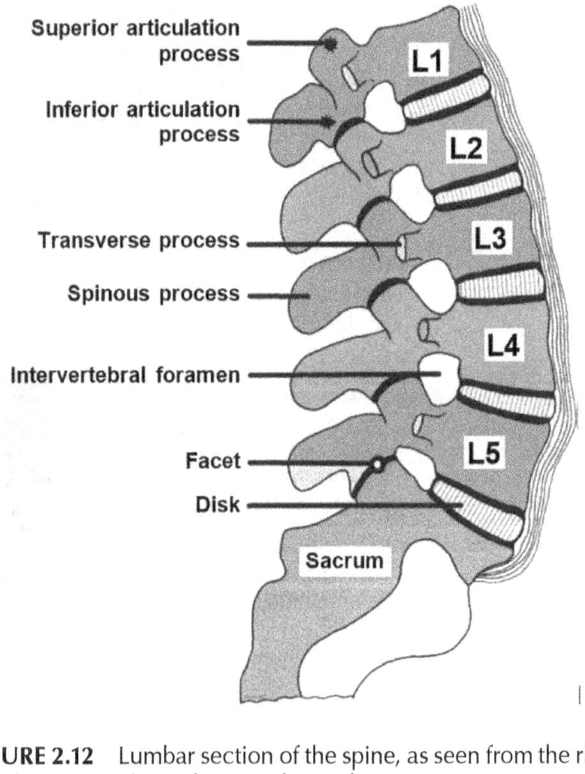

FIGURE 2.12 Lumbar section of the spine, as seen from the right, with bearing surfaces drawn in heavy lines.

the trunk, both in bending and twisting, occur primarily in the lower parts of the spine, especially in the lumbar section. The spine's ability to transmit large forces, primarily in compression, is remarkable. However, overloading and/or repetitive motions can cause damage, particularly to the spinal discs. The lumbar region* frequently experiences discomfort, pain, and injury because it must transmit substantial forces and torques to and from the upper body. Low back pain is common, especially among people who handle loads (see Chapter 21).

Skeletal adaptation for pregnancy

In pregnant women, increasing bulk in the abdomen shifts the center of mass forward; to counteract this, the mother-to-be may lean back slightly, which further curves the lumbar spine. Increasing the lumbar curve can create complications: vertebrae are more likely to slip against each other, potentially causing back pain or injury. The female spine* has several features that help reduce the risk of problems by distributing strain over a wider area. For example, in women, the curve in the lower back mainly includes three vertebrae, whereas it spans only two vertebrae in men. Furthermore, the facet joints are larger in women than in men, and the joints are oriented at a slightly different angle, allowing them to resist higher forces and better brace the vertebrae against slipping.

2.3 Designing for mobility

Given the many ways we can move, it is convenient to describe the rotations in each body articulation separately. Figure 2.13 depicts body joint displacements and their traditional descriptions for the arm and wrist.

Actual mobility

As we move about, we usually combine movements across several body joints to generate the flexibility needed. Table 2.1 lists mobility measurements taken from a group of physical education students, including 100 females and 100 males. Many ordinary people are likely to be less flexible, but, of course, actual ranges of motion depend greatly on health, fitness, training, skill, age, and any associated disabilities. Furthermore, differing measuring techniques and instructions provided to clients can cause significant diversity in the reported results of mobility ranges among groups of people*.

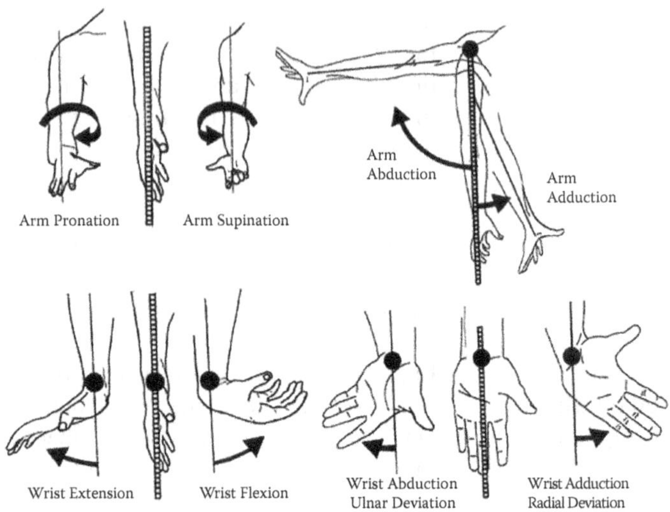

FIGURE 2.13 Arm and wrist displacements.

2.4 Designing workspaces

In our everyday activities, whether at work or during leisure, we like to move and walk about; changes in posture are essential for our well-being. Being forced to stand, sit, or lie still for long periods of time is difficult to tolerate.

Preferred motions

Preferred motion ranges of the feet or hands depend on habits and skills, on workplace layout, and on the dominant task requirements* such as strength, speed, accuracy, or vision. Clearly, there is not just one work envelope for every person or task; rather, different people prefer to do different tasks in different workspaces.

Everyday motion ranges

"Convenient" mobility is somewhere within the overall motion range, as listed in Table 2.2, but is not always in the middle of the range of motion; often, motions take place close to a joint's flexibility limits. For example, a person walking around on a job site or standing often has the hips and knees almost fully extended, near 180°. However, when sitting in a chair, the angles of both hips and knees cluster around 90°.

Who uses chairs?

In many parts of the world, working habits differ. For example, work chairs are widely used in so-called Western countries, such as in Europe, North America, and Australia/New

Table 2.1 Mobility of 100 female and 100 male students of physical education

Joint	Movement (in degrees)	5th percentile		50th percentile		95th percentile		Difference* at 50th percentile
		Females	Males	Females	Males	Females	Males	
Neck	Ventral flexion	34.0	25.0	51.5	43.0	69.0	60.0	Fem+8.5
	Dorsal flexion	47.5	38.0	70.5	56.5	93.5	74.0	Fem+14.0
	Right rotation	67.0	56.0	81.0	74.0	95.0	85.0	Fem+7.0
	Left rotation	64.0	67.5	77.0	77.0	90.0	85.0	None
Shoulder	Flexion	169.5	161.0	184.5	178.0	199.5	193.5	Fem+6.5
	Extension	47.0	41.5	66.0	57.5	85.0	76.0	Fem+8.5
	Adduction	37.5	36.0	52.5	50.5	67.5	63.0	NS
	Abduction	106.0	106.0	122.5	123.5	139.0	140.0	NS
	Medial rotation	94.0	68.5	110.5	95.0	127.0	114.0	Fem+15.5
	Lateral rotation 1	9.5	16.0	37.0	31.5	54.5	46.0	Fem+5.5
Elbow	Flexion	135.5	122.5	148.0	138.0	160.5	150.0	Fem+10.0
Wrist	Supination	87.0	86.0	108.5	107.5	130.0	135.0	NS
	Pronation	63.0	42.5	81.0	65.0	99.0	86.5	Fem+16.0
	Extension	56.5	47.0	72.0	62.0	87.5	76.0	Fem+10.0
	Flexion	53.5	50.5	71.5	67.5	89.5	85.0	Fem+4.0
	Adduction	16.5	14.0	26.5	22.0	36.5	30.0	Fem+4.5
	Abduction	19.0	22.0	28.0	30.5	37.0	40.0	Male+2.5
Hip	Flexion	103.0	95.0	125.0	109.5	147.0	130.0	Fem+15.5
	Flexion	27.0	15.5	38.5	26.0	50.0	39.0	Fem+12.5

(Continued)

Table 2.1 (Continued) Mobility of 100 female and 100 male students of physical education

Joint	Movement (in degrees)	5th percentile		50th percentile		95th percentile		Difference* at 50th percentile
		Females	Males	Females	Males	Females	Males	
	Adduction	47.0	38.0	66.0	59.0	85.0	81.0	Fem+7.0
	Abduction	30.5	30.5	44.5	46.0	58.5	62.5	Ns
	Medial rotation (prone)	29.0	21.5	45.5	33.0	62.0	46.0	Fem+12.5
	Lateral rotation (prone)	20.5	18.0	32.0	28.0	43.5	43.0	Fem+4.0
	Medial rotation (sitting)	20.5	18.0	33.0	26.5	45.5	37.0	Fem+6.5
	Lateral rotation (sitting)	99.5	87.0	113.5	103.5	127.5	122.0	Fem+10.0
Knee	Flexion (standing)	116.0	99.5	130.0	117.0	144.0	130.0	Fem+13.0
	Flexion (prone)	18.5	14.5	31.5	23.0	44.5	35.0	Fem+8.5
	Medial rotation	28.5	21.0	43.5	33.5	58.5	48.0	Fem+10.0
	Lateral rotation	13.0	18.0	23.0	29.0	33.0	34.0	Male+6.0
Ankle	Flexion	30.5	21.0	41.0	35.5	51.5	51.5	Fem+5.5
	Extension	13.0	15.0	23.5	25.0	34.0	38.0	Ns
	Adduction	11.5	11.0	24.0	19.0	36.5	30.0	Fem+5.0
	Abduction	103.0	95.0	125.0	109.5	147.0	130.0	Fem+15.5

Source: Adapted from data from Houy (1983) and Staff (1983).

* Differences are listed only if statistically significant ($\alpha<0.5$).

ns, not statistically significant.

Table 2.2 Mobility ranges at work

Body joints/parts	Angles/positions
Shoulder	Mostly mid-range; upper arm often hanging down
Elbow	Mostly mid-range, at about 90°
Wrist	Mostly mid-range; about straight (unflexed) position
Neck/head	Mostly mid-range; about straight position
Back	Near complete stretch; about erect position
Hip (side view)	When walking or standing: near complete stretch, at about 180°
	When sitting: mostly mid-range, at about 90°
Knee	When walking or standing: near-extreme stretch, near 180°
	When sitting: mostly mid-range, at about 90°

Source: Adapted from Table 11.13 of Kroemer, Kroemer, and Kroemer-Elbert (2020).

FIGURE 2.14 Work postures without a chair.

Zealand. In other regions, working people often hover close to the ground, as sketched in Figure 2.14. Furthermore, work tasks may vary significantly by region and require special skills and motions. Unfortunately, few ergonomic recommendations for "non-Western" work conditions are readily available.

Reach envelopes In general, preferred work areas for the hands and feet are in front of the trunk of the body, within curved envelopes that reflect the mobility of the arms and legs. Thus, it is convenient to describe these reach envelopes for hands and feet as partial spheres around the temporary locations of body segments.

In reality, however, several parts of the body may move simultaneously, and force or vision requirements may impose specific constraints.

Hand workspace Movement of the forearm at the elbow joint, often close to a 90° bend, combined with rotation at the shoulder, determines the envelope for fine manipulation and other light work; this is the shaded space in Figure 2.15 (and in Figure 11.11). Elbow supports (as shown in Figure 11.12) may facilitate precision work. Full movements of the upper and forearms allow handwork from about hip to well above shoulder height in a wide space, mostly in front of the body. The main determinant of the proper height of the workspace is the height of the person's elbow, whether standing or seated; see Figure 2.16 (same as Figure 20.2).

Easy foot actions When a person sits, foot activities usually occur in a shallow area below the knees, as shaded in Figure 2.17. The operator can cover this area with small displacements in the knee and ankle joints, both usually flexed near 90°. This is the best space for foot activities that require exertion of only small force and energy, usually by pushing down with the whole foot, or just the heel or toes.

Strong foot push In contrast, foot controls that require substantial forward-push force from the seated operator (such as the brake pedal in a motorized vehicle) must be positioned higher and farther away

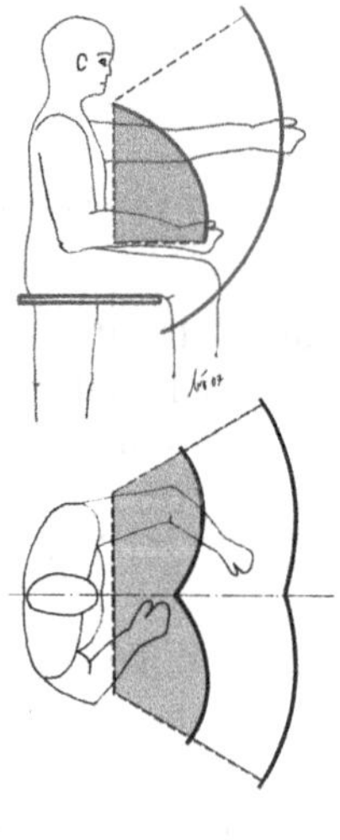

FIGURE 2.15 Workspace of the hands, from side and top views.

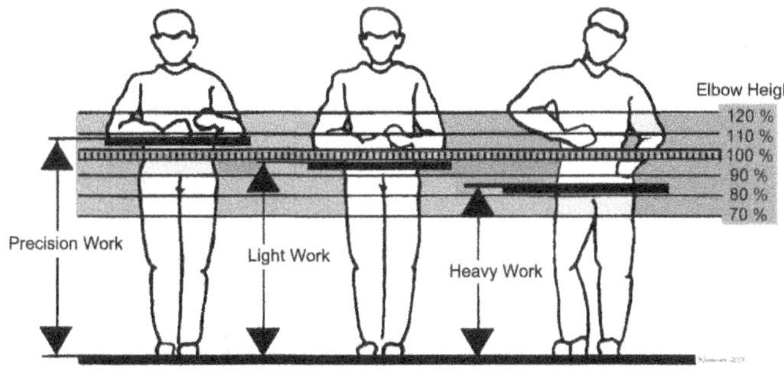

FIGURE 2.16 Height of work surfaces for handwork, as percentages of elbow height.

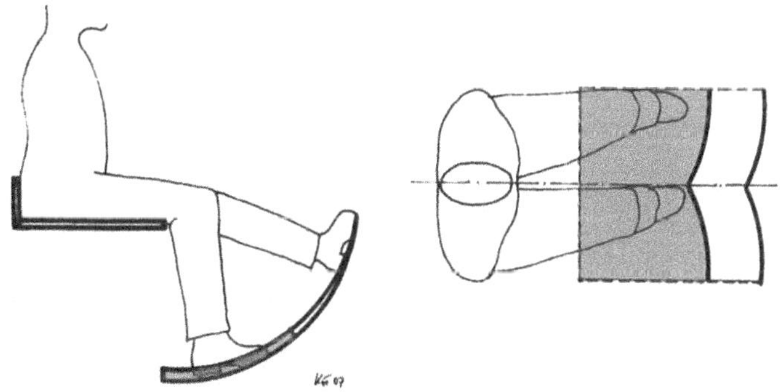

FIGURE 2.17 Workspace of the feet, from side and top views.

so that the knee angle is around 150°–160°. However, the designer needs to pay attention to two factors: the nearly fully extended leg does not permit much displacement of the foot, and a strong backrest is necessary to counteract the pushing force exerted on the pedal.

No standing on one foot

A seated operator may operate pedals or other foot controls, even with both feet simultaneously, in a suitable workspace. A standing person can only use one foot at a time (as shown in Figure 2.18), which requires balancing the entire body on the other foot. (Note also that while birds can stand on one foot for a long time, humans cannot.) The use of both feet to operate foot controls necessitates appropriate seating.

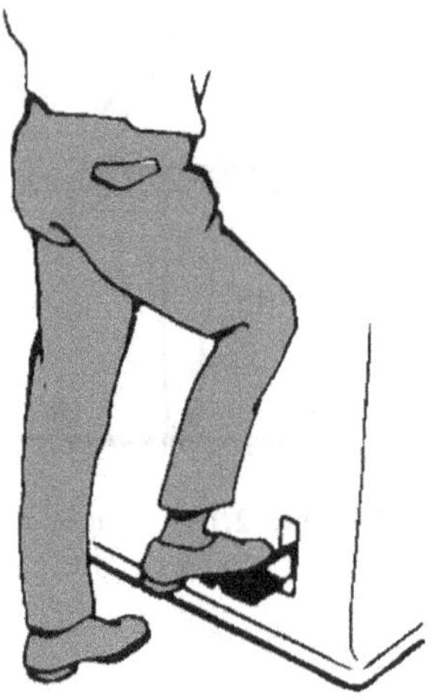

FIGURE 2.18 Forced to stand on one foot.

Summary

Humans cannot maintain any given body position for long periods; we need to move our bodies. Abolishing the idea of fixed postures and instead designing for suitable ranges of motion is not difficult and is often necessary. Our bodies' joints have known movement limits within which we prefer certain workspaces for tasks that require strength, fast motions, or precise and enduring work.

Steps to fit the human

Step 1: Determine which specific task must be done.

Step 2: Decide how to perform the work, using the hand or foot, seated or standing, in a fixed position or in motion, and where within the movement space.

Step 3: Design, then test. Modify and iterate as necessary.

Notes

The text contains markers, *, to indicate specific references and comments, which follow.

2.1 Basics of work and motion

Design for movement: Delleman, Haslegrave, and Chaffin (2004), Freivalds (2011), Kroemer (2006), Kroemer, Kroemer, and Kroemer (2018, 2020), Marras and Karwowski (2006a, b).

Repetitive movement requirements: Armstrong (2006), Arndt and Putz-Anderson (2006), Kroemer (2001), Violante, Armstrong, Kilbom (2000), Walji (2007), Ramazzini (1713) (with translation from 2001).

Low back problems: Marras (2008), Snook (2000).

2.2 Body joints and appendages

Articulations of the body: Chaffin, Andersson, and Martin (2006), Kapandji (2008); Oezkaya, Nordin, Goldsheyder et al. (2012), Walji (2007).

The hand, carpal tunnel: Freivalds (2006), Hughes and An (2007), Kroemer (2008); Walji (2007).

Vertebra: there are 33 vertebrae in total (7 cervical, 12 thoracic, 5 lumbar, 5 sacral and 4 coccygeal), however since the sacral and coccygeal (also called tailbone) vertebrae are fused together, they can also each be counted as one.

The spine, low back problems: Deyo and Weinstein (2001), Marras (2008), Violante, Armstrong and Kilbom (2000), Snook (2000).

Features of female and male spines: already present in Australopithecus fossils, as recorded in Whitcome, Shapiro, and Lieberman (2007).

2.3 Designing for mobility

Diversity in mobility measurements: Wu, Siegler, Allard et al. (2002); data on NASA astronauts at http://msis.jsc.nasa.gov/.

2.4 Designing workspaces: *see also Chapter 20*

Dominant strength, speed, accuracy or vision requirements: Kroemer, Kroemer, and Kroemer (2018).

References and further reading

Armstrong TJ (2006). The ACGIH TLV for hand activity work. Chapter 41 in Marras WS, Karwowski W (eds), *The Occupational Ergonomics Handbook: Fundamentals and Assessment Tools for Occupational Ergonomics*, 2nd ed. Boca Raton, FL, CRC Press. https://doi.org/10.1201/9781420003635

Arndt R, Putz-Anderson V (2014). *Cumulative Trauma Disorders.* Boca Raton, FL, CRC Press.

Chaffin DB, Andersson GBJ, Martin BJ (2006). *Occupational Biomechanics*, 4th ed. New York, NY, Wiley.

Delleman NJ, Haslegrave CM, Chaffin DB (2004). *Working Postures and Movements.* Boca Raton, FL, CRC Press. https://doi.org/10.1201/9781482265095

Deyo RA, Weinstein JN (2001). Low back pain. *New England Journal of Medicine*, 344:363–370. https://doi.org/10.1056/NEJM200102013440508

Freivalds A (2006). Upper extremity analysis of the wrist. Chapter 45 in Marras WS, Karwowski W (eds), *The Occupational Ergonomics Handbook: Fundamentals and Assessment Tools for Occupational Ergonomics*, 2nd ed. Boca Raton, FL, CRC Press. https://doi.org/10.1201/9781420003635

Freivalds A (2011), *Biomechanics of the Upper Limbs: Mechanics, Modelling and Musculoskeletal Injuries*, 2nd ed. Boca Raton, FL, CRC Press. https://doi.org/10.1201/b11547

Houy DA (1983) Range of joint motion in college males. *Proceedings of the 27th annual meeting Human Factors Society*, pp. 374–378. Santa Monica, CA, Human Factors Society.

Hughes RE, An KN (2007). Biomechanical models of the hand, wrist, and elbow in ergonomics. Chapter 14 in Kumar S (ed), *Biomechanics in Ergonomics*, 2nd ed. Boca Raton, FL, CRC Press.

Kapandji IA (2007). *The Physiology of the Joints*, 6th ed. Edinburgh, UK, Churchill Livingstone.

Kroemer KHE (2001). Keyboards and keying: an annotated bibliography of the literature from 1878 to 1999. *Universal Access in the Information Society UAIS*, 1(2):99–160. https://doi.org/10.1007/s102090100012

Kroemer KHE (2006). *"Extra-Ordinary" Ergonomics: How to Accommodate Small and Big Persons, the Disabled and Elderly, Expectant Mothers and Children.* Boca Raton, FL, CRC Press. https://doi.org/10.1201/9780203025246

Kroemer KHE (2008). Anthropometry and biomechanics: anthromechanics. Chapter 2 in Kumar S (eds), *Biomechanics in Ergonomics*, 2nd ed. Boca Raton, FL, CRC Press. https://doi.org/10.1201/9780849379093

Kroemer AD, Kroemer KHE (2017). *Office Ergonomics: Ease and Efficiency at Work*, 2nd ed. Boca Raton, FL, CRC Press. https://doi.org/10.1201/9781315368603

Kroemer Elbert K, Kroemer HB, Kroemer Hoffman AD (2018). *Ergonomics: How to Design for Ease and Efficiency*, 3rd ed. London, UK, Elsevier. https://doi.org/10.1016/C2016-0-02516-0

Kroemer KHE, Kroemer HJ, Kroemer-Elbert KE (2020). *Engineering Physiology: Bases of Human Factors Engineering/Ergonomics*, 5th ed. Heidelberg, Springer. https://doi.org/10.1007/978-3-030-40627-1

Marras WS (2008). *The Working Back: A Systems View.* New York, Wiley & Sons. https://doi.org/10.1002/9780470258576

Marras WS, Karwowski W (2006a). *The Occupational Ergonomics Handbook: Fundamentals and Assessment Tools for Occupational Ergonomics*, 2nd ed. Boca Raton, FL, CRC Press. https://doi.org/10.1201/9781420003635

Marras WS, Karwowski W (2006b). *The Occupational Ergonomics Handbook: Interventions, Controls, and Applications in Occupational Ergonomics*, 2nd ed. Boca Raton, FL, CRC Press. https://doi.org/10.1201/9781420003642

NASA-STD-3000 (2023). The Man-System Integration Standards. Archived at https://msis.jsc.nasa.gov/

Oezkaya N, Leger D, Goldsheyder D, Nordin M (2017). *Fundamentals of Biomechanics: Equilibrium, Motion, and Deformation*, 4th ed. Switzerland, Springer. https://doi.org/10.1007/978-3-319-44738-4

Putz-Anderson V (1988). *Cumulative Trauma Disorders*. Boca Raton, FL, CRC Press. https://doi.org/10.1201/9781315140704

Ramazzini B (1713). *De morbis artificum diatriba* [diseases of workers]. Translated in *American Journal of Public Health*, 2001, 91(9):1380–1382. https://doi.org/10.2105/AJPH.91.9.1380

Snook SH (2000). Back risk factors: an overview. Chapter 11 in Violante F, Armstrong T, Kilbom A (eds), *Occupational Ergonomics. Work Related Musculoskeletal Disorders of the Upper Limb and Back*. London, UK, Taylor & Francis. https://doi.org/10.1201/9781482268287

Staff KR (1983). *A comparison of range of joint mobility in college females and males*. Master's Thesis, Industrial Engineering, A&M University, College Station, TX

Violante F, Armstrong T, Kilbom A (2000). *Occupational Ergonomics. Work Related Musculoskeletal Disorders of the Upper Limb and Back*. London, UK, Taylor & Francis. https://doi.org/10.1201/9781482268287

Walji AH (2007). Functional anatomy of the upper limb (extremity). Chapter 8 in Kumar S (eds), *Biomechanics in Ergonomics*, 2nd ed. Boca Raton, FL, CRC Press. https://doi.org/10.1201/9780849379093

Whitcome KK, Shapiro LJ, Lieberman DE (2007). Fetal load and the evolution of lumbar lordosis in bipedal hominins. *Nature*, 450(7172):1075–1078. https://doi.org/10.1038/nature06342. Corrigendum in *Nature* 487, 128 (2012). https://doi.org/10.1038/nature11168

Wu G, Siegler S, Allard P, Kirtley C, Leardini A, Rosenbaum D, Whittle M, D'Lima DD, Cristofolini L, Witte H, Schmid O, Stokes I (2002). Standardization and terminology committee of the international society of biomechanics. ISB recommendation on definitions of joint coordinate system of various joints for the reporting of human joint motion – part I: ankle, hip, and spine. *Journal of Biomechanics*, 35(4):543–548. https://doi.org/10.1016/s0021-9290(01)00222-6

CHAPTER THREE

Muscular work

Muscles

Muscles are our body's natural engines. A muscle is attached to a bone and extends across one or two body joints to another bone. When contracting, the muscle pulls on the body's internal bone framework. That pull can change the angle between bones, setting body segments into motion or stabilizing their positions. We need to understand how our muscles work to fit the task to the human.

Basic physics science explains the relationships between common strength measurements:

- *Force* (in Newton, *N*) divided by the transmitting cross-sectional area is *stress* (usually in N/mm^2), which can be a *tension stress* when being pulled or *compression stress* or *pressure* when being pushed.
- Force multiplied by lever arm is *moment* or *torque* (usually in Newton-meter, *Nm*).
- The integral of force and distance traveled (meaning, the area under the curve when force is considered to act over a distance) is *work* (usually in joule, *J*, or *Nm*)
- Work divided by time is *power* (usually in *watt W* or joule per second *J/s* or *Nm/s*).

Kinds of muscles

The human body has three kinds of muscle: smooth, cardiac, and skeletal. Smooth muscle is found in the walls of hollow organs such as blood vessels (also functioning as sphincters that control the opening of blood vessels), the bladder, and the uterus. Cardiac muscle is located in the heart. Skeletal muscle is what enables our bodies to move. Smooth and cardiac muscles have characteristics that differ from those of skeletal muscle*, which is the focus of the following text.

DOI: 10.1201/9781032674575-5

3.1 Basics of physiology

Muscle components

The human body has several hundred skeletal muscles, known by their Latin names; for example, the "biceps muscle" (*m. biceps brachii*) helps flex the elbow by pulling the upper and lower arms together. Every muscle consists of bundles of fibers, as illustrated in Figure 3.1. Tissue embedded with blood vessels and nerves wraps around and permeates the muscle. At the muscle ends, the wrapping tissues combine to form tendons, which act like cables that extend and attach to bones*.

Power to the muscle

Thousands of individual muscle fibers (cells) run in bundles, essentially parallel, along the length of the muscle. Inside the bundles are hundreds of mitochondria, which serve as the muscle's "power factories." These cellular organelles are specialized to liberate chemically stored energy (in the form of adenosine triphosphate, ATP, and creatine phosphate, CP)*. This energy enables the muscle to contract. Blood supply is essential for muscle function because it provides energy and oxygen and removes the byproducts of metabolic processes (see Chapter 10) such as heat, water, and carbon dioxide. Nervous signals trigger the actions of the muscle and control their intensity.

Contractile microstructure

Contraction is the only *active* action that a muscle can undertake. Elongated protein molecules, arranged side by side as thread-like filaments, temporarily adhere to each other when triggered by nervous signals. One kind of filament, called myosin, has projections that extend (see Figure 3.2) toward the surrounding

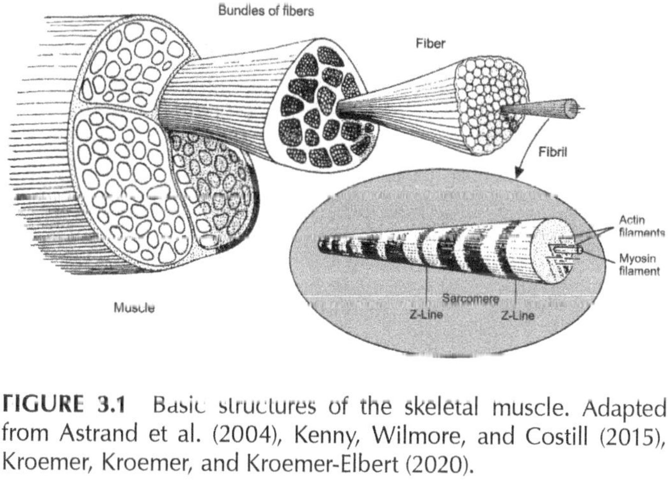

FIGURE 3.1 Basic structures of the skeletal muscle. Adapted from Astrand et al. (2004), Kenny, Wilmore, and Costill (2015), Kroemer, Kroemer, and Kroemer-Elbert (2020).

FIGURE 3.2 Diagram of the club-like myosin with its projections that can form cross-bridges with actin filaments. Adapted from Herzog (2007).

type of filaments, called actin. When the muscle is at rest, the projections remain relaxed and do not transmit forces between the actin and myosin filaments. However, when the muscle receives a signal to contract, the myosin's projections activate and attach to the adjacent actins (forming cross-bridges) and, in ratcheting motions, attempt to pull the actin rods along the myosin. If that pull is stronger than any opposing force, the ends of the actin filaments slide. This shortens the sarcomere (see the following section) and thus shortens the muscle. If the pull is equal to an opposing force, the length of the muscle remains unchanged; this is called an *isometric contraction*. If the contraction pull is weaker than an opposing force, the muscle can lengthen (despite contracting).

Sarcomere

The sarcomere is the smallest functional unit of the muscle and appears as alternating bands of light and dark. The dark bands are called z-lines (or z-discs, where the letter z stands for the German word *zwischen*, between), and define the length of the sarcomere. At rest, a sarcomere has a length of approximately 250 angstroms (also spelled ångström, Å). Thus, since 1 Å$=10^{-10}$m, about 40,000 sarcomeres can lie in series within 1 mm of muscle fiber length. In full contraction, the length of a sarcomere—and hence the length of a muscle—can shorten to about 60% of its resting length, as depicted schematically in Figure 3.3.

Striated skeletal muscle

This shortening (or stretching) of muscle is visible under the microscope when examining the z-lines. These lines appear as dark stripes that run across the muscle fibers, where the fixed ends of actin filaments connect. When a contraction shortens the muscle, the z-lines are pulled closer to each other. The striping, caused by the z-lines and by other bands crossing skeletal muscle, provides an alternate name for skeletal muscle: striated (striped) muscle.

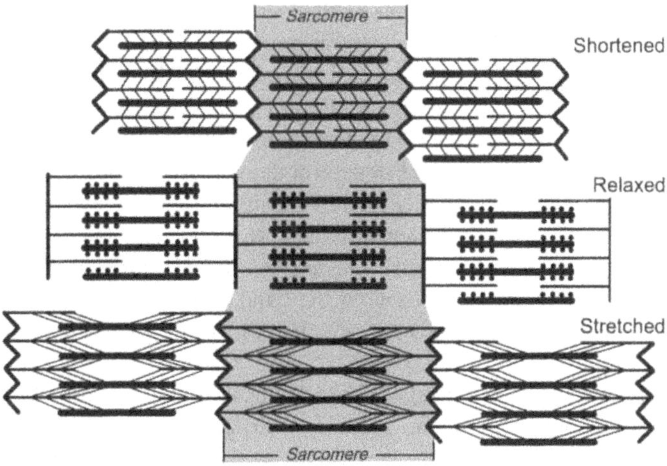

FIGURE 3.3 Patterns of a sarcomere when shortened, relaxed (resting length), or stretched. Adapted from Kroemer, Kroemer, and Kroemer-Elbert (2020).

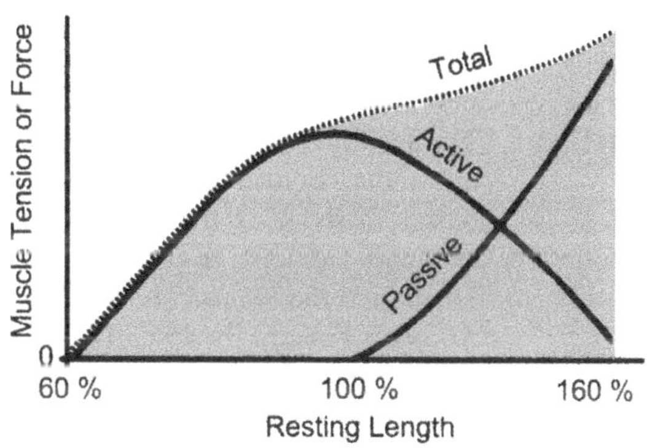

FIGURE 3.4 Active, passive, and combined tensions at different muscle lengths. Adapted from Kroemer, Kroemer, and Kroemer-Elbert (2020).

Strength of skeletal muscle The greater the number of actin–myosin bundles that exist side by side in a muscle, the more strength it can exert. Hence, the strength of a muscle is proportional to its cross-sectional area. The muscle's contractile force is largest at about resting length (around 100% in Figure 3.4) and decreases as the muscle shortens (60%–100% of resting length in Figure 3.4).

Muscle tension A strong force opposing muscle contraction (such as due to gravity or stemming from the action of another muscle) can cause the muscle to lengthen beyond its relaxed length. The farther the stretch, the more the muscle resists: the overall tension inside the muscle is the result of adding the active and passive tensions (as illustrated in Figure 3.4 beyond 100% resting length). That elongation can reach about 160% of resting length, after which point structural damage to the muscle occurs, which may eventually lead to tearing or breaking of the muscle. By experience, we learn to use the "pre-stretch trick" of combining active and passive tension to increase our muscular power; for example, by moving the hand and arm back behind the shoulder before throwing an object forward to obtain greater force.

Muscle pull around the elbow

Skeletal muscles usually appear in pairs: one muscle moves a bone around an articulation in one direction while the other muscle moves it in the opposite direction. The elbow provides a good example: here, the pull of the biceps muscle reduces the elbow angle, while the pull of the triceps muscle increases it, as illustrated in Figure 3.5. As the name triceps indicates, this muscle has three "heads" for attachment: one reaches across the shoulder joint to the shoulder blade, while the other two heads attach to the humerus (the long bone in the upper arm). The biceps has two heads for attachment.

Co-contraction

The seemingly simple muscle system around the elbow is actually quite complex because two additional muscles assist the biceps in reducing the elbow angle: the *m. brachialis*, which connects the upper arm bone (humerus) to one in the forearm (ulna), and the *m. brachioradialis*, which links the humerus to the other bone in the forearm (radius). When muscles contract simultaneously to pull in the same direction, their combined effort exerts an overall force on the attached bone(s) and across the affected joint. Co-contraction of two opposing muscles (such as biceps and triceps) generates a coalition of forces that helps control and stabilize segment movement, even if there is no net movement across the joint.

3.2 Designing for dynamic and static efforts and strength

As discussed in Chapter 2, the human body functions best in motion rather than when immobile. However, it is easier to assess certain traits of the body when it remains motionless in a defined and controlled position. Thus, for individuals holding still, body sizes are simple to measure (see Chapter 1); likewise,

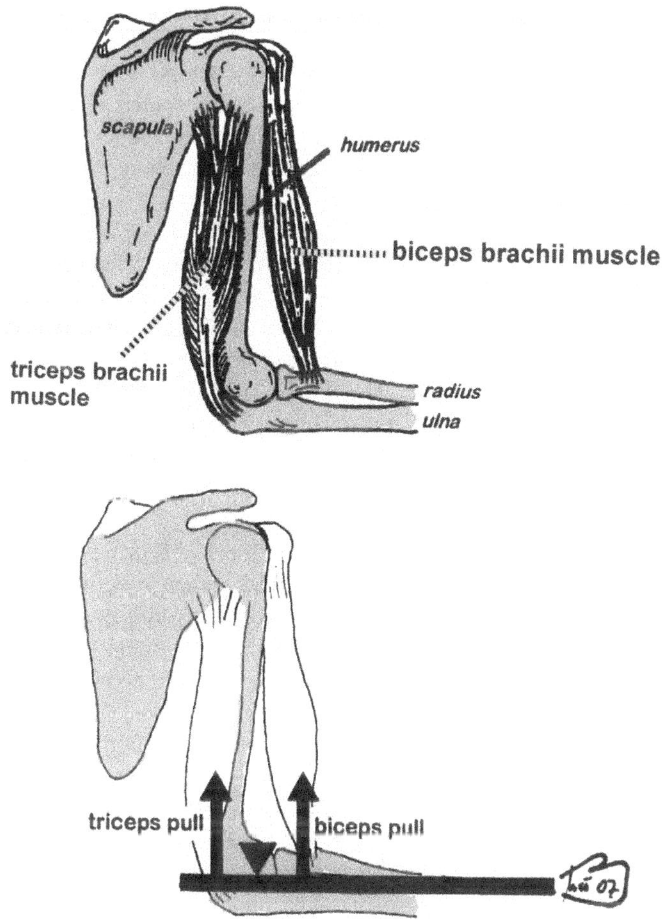

FIGURE 3.5 Triceps and biceps muscles control elbow extension and flexion. Adapted from Kroemer, Kroemer, and Kroemer-Elbert (2020).

it is relatively easy to assess muscular strength when muscle lengths remain constant.

Contraction and motion

There is much interplay between the tension (force, energy, power) a muscle can develop and how it moves while contracting. The simplest case is no net motion: while its myofilaments attempt to contract, the overall length of the muscle does not change (and the involved body segments stay in place). While physiologists call this condition *isometric* (no change in length), to physicists, it is *static* (not moving). If the muscle succeeds in shortening its length, this is a concentric contraction.

Such shortening can occur only if the muscle is stronger than any opposing force. However, if the opposing force overwhelms the muscle while it attempts to contract, the muscle becomes lengthened. This is an eccentric contraction. To reduce confusion, it is often better to use the terms activation, effort, or exertion, since the term contraction does not necessarily mean that the muscle shortens.

Dynamics

Dynamic muscular efforts are more complicated to describe than static contractions, and they are more difficult to control in experiments. In dynamic activities, muscle length changes, and therefore involved body segments move, resulting in displacement. The time derivatives of displacement (velocity, acceleration, and jerk) are important* for both the internal muscular effort and its external effect. For example, change in velocity determines impact and force, as per Newton's second law.

Anthromechanics

The application of physics principles to biological events is called biomechanics; when specifically applied to the human body, it is referred to as *anthromechanics*. Using knowledge gained in overlapping traditional disciplines is typical in ergonomics/human factors engineering, as exemplified in this book.

Athletics and sports

We can measure muscle strength in different ways. One kind of assessment relies on measurements of athletic performance, which is common in sports events characterized by immense and often explosive efforts, usually conducted in competition with other people selected for their abilities and competitive spirits. The outcome is often scored in terms of records achieved, weights lifted, distances covered, and matches won. However, these "experimental conditions" are impossible to control and reproduce in the laboratory. Furthermore, these performances have little bearing on everyday work requirements.

Work requirements

At work, extreme efforts are rarely required (or recommended, due to the risk of injury). Instead, moderate muscular labor is usually in demand throughout the working time, such as 8 hours a day. These efforts are often intermittent and may be repeated several times during the work shift. The individuals performing such work commonly include people of different sizes, physical capabilities, and motivations. For various reasons, the required muscular exertions are usually at a fairly low level, enabling all workers to manage them over the duration of the work shift.

Controlled strength tests

For jobs that require physical strength, it may be important to determine work-related muscular capabilities. Strength tests are routinely conducted in a controlled laboratory environment, where subjects are asked to perform specific exertions. This may involve elbow flexion, where the biceps muscle is predominant, or a whole-body lift that engages multiple muscles. These tests follow carefully planned experimental protocols*. The peak force (or torque, moment) observed during the test performance, which typically 3–5 seconds, is commonly recorded as the strength measurement. While this laboratory approach does not fully reflect actual work conditions, such as those found in industry or agriculture, the controlled procedure allows for standardization and provides expected test–retest reliability.

Testing muscle strength

The strength of a muscle can be measured objectively and reliably when removed from the body, as has been done with animal models. A less invasive approach is to instrument an intact muscle *in situ* (in its natural position within the body) with electrodes to capture excitation signals from the nerves. Such tests have demonstrated that the strength of exertion depends on the muscle length at which the exertion occurs. Figure 3.4 illustrates how active and passive tension within the muscle changes with its length. The actual length at which a muscle performs depends on the body posture at that moment; for example, the strength of elbow extension depends on the angle between the upper arm and forearm, which determines the length of the triceps muscle. Additionally, the strength capabilities of muscles also depend on physiological factors, including muscle size (thickness), shape (pennation), and fiber composition (fast or slow twitch), as well as psychological factors such as mental fatigue and motivation.

Maximum voluntary contraction

Clearly, it is not permissible to measure human muscle strength by forcing muscle fibers to perform a maximal effort, as this is the point at which damage might occur. Therefore, all traditional strength measures rely on subjects "giving their best effort"; the literature refers to this as MVC, or maximum voluntary contraction. Subjects are regularly reminded not to hurt themselves during the test. This means that all maximal strength data reported in the literature depend on the subjects' motivation, which, in turn, is influenced by various physiological, psychological, and environmental conditions present at the time of measurement.

Influencing test scores

The actual results of strength tests depend on many possible variables. These include the age and gender of the subject; health, nutritional and hydration status; fitness, training, skill, and experience; rewards and possible drawbacks such as pain or injury; and, of course, the test subject's willingness to participate and perform truly maximal voluntary contractions. The actions of the experimenters, such as instructions, encouragement, and even feedback of results, can easily influence the subject's ego involvement and hence the test scores*. Table 3.1 lists conditions that may increase or decrease muscular performance.

Situational factors

As just mentioned, muscle length, and hence muscle strength, depend on so-called situational variables*: one is obviously posture, and another is external body support. (More on situational variables and factors in Chapter 4.) For example, when you stand upright to exert a horizontal pull with one hand, that pull is weakest when you keep your feet side by side; the pull becomes stronger with one foot placed forward, and strongest if you can grasp an external structure with your free hand that provides support. Another example is pushing with a foot on a pedal while sitting, such as when operating the brake pedal in a vehicle: that push becomes strongest if we can extend the leg almost fully while bracing ourselves against the backrest of the seat.

Test instruments

To date, there exists no practical non-invasive technique for measuring the strength of a single muscle in the living human body. Instead, external gauges are used which, as a rule, record the combined output of several muscles. Their designs and usage

Table 3.1 Circumstances likely to increase (+) or decrease (–) maximal muscular performance

Circumstances	Likely effect
Feedback of results to subject	+
Instructions on how to exert	+
Arousal of ego involvement/aspiration	+
Medications/drugs	+
Subject's outcry, startling noise	+
Hypnotic commands to perform strongly	+
Setting of goals; incentives	+
Competition; contest	?
Spectators	?
Fear of injury	–

Source: Adapted from Kroemer (1974).

practices generate a set of situational conditions that affect the strength test score. By far, most test instruments are of the static type, not of the dynamic sort: the handles of a grip strength tester or of other so-called dynamometers do not flex appreciably while the test subject squeezes, pushes, or pulls. With no motion, muscular exertion is isometric. True dynamometers, which measure while muscle length changes, are much less common; most are expensive, complex, and difficult to use, and hence are seldom employed. The only simple yet effective means for dynamic measurements are objects of constant mass: moving an object, called *isoinertial* testing, supplies information on the ability to push and pull, lift and lower loads. Otherwise, there continues to be a general need for ergonomic research on practical measurements of dynamic strength exertions*.

3.3 Addressing fatigue and recovery

Muscle fatigue

Muscle fatigue* is a subjective experience signaling that one is becoming unable (or unwilling) to continue or repeat a muscular effort. The onset of fatigue depends on the muscle fiber types involved in different kinds of muscular exertions, such as dynamic or static, sustained or brief. The underlying physiological reasons for fatigue are manifold: they include restriction of blood flow through muscle tissue due to intramuscular pressure, depletion of available energy sources (especially ATP and CP; see Chapter 10) in the muscle, formation of lactic acid (a byproduct of the energy conversion process), and other events caused by prior muscular effort. These events especially impede the processes associated with the formation and detachment of cross-bridges within the sarcomere. Furthermore, so-called central fatigue may occur in the nervous control system, one's sense of effort and motivation; however, as often observed in competitive sports, a new emotional drive to perform can temporarily overcome muscular fatigue.

Fatigue ensues when an individual tries to exert a larger effort than she or he is capable of producing. The required exertion can be excessive due to its extreme magnitude, unwarranted duration, or an undue combination of amount and time. The two variables interact with each other: the longer the effort, the smaller the possible amount of exertion, and vice versa. The limiting organ may not be the one doing the actual task, but could be in an auxiliary part of the body: a typical situation is working overhead, requiring the arms to extend upward to reach the work site, with the neck bent backward to see the task,

FIGURE 3.6 Fatiguing overhead work.

as sketched in Figure 3.6. In this case, the shoulder muscles that keep the arms up and the muscles that bend the neck are likely to fatigue earlier than the hand muscles performing the actual intended work task.

Avoiding fatigue

Fortunately, complete recovery from fatiguing work can result from simply discontinuing what caused muscular fatigue and resting regularly. The benefit of the feeling of fatigue is the prevention of serious damage to muscles. Muscle training and skill development can be (to some extent) effective as subjective counteractions to fatigue. However, the incidence of fatigue when performing tasks indicates the need to improve the conditions of the task: the proper ergonomic approach is to eliminate any work requirements that generate fatigue. The human engineering solution is to fit the task to the human, not the other way around.

3.4 Designing for muscle strength

Factors affecting strength

As mentioned, the actual strength (usually defined in terms of amount, direction, and duration) that people can generate depends significantly on individual factors including gender, age, training, fitness, skill, experience, and motivation. Employing muscles in suitable body positions or motions while bracing the body against a support structure are examples of situational factors that also affect the strength exerted by individuals. Figures 3.7 and 3.8 illustrate such effects. The next

	Force plate height(a)	Distance(b)	Force (N) Mean	SD
	Percent of shoulder height		With both hands	
	50	80	664	177
	50	100	772	216
	50	120	780	165
	70	80	716	162
	70	100	731	233
	70	120	820	138
	90	80	625	147
	90	100	678	195
	90	120	863	141
	Percent of shoulder height		With one shoulder	
	60	70	761	172
	60	80	854	177
	60	90	792	141
	70	70	580	110
	70	80	698	124
	70	90	729	140
	80	70	521	130
	80	80	620	129
	80	90	636	133
	Percent of shoulder height		With both hands	
	70	70	623	147
	70	80	688	154
	70	90	586	132
	80	70	545	127
	80	80	543	123
	80	90	533	81
	90	70	433	95
	90	80	448	93
	90	90	485	80

	Force plate height(a)	Distance(b)	Force (N) Mean	SD
	100% of shoulder height	Percent of thumb-tip reach*	With both hands	
		50	581	143
		60	667	160
		70	981	271
		80	1285	398
		90	980	302
		100	646	254
			With the preferred hand	
		50	262	67
		60	298	71
		70	360	98
		80	520	142
		90	494	169
		100	427	173
	100% of shoulder height	Percent of span*		
		50	367	136
		60	343	125
		70	519	164
		80	707	190
		90	325	132

(a)Height of the center of the force plate (20cm high by 25cm wide) upon which the force is applied.

(b)Horizontal distance between the vertical surface of the force place and the opposing vertical surface (wall or footrest, respectively) against which the subjects brace themselves.

SD: Standard deviation of force (reported in Newtons)

*Thumb-tip reach: distance from backrest to tip of subject's thumb as arm and hand are extended forward.

*Span: maximal distance between the subject's fingertips when arms and hands are extended to each side.

FIGURE 3.7 Average maximal static horizontal push forces, in N, exerted by standing military men with hand, shoulder, or back. Adapted from Kroemer and Robinson (1971).

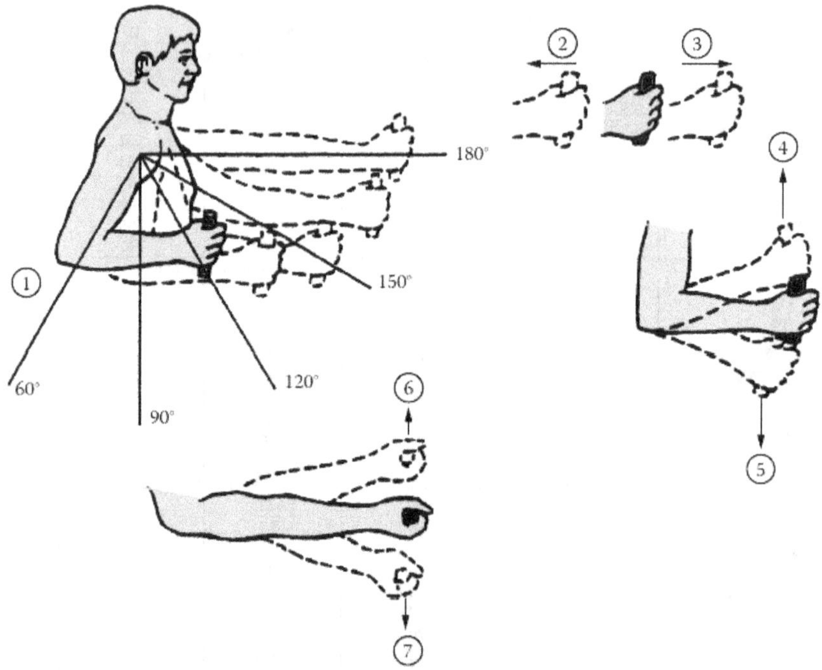

1 Elbow angle (degrees)	2 Pull		3 Push		4 Up		5 Down		6 In		7 Out	
	Left	Right	Left	Right	Left	Right	Left	Right	Left	Right	Left	Right
180	222	231	187	222	40	62	58	76	58	89	36	62
150	187	249	133	187	67	80	80	89	67	89	36	67
120	151	187	116	160	76	107	93	116	89	98	45	67
90	142	165	98	160	76	89	93	116	71	80	45	71
60	116	107	96	151	67	89	80	89	76	89	53	71

FIGURE 3.8 Fifth percentile arm forces, in N, exerted by sitting military men with their arms in various positions. Adapted from MIL-HDBK-759 (1981).

chapter (Chapter 4) contains more data and discusses their use in design; Chapter 21 provides information on material handling, lifting and lowering, and pushing and pulling.

Sources of data

The ergonomic literature provides a wealth of static muscle strength data*. Sources of such information are most often journals because they contain the latest information on human factors engineering and ergonomics. International and national agencies, insurance companies, and specialized research groups can provide answers to specific inquiries. Textbooks, handbooks, encyclopedias, and military and industry standards* can provide comprehensive overviews of accumulated data as well as historical data.

Naturally, all data published in the literature describe the measurements of strength capabilities of a selected population under specific situational conditions. However, the specific users of tools and equipment whom the designer has in mind may differ from previously measured groups. If differences are expected, in some cases it may be possible to extrapolate information from existing data; however, it is often necessary to conduct specialized tests on a representative sample of the actual users under actual use conditions.

Strong or weak persons

For design purposes, "average user strength" typically has no practical importance. (This statement resembles the one made in Chapter 1 regarding the "average person" phantom.) Instead, the ergonomist must usually determine the strongest and weakest strength exertions of relevance. When checking a (normal, Gaussian, bell-shaped) distribution of strength data for distinct numbers that delineate the lower and upper limits of the design range, the average (mean, m) is useful as a *statistical tool*, along with the standard deviation (SD). With these two descriptors of the data distribution, we can calculate the values of key design points. For example, multiplying the standard deviation (SD) of a normal distribution by, say, 4.25 (the value of k from Table 1.5 in Chapter 1 used to calculate the 99.999th percentile) and add ing that product to the mean (average) leads to the strength value that represents the strongest tabulated person.

$$\text{Percentile value} = m + k \times \text{SD}\left(\text{for percentiles above the 50th}\right)$$

Subtracting 1.65 SD from the mean leads to the fifth percentile operator, a statistically weak person who is stronger than only 5% of his fellow users.

$$\text{Percentile value} = m - k \times \text{SD}\left(\text{for percentiles below the 50th}\right)$$

This is the statistical technique used to establish minimal and maximal design limits, as discussed in Chapter 1.

Identifying critical strength values

Figures 3.7 and 3.8 illustrate such statistical exercises: Figure 3.7 contains mean values with their standard deviations, whereas Figure 3.8 lists 5th percentile strength values. Note that we can calculate an unknown value of SD from a table that lists percentiles for normal data; for example, all the fifth percentiles in Figure 3.8 lie 1.65 SDs below the mean. The mean lies, of course, halfway between symmetrical percentiles, such as between the 5th and 95th, or the 10th and 90th.

Non-normal data sets

Calculations involving average and standard deviation are straightforward with normally distributed data; but such procedures lead to misleading results if the data set is severely skewed or otherwise non-Gaussian. Unfortunately, many collections of strength data exhibit non-normal distributions. To determine specific points in a non-normal data distribution, non-parametric procedures are appropriate; these are explained in statistics books*. In some cases, it may be sufficiently accurate for designers to estimate the percentile values of interest.

The MiniMax design rule

Any device requiring strength for operation should be designed with both minimum and maximum strengths in mind. The MiniMax rule for design states that (1) the device can be operated by applying only very small strength but (2) that even the largest strength applied will not damage the device.

Summary

The literature contains many sources of information on maximal human muscle strength. Most strength statistics concern static exertions. The ergonomist must be cautious when applying the data contained in textbooks, handbooks, design manuals, and data compilations, even national or international standards, because they may rely on test conditions that are different from those encountered during the actual work being considered. Sometimes, design data can be extrapolated from existing information. Often, it is necessary to obtain relevant data by testing a representative sample of actual users under representative or actual use conditions.

Steps to fit the human

Step 1: Assess which kind and magnitude of body strength is necessary to perform a specific task.

Step 2: Determine whether people can safely perform the required exertion. Consider whether the task is at the extremes of strength and/or muscle abilities, and if the task is repetitive and/or fatiguing. If the task is outside of safe or appropriate operation, redesign the task.

Step 3: Design, then test. Modify and iterate as necessary.

Notes

The text contains markers, *, to indicate specific references and comments, which follow.

3.1 Basics of physiology

Smooth, cardiac and skeletal muscle: Astrand, Rodahl, Dahl et al. (2004).

Tendons attach like cables to bones: Chaffin, Andersson, and Martin (2006), Kroemer (1998), Kumar (2004, 2007), Nordin and Frankel (1989), Oezkaya, Nordin, and Goldsheyder (2017).

Chemically stored energy: Astrand, Rodahl, Dahl et al. (2004), Hall (2015), Kroemer, Kroemer, and Kroemer (2010), Rodahl (1989), Kenny, Wilmore, and Costill (2015).

3.2 Designing for dynamic and static efforts and strength

Time derivatives of displacement are important: Marras, McGlothlin, McIntyre et al. (1993).

Work-related strength tests and strength test protocols: Chaffin, Andersson, and Martin (2006), Kroemer (1999), Kroemer, Kroemer, and Kroemer (2018), Marras, McGlothlin, McIntyre et al. (1993).

Experimenters can influence test scores: Kroemer (1999).

Situational variables: Daams (1993, 1994, 2001), Kroemer and Robinson (1971).

Dynamic strength tests: Dempsey (2006); Kumar (2004, 2007), Marras, McGlothlin, McIntyre et al. (1993), Winter (2009).

3.3 Designing for fatigue and recovery

Muscle fatigue: Astrand, Rodahl, Dahl et al. (2004), Chaffin, Andersson, and Martin (2006), Kroemer, Kroemer, Kroemer (2020), Kumar (2004, 2007), Toomingas, Mathijssen, and Tornquist (2011), Winter (2009), Beltran Martinez, Nazarahari, and Rouhani (2023). Note that fatigue can be physically in the muscle (metabolic and neural) as well as psychologically perceived by the individual.

3.4 Designing for muscle strength

Strength data in textbooks … and standards: Chengular, Rodgers, and Bernard (2003), Konz and Johnson

(2007), Karwowski (2006), Kroemer Elbert, Kroemer, and Kroemer Hoffman (2018), Marras and Karwowski (2006), Peebles and Norris (1998, 2000, 2003), Univ. Nottingham (2002), Salvendy (2012).

Standards: U. S. military standards such as 759, 1472; NASA 3001; ISO standards.

Statistics books: Hinkelmann and Kempthorne (1994, 2005), Sprent and Smeeton (2007), Tamhane (2009), Williges (2007).

References and further reading

Astrand PO, Rodahl K, Dahl HA, Stromme SB (2004). *Textbook of Work Physiology: Physiological Bases of Exercise*, 4th ed. Champaign, IL, Human Kinetics.

Beltran Martinez K, Nazarahari M, Rouhani H (2023). Breaking the fatigue cycle: investigating the effect of work-rest schedules on muscle fatigue in material handling jobs. *Sensors (Basel)*, 23(24):9670. https://doi.org/10.3390/s23249670

Chaffin DB, Andersson GBJ, Martin BJ (2006). *Occupational Biomechanics*, 4th ed. New York, Wiley.

Chengular SN, Rodgers SH, Bernard TE (2003). *Kodak's Ergonomic Design for People at Work*, 2nd ed. New York, Wiley.

Daams BJ (1993). Static force exertion in postures with different degrees of freedom. *Ergonomics*, 36:397–406.

Daams BJ (1994). *Human Force Exertion in User-Product Interaction: Background for Design.* Delft, NL, Delft University Press – IOS Press.

Daams BJ (2001). Push and pull data and torque data. In Karwowski W (ed), *International Encyclopaedia of Ergonomics and Human Factors.* London, UK, Taylor & Francis, pp. 299–316, pp. 334–342.

Dempsey PG (2006). Psychophysical approach to task analysis. Chapter 47 in Marras WS, Karwowski W (eds), *The Occupational Ergonomics Handbook: Fundamentals and Assessment Tools for Occupational Ergonomics*, 2nd ed. Boca Raton, FL, CRC Press. https://doi.org/10.1201/9781420003635

Hall, JE (2015). *Guyton and Hall Textbook of Medical Physiology*, 13th ed. Amsterdam, NL, Elsevier.

Herzog W (2007). Determinants of muscle strength. Chapter 7 in Kumar S (ed), *Biomechanics in Ergonomics*, 2nd ed. Boca Raton, FL, CRC Press.

Hinkelmann K, Kempthorne O (1994). *Design and Analysis of Experiments, Vol. 1: Introduction to Experimental Design.* New York, Wiley.

Hinkelmann K, Kempthorne O (2005). *Design and Analysis of Experiments, Vol. 2: Advanced Experimental Design.* New York, Wiley-Interscience. https://doi.org/10.1002/0471709948

Karwowski W (2006b). *International Encyclopedia of Ergonomics and Human Factors*, 2nd ed. Boca Raton, FL, CRC Press. https://doi.org/10.1201/9780849375477

Kenny WL, Wilmore J, Costill D (2015). *Physiology of Sport and Exercise*, 6th ed. Champaign, IL, Human Kinetics.

Konz S, Johnson S (2007). *Work Design: Industrial Ergonomics*, 6th ed. Scottsdale, AR, Holcomb Hataway.

Kroemer Elbert K, Kroemer HB, Kroemer Hoffman AD (2018). *Ergonomics: How to Design for Ease and Efficiency*, 3rd ed. London, UK, Elsevier. https://doi.org/10.1016/C2016-0-02516-0

Kroemer KHE (1974). *Designing for muscular strength of various populations*. AMRL-Technical Report 72-46, Wright-Patterson AFB, OH, Aerospace Medical Research Laboratory.

Kroemer KHE (1998). Relating muscle strength and its internal transmission to design data. In Kumar S (ed), *Advances in Occupational Ergonomics and Safety.* Amsterdam, NL, IOS, pp. 349–352.

Kroemer KHE (1999). Assessment of human muscle strength for engineering purposes: a review of the basics. *Ergonomics*, 42(1):74–93. https://doi.org/10.1080/001401399185810

Kroemer KHE, Kroemer HJ, Kroemer-Elbert KE (2020). *Engineering Physiology: Bases of Human Factors Engineering/Ergonomics*, 5th ed. Heidelberg, Springer. https://doi.org/10.1007/978-3-030-40627-1

Kroemer KHE, Robinson DE (1971). *Horizontal static forces exerted by men standing in common working positions on surfaces of various tractions*. Technical Report AMRL-TR-70-114, Wright-Patterson Air Force Base, OH, Aerospace Medical Research Laboratory.

Kumar S (2004). *Muscle Strength*. Boca Raton, FL, CRC Press. https://doi.org/10.1201/9780203503591

Kumar S (2008). *Biomechanics in Ergonomics*, 2nd ed. Boca Raton, FL, CRC Press. https://doi.org/10.1201/9780849379093

Marras WS, Karwowski W (2006). *The Occupational Ergonomics Handbook*, 2 volumes, 2nd ed. Boca Raton, FL, CRC Press. https://doi.org/10.1201/9781420003635 and https://doi.org/10.1201/9781420003642

Marras WS, McGlothlin JD, McIntyre DR, Nordin M, Kroemer KHE (1993). *Dynamic Measures of Low Back Performance*. Fairfax, VA, American Industrial Hygiene Association.

MIL-HDBK 759, 1472 (and others). DLA Document Services, Building 4/D, 700 Robbins Avenue, Philadelphia, PA 19111–5094.

NASA/SP-2010-3407/REV1 (2014). Human Integration Design Handbook (HIDH). https://www.nasa.gov/wp-content/uploads/2023/03/human-integration-design-handbook-revision-1.pdf?emrc=664e5339d33ee

NASA-STD-3001 (2023). Space Flight Human-System Standard Volume 1 (Crew Health); Volume 2 (Human Factors, Habitability and Environmental Health) https://standards.nasa.gov/standard/NASA/NASA-STD-3001_VOL_1, https://standards.nasa.gov/standard/NASA/NASA-STD-3001_VOL_2

Nordin M, Frankel VH (1989). *Basic Biomechanics of the Musculoskeletal System*, 2nd ed. Philadelphia, PA, Lea & Febiger. https://doi.org/10.1002/art.1790030209

Oezkaya N, Leger D, Goldsheyder D, Nordin M (2017). *Fundamentals of Biomechanics: Equilibrium, Motion, and Deformation*, 4th ed. Switzerland, Springer. https://doi.org/10.1007/978-3-319-44738-4

Peebles L, Norris B (1998). *Adultdata. The Handbook of Adult Anthropometric and Strength Measurements – Data for Design Safety.* DTI/Pub 2917/3k/6/98/NP, London, UK, Department of Trade and Industry.

Peebles L, Norris B (2000). *Strength Data.* DTI/URN 00/1070, London, UK, Department of Trade and Industry.

Peebles L, Norris B (2003). Filling "gaps" in strength data for design. *Applied Ergonomics*, 34(1):73–88. https://doi.org/10.1016/s0003-6870(02)00073-x

Rodahl K (1989). *The Physiology of Work.* London, UK, CRC Press. https://doi.org/10.1201/b12599

Salvendy G (2012). *Handbook of Human Factors and Ergonomics*, 4th ed. New York, Wiley. https://doi.org/10.1002/9781118131350

Sprent P, Smeeton NC (2007). *Applied Nonparametric Statistical Methods.* New York, Chapman & Hall/CRC. https://doi.org/10.1201/b15842

Tamhane AC (2009). *Statistical Analysis of Designed Experiments: Theory and Applications.* Hoboken, NJ, Wiley. https://doi.org/10.1111/j.1751-5823.2010.00109_3.x

Toomingas A, Mathijssen SE, Tornquist EW (2011). *Occupational Physiology.* Boca Raton, FL, CRC Press. https://doi.org/10.1201/b11474

University of Nottingham (2002). *Strength Data for Design Safety, Phase 2.* DTI URN 01/1433, London, UK, Department of Trade and Industry.

Williges RC (2007). *CADRE: Computer-Aided Design Reference for Experiments.* Electronic Book CD-ROM-07-01, Blacksburg, VA, Virginia Tech.

Wilmore JH, Costill D, Kenney WL (2008). *Physiology of Sport and Exercise*, 4th ed. Champaign, IL, Human Kinetics.

Winter DA (2009). *Biomechanics and Motor Control of Human Movement*, 4th ed. New York, Wiley. https://doi.org/10.1002/9780470549148

CHAPTER FOUR

Body strength

The actual strength that people generate depends primarily on two clusters of factors (also mentioned in Chapter 3) whose effects overlap and influence each other.

Individual strength factors

The first group of strength factors consists of *individual factors* generally related to body build, sex, age, training, and fitness. Specific muscle properties related to an individual's strength are as follows:

- Muscle cross-sectional area and composition,
- The length (stretch) of the muscle,
- Speed of motion (static or dynamic exertion),
- Metabolic fatigue, and
- Neurophysiological excitation.

Situational strength factors

The second group of strength factors consists of *situational factors*, which are conditions that affect—often strongly—the amount of body strength that a person can exert. Situational factors related to strength are given as follows:

- Motivation, ego involvement, the will to succeed, or fear of injury,
- Skill and experience,
- Body motion, or lack thereof,
- Ability to brace the body against a support structure,
- Body posture, for example, allowing the use of strong muscles at advantageous leverage,
- The body site used for exertion, such as one foot or both feet, shoulder, hand or hands, and
- Coupling (such as hand-held grip) between body and object.

DOI: 10.1201/9781032674575-6

Both groups of strength factors* are of great importance for the human factors engineer who wants to make muscle use easy and fit the task to the human. The last five situational factors are of particular concern to, and are under the control of, the designer of work equipment and tools.

4.1 Designing for static and dynamic strength exertion

Body motion, or lack thereof, is a major factor that determines the force, torque, work, power, or impulse transmitted from the body to an external object*. When the body does not move, the muscles involved in this "static" strength exertion do not change in length. In physiological terms, this is called an *isometric* muscle contraction. In physics terms, all forces acting within the system are in equilibrium, as Newton's first law requires.

"Iso"

For characterizing strength exertions, the prefix "iso" has been in use for decades; derived from the Greek word for equal, it indicates something that is the same, not varying, or constant.

- *Isometric*: at the same length, no change in (muscle) length.
- *Isoinertial*: with the same mass (weight), no change in mass (weight). This term does not describe the kinematics or the kinetics of the muscle.
- *Isotonic*: at the same tension (force), no change in (muscle) tension (force), the same as *isoforce*.
- *Isokinematic*: at the same speed of the handle, pedal, or control, no change in speed; the same as *isovelocity*. This term does not describe the kinematics or the kinetics of the muscle.
- *Isokinetic*: at the same force, no change in (muscle) tension.

Static exertion

The static condition is theoretically simple and experimentally easy to control. It allows for straightforward measurements of muscular effort. Therefore, much of the information available on body strength describes outcomes of static (isometric) testing, as discussed in Chapter 3. Figure 4.1 shows, schematically, measurements of static torque about the elbow, taken every few degrees. Although the results are shown as being connected with a curve, the torques were each measured in a static condition.

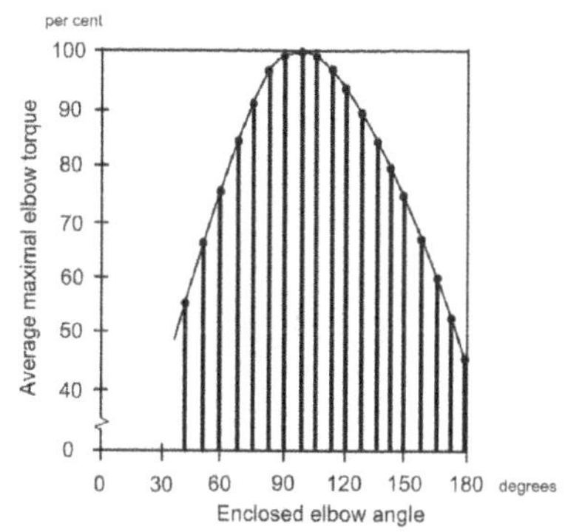

FIGURE 4.1 Schematic representation of static elbow torque measurements. Individual measurements of torque were taken at 10° increments of elbow angle. A curve connects the results of single isometric efforts.

Note of caution

Connecting the measured values with a continuous line, as is habitually done in Figures 4.1 and 4.5, can be misleading: the connecting curve seems to imply motion when, in fact, strength was exerted and measured in separate static positions, independently, one after the other, with a break after each exertion.

Isometric, isokine(ma)tic, isoinertial strength tests

The ease of measurement explains why, for many years, assessments of human muscle strength routinely used isometric muscle contractions. Accordingly, most tables on body segment strength in the human engineering and physiologic literature contain static strength data. Occasionally, strength is measured when exerted on a handle or pedal that moves at a constant speed (often called isokinetic or, more properly, isokinematic). The other exception involves isoinertial tests, where a person moves a constant external mass. This procedure often serves to determine "lifting and lowering" abilities, briefly mentioned in Chapter 3 and below in Section 4.5 and, when applied to material handling, extensively discussed in Chapter 21.

Static or dynamic?

Does the real-world task require static or dynamic strength exertion? This question is of great importance to the human factors engineer who intends to design for the easy and efficient use of muscles.

- If the exertion is *static*, information about isometric strength capabilities is obviously appropriate.
- If the exertion is *dynamic*, then, as a general rule, strength exerted in movement is of lower magnitude than the static strength values measured at positions located on the path of motion.

- If the motion is *very slow*, especially when eccentric (meaning, the muscle is actually lengthening during the exertion), data on isometric strength measured at points along the path of motion usually yield a reasonable estimate of the maximally possible exertion, although often a bit high.
- If the motion is *very fast*, then isometric data do not estimate fast exertions well. This is especially true if the exertions are concentric (meaning, the muscle is shortening during the exertion) and of the ballistic or impulse type, such as throwing or hammering.
- If the dynamic motion is *fast or variable*, special experiments* may be necessary to measure the strength capability that is specific to the demanded kind of exertion.

4.2 Designing for maximal or minimal strength exertion

Strong and weak operators

A designer who wants to consider operator strength must make a series of decisions when developing a new process or device. The first decision depends on whether strong or weak strength exertion is the critical design factor. The structural integrity of the object must exceed the maximal strength that any user might apply, because even the strongest operator should not break a handle or a pedal. Therefore, the design value should be above the highest strength application, with a safety margin. The other extreme is the minimal strength expected from the weakest operator that still yields the desired result; for example, successfully operating a brake pedal in a vehicle under the worst foreseeable circumstance, such as the failure of the power booster. This illustrates the MiniMax rule: design the device so it can be operated by applying minimal strength and will not be damaged by the largest strength.

No average user

Consequently, the designer must carefully determine both the minimum and maximum expected operator strength exertions. These values usually bracket the design range because "average user" strength typically has no applicability for ergonomic design (as already mentioned in Chapters 1 and 3). Determining the upper and lower design limits requires knowledge of the actual distribution of strength data. With this knowledge, we set the lower cutoff so that everybody, or nearly everybody, can operate the equipment; a common cutoff is the 5th percentile strength, meaning that only 5% of users are weaker than anticipated by the design. Correspondingly, the upper cutoff is commonly selected at the 95th percentile strength; however, this may not be sufficiently high (safe), as the strongest 5% of users could still break the equipment.

MAX and MIN values

It is easy to calculate percentile points when the data follows a normal (Gaussian, symmetrical, and bell-shaped) distribution, often encountered with anthropometric data; see, for example, Figure 1.2 in Chapter 1. Many strength data, however, are not evenly distributed but show irregular (non-parametric) accumulations. In these cases, the usual parametric statistical procedures should not be used. However, we can still determine—sometimes by rough estimates or more properly by formal non-parametric procedures—where we want to set our minimal and maximal design limits. Appropriate non-parametric procedures depend on the particular data distribution and are explained in non-parametric statistics books*.

4.3 Designing for hand strength

Most tasks, both at work and in our private lives, involve grasping, holding, turning, pressing, pushing, pulling, and other manipulations done with our hands on objects held between the thumb, fingers, and/or palm. Given the importance of our ability to "handle" objects, a large body of ergonomic literature discusses hand capabilities*, especially hand forces.

Intrinsic and extrinsic hand muscles

Hand force depends on the combined effect of intrinsic and extrinsic muscles. Intrinsic muscles and their tendons are located within the hand; extrinsic muscles are in the forearm, connected to the hand's digits with tendons that cross the wrist joint (more on this in Chapter 2). Most intrinsic muscles contribute to the finely controlled actions of the digits, while the extrinsic muscles primarily generate large forces for moving the whole hand about the wrist joint and for exercising the individual digits. Like all other skeletal muscles of the human body, they can be strengthened by use, as well as weakened by disuse, and they can suffer and recover from fatigue.

Types of hand tasks

There are three major types of requirements in hand tasks: accuracy, strength exertion, and displacement. Hand tasks may be further divided as follows:

- *Fine manipulation of objects*, with little displacement and force. Examples include writing by hand, assembling small parts, or adjusting fine controls.
- *Fast movements to an object*, requiring moderate accuracy to reach the target but relatively small force exertion once at the target. An example is the movement to a switch and its operation.
- *Frequent movements between targets*, usually with some accuracy but little force. Examples include typing on a keyboard, or an assembly task where small parts must be taken from bins and assembled.
- *Forceful activities with moderate displacement*, such as in many fabrication or repair activities. Examples include turning hand tools, such as wrenches or screwdrivers, against resistance.
- *Forceful activities with large displacements*, such as hammering.

Grips and grasps

Tools serve as extensions of the hands, fortifying and also protecting them. Depending on the nature of the job, the tools are held in different ways between the digits and the palm of the hand. Some hand tools require fairly small force but precise handling, such as surgical instruments, screwdrivers used by optometrists, or writing utensils. Commonly, the manner of holding these tools is called a *precision grip*. Other gadgets must be held strongly between the larger surfaces of the fingers, thumb, and palm in what is often called a *power grasp*. There are many transitions from merely touching an object with a finger (such as pushing a button) to pulling on a hook-like handle, from holding small objects between the fingertips to transmitting large energy from the hand to the handle. Figure 4.2 classifies couplings between the hand and the object.

Designing hand tools

Shaping the tool handle* to provide high friction (often aided by wearing a glove) or even to achieve mechanical interlocking between hand and handle (with suitable grooves and/or protrusions) facilitates a secure grip and effective transfer of energy. Proper tool design and use should keep the wrist straight, not bent, to avoid overexertion of connective tissues (muscles, tendons, tendon sheaths) and especially to prevent compression of the median nerve in the carpal tunnel at the base of the hand. This becomes an even more important design consideration for hand tool tasks that are repetitive, forceful, vibrating, and/or

1. Digit (finger) touch: one finger touches an object without holding it.	
2. Palm touch: some part of the inner surface of the hand touches the object without holding it.	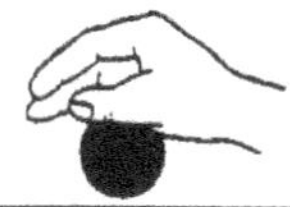
3. Finger palmar grip (hook grip): one or several fingers hook onto a ridge or a handle. This type of finger action is used when thumb counterforce is not needed.	
4. Thumb-fingertip grip (tip grip): the thumb tip opposes one fingertip.	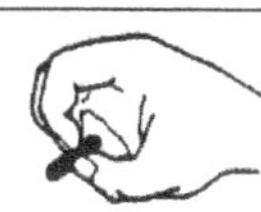
5. Thumb-finger palmar grip (pinch grip or plier grip): the thumb pad opposes the palmar pad of one (or several) finger(s) near the tips. This grip evolves easily from coupling #4.	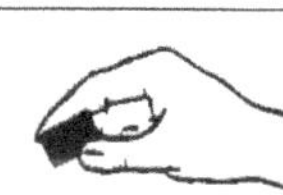
6. Thumb-forefinger side grip (lateral grip or side pinch): the thumb opposes the (radial) side of the forefinger.	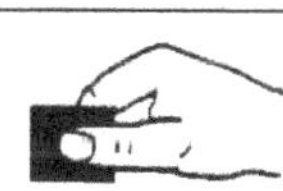
7. Thumb-two-finger grip (writing grip): the thumb and two fingers (often index and middle fingers) oppose each other at or near the tips.	
8. Thumb-fingertips enclosure (disc grip): the thumb pad and the pads of three or four fingers oppose each other near the tips (object grasped does not touch the palm). This grip evolves easily from coupling #7.	
9. Finger-palm enclosure (collet enclosure): most, or all, of the inner surface of the hand is in contact with the object while enclosing it.	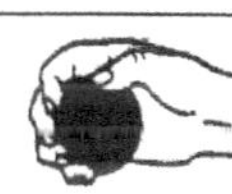
10. Power grasp: the total inner hand surface is grasping the (often cylindrical) handle which runs parallel to the knuckles and generally protrudes on one or both sides from the hand.	

FIGURE 4.2 Couplings between hand and handle. Adapted from Kroemer (1986).

sustained. Carpal tunnel syndrome (Chapter 2) is a painful and disabling condition that often results from repetitive motions of the hand, such as in assembly work or keyboard use.

Avoiding wrist bending

The natural grasp centerline of a straight handle is not perpendicular to the forearm axis, but at an angle of about 60°–70°, as shown in Figure 4.3. For example, using common straight-nose

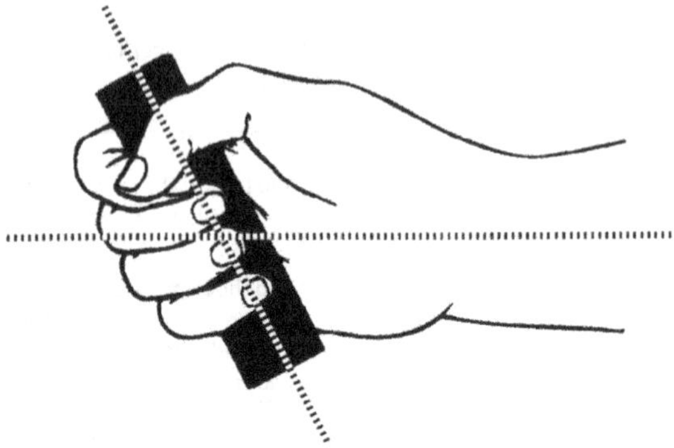

FIGURE 4.3 Natural angle between the forearm and the hand's grasp centerline.

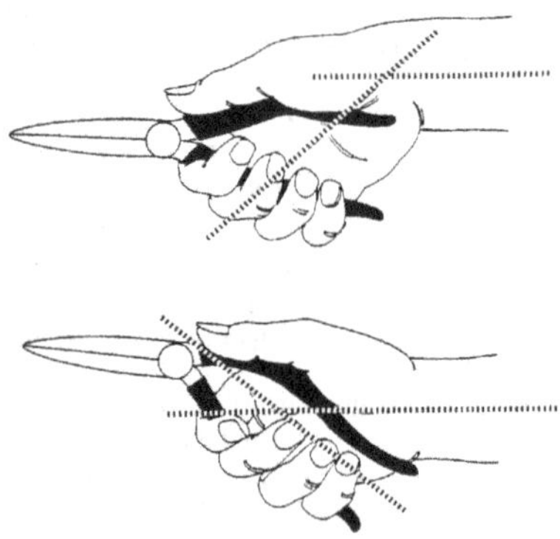

FIGURE 4.4 Don't bend the wrist, instead bend the tool.

pliers often requires a strong bend in the wrist, and neither the direction of thrust nor the axis of rotation corresponds with those of the hand and arm. This results in wrist bending, which reduces the force that can be applied. Bending the tool, not the wrist, improves this situation, as shown in the lower image in Figure 4.4.

"Lefties"

Most women and men are right-handed, but about one in ten people prefers to use the left hand for many tasks and has better skills and more strength available there. While many tools can be used with either the left or right hand, some tools are shaped to fit only the right hand. Hence, it is generally advisable to provide left-handers with tools specifically designed to accommodate their preferred dominant hand, or better yet, to design tasks to allow for use with either hand.

4.4 Designing for foot strength

Foot controls require careful design. If a person must stand at work, foot controls should only be used if relatively little force and infrequent foot control operation are required. This is because, during these exertions, the operator must support the body solely on the other leg (as shown in Figure 2.18).

Bicycle pedals

For a seated operator, however, operating foot controls is much easier because the seat supports the weight of the body. Thus, the feet can move more freely (as in Figure 2.17) and, given suitable conditions, can exert large forces and energies, such as when pedaling a bicycle. Bicycle pedals are most effective when positioned directly beneath the body, allowing the body weight above them to contribute to the reactive force generated by the foot force applied to the pedal. Placing the pedals higher and forward reduces the effectiveness of body weight in generating reaction force; however, if a suitable backrest is present against the buttocks and lower back, the feet can effectively push forward on the pedals. This design principle is applied in so-called recumbent bicycles and, similarly, in automobiles.

Foot controls

Most of us can walk and run well on our feet. However, compared to hand motions, foot movements are relatively slow and clumsy, especially when involving the rather large leg masses and potentially awkward footwear (such as heavy work boots). Yet with proper seat support, our feet can exert significant forces in both the downward and forward directions. This fact was utilized in the early years of automotive design (and, surprisingly, is still used today) when drivers had to stomp hard on pedals to operate the brake and clutch—actions that are now, in modern vehicles with power assistance, much less demanding.

Foot thrust

A nearly fully extended leg can generate a very large foot thrust force, but only in the direction of the lower leg. If there is no seat providing support to the operator's back and backside, or if the angle at the knee becomes smaller or larger than about 160°, the pedal force diminishes dramatically, as shown in Figure 4.5. Thus, there is very little space available for the designer to effectively position the pedal. This case is a striking example of how the strength of a body exertion depends not only on nominal muscle strength but also on how the seat provides necessary reactive support, as well as how muscle strength is transmitted across the hip, knee, and ankle joints. An improper seat, a weak hip, a painful knee, or a frail ankle—even a poor coupling at the shoe—can all lead to a weakened push.

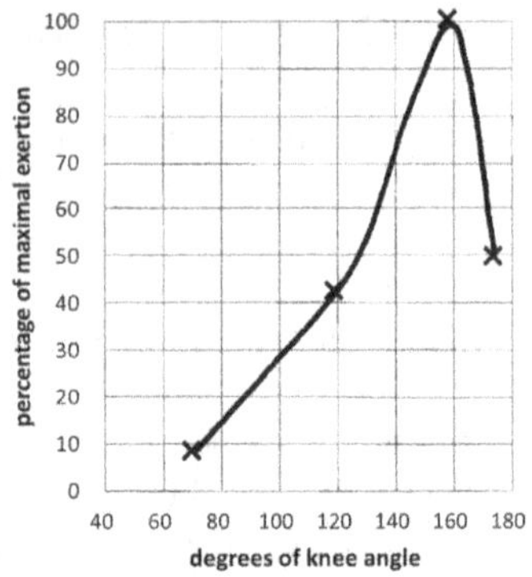

FIGURE 4.5 Effect of knee angle on static pedal push force. A curve connects the results of single isometric efforts. Adapted from diagram in Kroemer-Elbert, Kroemer, and Kroemer Hoffman (2018).

4.5 Designing for whole body strength

Just a few jobs require the use of only one finger; most hand tasks also involve the hand, arm, and shoulder. Likewise, foot use typically entails leg activity as well. The other body parts involved, such as the shoulder, legs, and back, also have characteristic strength capabilities, both individually and situationally.

Chains of strength vectors

Pushing and pulling, lifting and lowering loads (discussed in detail in Chapter 21) are examples of isoinertial strength exertions that involve more than one body segment engaged to perform the task; often, nearly the whole body participates. A "chain" of strength-transmitting segments starts at the hand on the steering wheel, or at the foot on the brake pedal, or at the shoulder pushing against a stuck car. These chains then run through all the body parts involved to the surface that provides the reaction force: usually the seat when sitting, or the ground when standing. Figure 4.6 depicts the chaining of strength-transmitting body segments. Every involved body segment must be able to perform its duty; if one fails, the task cannot be completed. The weakest segment determines the strength of the whole chain. Furthermore, according to Newton's third law, reaction forces must be strong enough to counteract the forces exerted by the hand, foot, or shoulder.

Pulling and pushing

In the last chapter, Figures 3.7 and 3.8 showed examples of one-time static push and pull forces measured under experimental conditions, where the subjects kept their bodies still

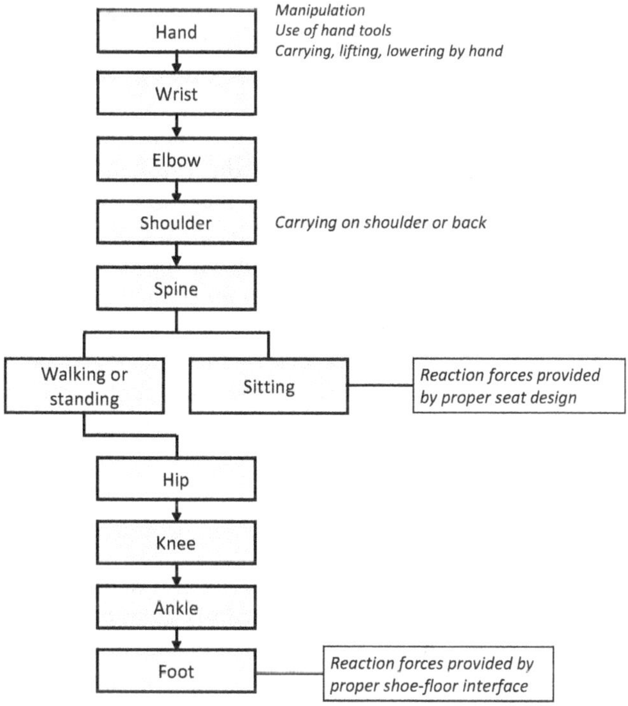

FIGURE 4.6 Chain of strength-transmitting body segments. Adapted from Kroemer, Kroemer, and Kroemer-Elbert (2020).

while they exerted one-time maximal efforts on non-moving measuring devices*. In reality, however, most people move objects by dynamically pushing and pulling on them; these activities involve body motions and are often repetitive. Psychophysical* experiments have been conducted to assess tasks performed by firefighters and soldiers and, for industrial work, while handling loads; material handling is discussed in some detail in Chapter 21.

4.6 Designing for use preferences

If muscle strength is critical for operating a hand or foot control or for task performance in general, that strength requirement must be considered during the design of the task, tools, and workplace. Once the need for forceful operation is met, other safety and usability criteria become important for design decisions. These include ease of performance, consideration of local customs, striving for similarity with other operations, utilizing skills and habits, accommodating individual preferences, and generating a pleasing appearance.

Carrying loads on the body

Load carrying can be done in many ways, not only with the hands but also on the shoulders, back, or chest; the load can be held, strapped on, or contained in some sort of pocket. Table 4.1 compiles the results of several studies on the ways of carrying weighty objects on the body. In contrast to the preceding discussions of designing for strength, the recommendations for carrying loads on the body do not primarily rely on muscle strength but reflect several other criteria, including energy consumption, discomfort, stability, and ease of movement. Chapter 21 deals with such assessments related to material handling.

Summary

The engineer or designer wanting to consider operator strength must ask a series of questions and, based on the answers, make a series of decisions. These include the following:

1. Is the *action static or dynamic*? If static, information about isometric capabilities can be used. If dynamic, additional considerations apply, such as the physical (circulatory, respiratory, metabolic) endurance capabilities of the operator, or prevailing environmental conditions. Sections I, II, and V of this book provide related information.

   Most body segment strength data are available for static (isometric) exertions. They provide reasonable guidance for slow motions as well. As a general rule, strength exerted in motion is less than that measured in the static positions located on the path of motion.

2. Is the *exertion by hand, foot*, or other body segments? For each, specific design information is often available.
3. Is *maximal or minimal exertion* the critical design factor? Maximal user strength determines the necessary structural strength of an object so that even the strongest operator cannot break it; therefore, the design value should be set, with a safety margin, above the highest perceivable strength application. However, normal use must also consider the ability of the weakest operator to perform the necessary task.

Table 4.1 Techniques of carrying weights of about 30 kg

Technique to carry 30 kg on a straight, flat path	Estimated energy expenditure (kcal/min)	Relative muscular fatigue	Localized pressure, risk of ischemia	Stability of loaded body	Special considerations or benefits
In one hand	?	Very high	Very high	Very poor	Load easily manipulated and released.
In both hands, equal weights	Very high: about 7	High	High	Poor	Suitable for quick pick-up and release, and for short-term carriage even of heavy loads.
Clasped between arms and trunk	?	?	?	?	Compromise between hand and trunk use.
On head, supported with one hand	Fairly low: about 5	High if hand guidance needed	?	Very poor	May free hand(s); strongly limits body mobility; determines posture; pad is needed. Suitable for heavy and bulky loads If accustomed to this technique.
On neck, often with strap around forehead	Medium: about 5.5	?	?	Poor	May free hand(s); affects posture. Suitable for heavy and bulky loads If accustomed to this technique.
On one shoulder	?	High	Very high	Very poor	May free hand(s); strongly affects posture. Suitable for short-term transport of heavy and bulky loads.
Across both shoulders by yoke, held with one hand	High: about 6.2.	?	High	Poor	May free hand(s); affects posture. Suitable for bulky and heavy loads; pads and means of attachment must be carefully provided.
On back	Medium: 5.3 with backpack; 5.9 with bag held in place with hands	Low	?	Poor	Usually frees hands; forces forward trunk bend; may have skin-cooling problem. Suitable for large loads and long duration carriage. Packaging must be done carefully, attachment means shall not generate areas of high pressure on body.

(Continued)

Table 4.1 (*Continued*) Techniques of carrying weights of about 30 kg

Technique to carry 30 kg on a straight, flat path	Estimated energy expenditure (kcal/min)	Relative muscular fatigue	Localized pressure, risk of ischemia	Stability of loaded body	Special considerations or benefits
On chest	?	Low	?	Poor	Frees hands; easily reached with hands; reduces trunk mobility; may have skin-cooling problem. Highly advantageous for several small loads that must be accessible.
Distributed on chest and back	4.8, lowest	Lowest	?	Good	Frees hands; may reduce trunk mobility; may have skin-cooling problem. Highly advantageous for loads that can be divided or distributed; suitable for long durations.
At waist, on buttocks	?	Low	?	Very good	Frees hands; may reduce trunk mobility. Around waist for smaller items, distributed in pockets or by special attachments; superior surface of buttocks may partially support backpacks.
On hip	?	Low	?	Very good	Frees hands; may affect mobility. Often used to prop up large loads temporarily.
On legs	?	High	?	Good	Easily reached with hands; may affect walking. Requires pockets in garments or special attachments.
On foot	Highest	Highest	?	Poor	Usually not useful

Source: Adapted from Kroemer Elbert, Kroemer, and Kroemer Hoffman (2018).

4. Is *muscle strength only one design criterion among other use criteria*, such as ease of movement, body stability, and avoidance of discomfort or fatigue? If the strength requirement is met, the actual design may be chosen to satisfy user preferences.

Steps to fit the human

Step 1: Determine which specific task must be completed. Can a machine perform it? If a human operator is needed, proceed to Step 2.

Step 2: Decide how the operator can best perform the work. Consider whether the task must be done in a fixed position or while in motion, using the hand or foot, and the specific area within the movement space. Evaluate whether the task is at the limits of the operator's or the task object's strength, as well as any other important criteria.

Step 3: Design, then test. Modify and iterate as necessary.

Notes

The text contains markers, *, to indicate specific references and comments, which follow.

Individual and situational strength factors: Daams (1993, 2001), Kumar (2004, 2008). Note that individual physiological factors are also related to metabolic and circulatory body functions, especially for hard physical work as discussed in Chapter 10 of this book.

4.1 Designing for static and dynamic strength exertions

Transmitted from the body to an external object: Astrand, Rodahl, Dahl et al. (2004), Kroemer (1974, 1999), Marras, McGlothlin, McIntyre et al. (1993), Winter (2009).

Special experiments: Hinkelmann and Kempthorne (1994, 2005), Williges (2007).

4.2 Designing for maximal or minimal strength exertions

Non-parametric statistics: Gibbons (1997), Sprent and Smeeton (2007).

4.3 Designing for hand strength

Hand capabilities: Fagarasanu and Kumar (2004), Feathers, D'Souza, and Paquet (2015), Kumar (2008).

Shaping the handle: Freivalds (1999), Konz and Johnson (2007), Kroemer Elbert, Kroemer, and Kroemer Hoffman (2018).

4.5 Designing for whole body strength, pulling and pushing

Static tests: Dempsey (1998), Stanton, Hedge, Brookhuis et al. (2005).

Psychophysical testing: Psychophysics (psychophysiology) is the study of the relationship between the intensity of the physical (physiological) stimuli and the strength of its (psychological) perception and sensory responses. Dempsey (1998), Snook and Ciriello (1991) and updated in Potvin, Ciriello, Snook et al. (2021). See also the discussion on rating scales, subjectivity, and (dis)agreements in ergonomics in the *J. Ergonomics* (2002), volume 45(14), including Annett (2002).

References and further reading

Annett J (2002). Subjective rating scales: science or art? *Ergonomics*, 45(14):966–987. https://doi.org/10.1080/00140130210166951

Astrand PO, Rodahl K, Dahl HA, Stromme SB (2004). *Textbook of Work Physiology: Physiological Bases of Exercise*, 4th ed. Champaign, IL, Human Kinetics.

Daams BJ (1993). Static force exertion in postures with different degrees of freedom. *Ergonomics*, 36:397–406.

Daams BJ (2001). Push and pull data and torque data. In Karwowski W (ed), *International Encyclopaedia of Ergonomics and Human Factors*. London, UK, Taylor & Francis, pp. 299–316, 334–342.

Dempsey PG (1998). A critical review of biomechanical, epidemiological, physiological and psychophysical criteria for designing manual materials handling tasks. *Ergonomics*, 41(1):73–88. https://doi.org/10.1080/001401398187332

Fagarasanu M, Kumar S (2004). Hand strength. Chapter 10 in Kumar S (ed), *Muscle Strength*. Boca Raton, FL, CRC Press. https://doi.org/10.1201/9780203503591

Feathers D, D'Souza C, Paquet V (2015). Anthropometry in ergonomic design. Chapter 27 in Wilson JR, Sharples S (eds), *Evaluation of Human Work*, 4th ed. Boca Raton, FL: CRC Press. https://doi.org/10.1201/b18362

Freivalds A (1999). Ergonomics of hand controls. Chapter 27 in Marras WS, Karwowski W (eds), *The Occupational Ergonomics Handbook: Fundamentals and Assessment Tools for Occupational Ergonomics*. Boca Raton, FL, CRC Press.

Gibbons JD (1997). *Nonparametric Methods for Quantitative Analysis*, 3rd ed. Columbus, OH, American Sciences Press.

Hinkelmann K, Kempthorne O (1994). *Design and Analysis of Experiments, Vol. 1: Introduction to Experimental Design*. New York, Wiley.

Hinkelmann K, Kempthorne O (2005). *Design and Analysis of Experiments, Vol. 2: Advanced Experimental Design*. New York, Wiley-Interscience. https://doi.org/10.1002/0471709948

Konz S, Johnson S (2007). *Work Design: Industrial Ergonomics*, 6th ed. Scottsdale, AR, Holcomb Hataway.

Kroemer Elbert K, Kroemer HB, Kroemer Hoffman AD (2018). *Ergonomics: How to Design for Ease and Efficiency*, 3rd ed. London, UK, Elsevier. https://doi.org/10.1016/C2016-0-02516-0

Kroemer KHE (1974). *Designing for muscular strength of various populations*. AMRL-Technical Report 72-46, Wright-Patterson AFB, OH, Aerospace Medical Research Laboratory.

Kroemer KHE (1986). Coupling the hand with the handle: an improved notation of touch, grip, and grasp. *Human Factors*, 28(3):337–339. https://doi.org/10.1177/001872088602800309

Kroemer KHE (1997). *Ergonomic Design for Material Handling Systems*. Boca Raton, FL, CRC Press. https://doi.org/10.1201/9780203756683

Kroemer KHE (1999). Assessment of human muscle strength for engineering purposes: a review of the basics. *Ergonomics*, 42(1):74–93. https://doi.org/10.1080/001401399185810

Kroemer KHE, Kroemer HJ, Kroemer-Elbert KE (2020). *Engineering Physiology: Bases of Human Factors Engineering/ Ergonomics*, 5th ed. Heidelberg, Springer. https://doi.org/10.1007/978-3-030-40627-1

Kumar S (2004). *Muscle Strength*. Boca Raton, FL, CRC Press. https://doi.org/10.1201/9780203503591

Kumar S (2008). *Biomechanics in Ergonomics*, 2nd ed. Boca Raton, FL, CRC Press. https://doi.org/10.1201/9780849379093

Marras WS, McGlothlin JD, McIntyre DR, Nordin M, Kroemer KHE (1993). *Dynamic Measures of Low Back Performance*. Fairfax, VA, American Industrial Hygiene Association.

Potvin JR, Ciriello VM, Snook SH, Maynard WS, Brogmus GE (2021). The liberty mutual manual materials handling (LM-MMH) equations. *Ergonomics*, 64(8):955–970. https://doi.org/10.1080/00140139.2021.1891297

Snook SH, Ciriello VM (1991). The design of manual handling tasks: revised tables of maximum acceptable weights and forces. *Ergonomics*, 34(9):1197–1213. https://doi.org/10.1080/00140139108964855

Sprent P, Smeeton NC (2007). *Applied Nonparametric Statistical Methods*. New York, Chapman & Hall/CRC. https://doi.org/10.1201/b15842

Stanton N, Hedge A, Brookhuis K, Salas E, Hendrick HW (2004). *Handbook of Human Factors and Ergonomics Methods*. Boca Raton, FL, CRC Press. https://doi.org/10.1201/9780203489925

Williges RC (2007). *CADRE: Computer-Aided Design Reference for Experiments*. Electronic Book CD-ROM-07-01, Blacksburg, VA, Virginia Tech.

Winter DA (2009). *Biomechanics and Motor Control of Human Movement*, 4th ed. New York, Wiley. https://doi.org/10.1002/9780470549148

SECTION II

The human mind

The second section of this book on fitting the human discusses how we receive, process, and respond to information about our environment. What deals with such information is generally called the *mind*, which consists primarily of our body's network of nerves and the brain. This command-and-control center has many complex functions that depend on inputs (such as what we see and hear, and how we sense objects) and is affected by the surrounding environment and climate. Our brain processes inputs and makes decisions about actions to be taken. Current knowledge about brain functions provides guidance to the ergonomist for designing human-controlled systems.

We recognize and consider differences among people so that we can design tasks, gear, equipment, and procedures to fit their specific needs. Therefore, the following chapters provide relevant information for ergonomic design:

- Chapter 5, How We See
- Chapter 6, How We Hear
- Chapter 7, How We Sense Objects and Energy
- Chapter 8, How We Experience Indoor and Outdoor Climates
- Chapter 9, How We Think

DOI: 10.1201/9781032674575-7

CHAPTER FIVE

How we see

Eyes

Our eyes sense energy from the outside world in the form of light rays and convert these into nerve impulses, which the brain then integrates into a visual picture of what is in front of us. However, this perceived picture is a subjective modification of what the eye reports. For example,

- We are subject to optical illusions, such as when a straight line appears distorted against a background of curved or radiating lines.
- We are subject to color biasing, such as when a color seems darker against a bright background than when seen on a dark background.
- We experience inter- and intra-personal variation, such as when the perceived hue and intensity of greenish-blue differ from person to person and may change as the beholder's eyes age.
- Our interpretation of visual data varies depending on age, experience, attitude, and preconceived ideas. People also differ in their abilities to recognize colors and focus clearly on visual targets, with significant changes in these abilities typically occurring with age.

Nevertheless, there are general similarities in how the human visual sense functions, which allows us to develop ergonomic recommendations. We need to understand how we see in order to fit the task to the human.

Seeing requires light

We cannot see objects without light. Some objects generate light, such as the sun, a lamp, or an electronic computer display. Other objects reflect light, such as the moon, the walls

DOI: 10.1201/9781032674575-8

of a room, or a printed page. Unless we look directly into a light source, the light energy reflected from a surface (its luminance) activates the eye and is hence the most important factor for human vision.

5.1 Our eyes

The eyes continually adjust the amount of light they let in, change their focus on objects far and near, and produce continuous images, which they instantly transmit to the brain.

The human eyeball is a roughly spherical organ, about 2.5 cm in diameter, surrounded by a layer of fibrous tissue called the *sclera*. Light enters through the *cornea*, a transparent dome on the front surface of the eye. The cornea serves as a protective covering and a weak lens that helps to focus light on the *retina* at the back of the eye (see Figure 5.1).

Pupil in the iris

After passing through the cornea, light enters the *pupil*, the small round black area in the middle of the iris. (The pupil appears black because no light emerges from inside the eye. However, it can appear red in photographs when a flash illuminates the inside and reflects off the retina, which contains many blood vessels.) The *iris* is the circular, colored area of the eye, containing nerves and muscles. The pupillary dilator and sphincter muscles open and close the pupil (much like a camera lens aperture) to control the amount of light entering the eye: allowing more light in when the environment is dim and less light when the surroundings are bright. (Pigments in the iris not only protect it but also provide a wide variety of eye colors due to genetic variation in melanin production, ranging from brown

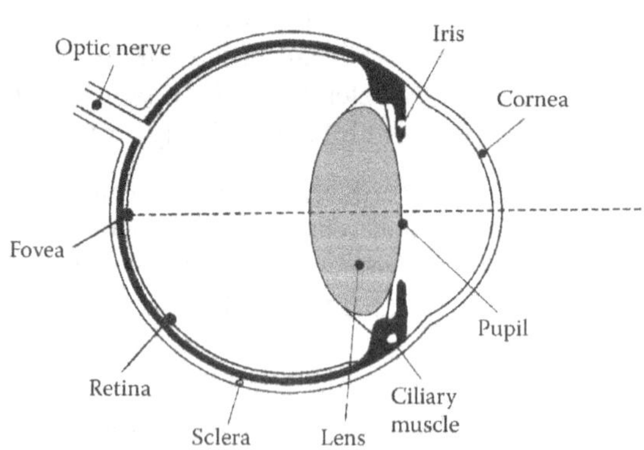

FIGURE 5.1 Right eye, as seen from above.

to green, gray, blue, or even being so pale as to appear as red in individuals with albinism.)

The lens focuses

The *lens* is located behind the iris. It focuses light onto the retina by changing its shape. When thin, it is a "weak" lens (in optics terminology*) that focuses on distant objects. The contracting *ciliary muscle* pulls on the lens, making it thicker and optically "stronger" so that it can focus on nearby objects. The eyes of a healthy young adult can usually focus on an object as close as 10 cm; however, with aging, the lens becomes less flexible and therefore less able to thicken, which diminishes its ability to focus on nearby objects. Consequently, the accommodation distance commonly increases with age. This condition is known as farsightedness or *presbyopia*.

Light, focused by the cornea and lens, travels through the vitreous humor, a gel-like fluid (with refractive properties similar to those of water) that fills the interior of the eyeball. Finally, the light reaches the retina, a thin tissue lining about three-quarters of the rear inner surface of the eyeball. The retina is well supplied with blood from numerous arteries and veins.

Rods for white/gray/black perception

The retina contains about 130 million light sensors. Their arrangement is densest in the center, the *fovea*, directly behind the lens. The two types of light-sensing cells are named for their shape: rods and cones. Rods are the most prevalent, numbering around 120 million. They contain only one pigment, which responds even to low-intensity light. Rods trigger electrical impulses that travel along the optic nerve to the brain, allowing perception of white, black, and shades of gray. They provide us with the most visual information, as well as the most important visual information.

Cones signal color

The retina also contains about 6 million cones, primarily located in the fovea. They respond best to relatively bright light and react differently to various wavelengths of light, enabling our ability to detect color. Each cone contains one pigment that is most sensitive to either blue, yellow-to-green, or red wavelengths. An incoming light beam, if intense enough, triggers chemical reactions in one of the three types of pigmented cones, creating electrical signals that pass along the optic nerve to the brain, which can compose and distinguish about 150 color hues.

Optic nerve

The *optic nerve* exits the eye at its rear, but not directly aligned with the centers of the cornea and lens. Instead, the nerve is

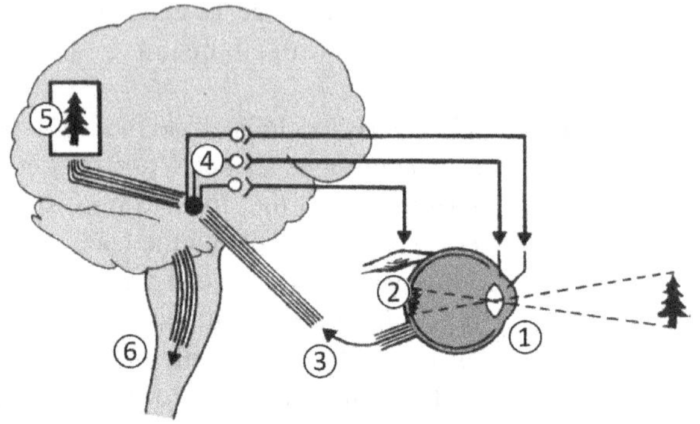

FIGURE 5.2 Diagram of the visual system.

offset from the fovea by about 15° toward the inside (medially). Since there are no light sensors in this area, we cannot see an image refracted in this so-called blind spot. However, as the blind spots of both eyes are each located medially, they do not overlap in our field of vision, and therefore we are usually unaware of their existence.

Visual control system

Proper vision requires the continuous operation of a complex control system, illustrated in Figure 5.2. It shows, first, the adjustment of the lens (1) behind the cornea to provide a sharply focused (and upside-down) image on the retina. The information (2) about that image is transmitted along the optic nerve (3) to the brain. Various nervous control mechanisms (4) continually readjust the position of the eyeball, the size of the pupil, and the shape (refractive property) of the lens. The visual perception of the external world takes place in the conscious sphere of the brain (5) and leads to command signals that travel through the spinal cord (6) to instigate appropriate actions of the body.

5.2 Seeing our environment

The visual field

The objects we can see are in the so-called visual field*, which is a roughly conical space in front of the eyes. Within this space, we can discern visual targets with high acuity only if they appear within a narrow cone at the center and at a suitable distance. For example, when we focus on written text, only a few adjacent and centrally located letters are clear, while the surrounding letters become increasingly blurred the farther away or to the side they are. The view to the sides is referred to as peripheral or indirect vision.

Fixated eyes

The areas visible to the two eyes do not overlap perfectly, partly because the nose is in the way. When the eye maintains fixation on a single spot directly ahead, researchers can measure the size of the visual field by presenting test objects away from this position. We can notice the presence of objects within about 90°–110° to the sides but colors are discernible only within the inner about 60°. Upward, we can perceive about 45° yet observe color only to about 30° upward; downward, we can see about 70°, but the area in which we see color extends downward to about 40°.

Rotating the eyes

Rotating the eyeball to the sides increases the visual area beyond the so-called field of visual fixation, but adds very little in the upward, downward, or inward directions since the eyebrows, cheeks, and nose obstruct the view. Six muscles attach to the outside of the eyeball, working together to move the eye. Rotational movements are most obvious, but some forward and backward motion of the eyeball also occurs. We can rotate our eyes up to about 50° in both pitch (up and down) and yaw (left and right). However, the eyes seldom rotate to these extreme angles.

Fixation on a visual target

When we try to look at an object at the periphery of our field of vision, we generally change the field of view by rotating our eyes, usually accompanied by a quick turn of the head in that direction. With a strong preference for body movement rather than extreme eye motion, we commonly adjust our body posture to fixate on a peripheral target so that our eyes can operate comfortably and precisely close to their normal resting positions.

Eye tracking

The eye can continuously track a visual target that moves left or right at less than 30° per second or cycles at less than 2 Hz (hertz, cycles per second). Above these rates, the eye is no longer able to follow continuously and must move in *saccades* (from the French for jerk): it lags behind and then jumps to catch up.

Avoiding eye fatigue

So-called eye fatigue is often the result of excessive demands on the muscles of the eye: those that move the eyeball and those that adjust the lens and iris. Eye fatigue is particularly problematic for older people whose lenses have become stiff and who often find it difficult to move their neck and trunk as easily as they could in their youth. Notably, the visual fatigue that computer users often complain about* is generally related to poor placement of the monitor, source documents, or other visual targets (see also Section V of this book) or to unsuitable lighting conditions at their workplace, which will be discussed later in this chapter (and in Chapter 18).

Locating visual targets

It is important to arrange visual targets properly when possible: in front, not to the side; at a distance to which the eyes can easily accommodate; and positioned slightly lower so that we can rotate the eyeballs down while keeping the trunk, neck, and head in comfortable positions. Makers of eyeglasses and optometrists have long known that looking down at small visual targets, such as printed text, is easier than looking up; therefore, they place the reading section of (bifocal or trifocal) corrective eyeglasses in the lower parts of the lenses. If several similar visual targets are present, they should be placed close together so that eye and head movements are minimal, as Figure 5.3 illustrates in a counter-example.

Line of sight

When we look at an object in front of us, we unconsciously adjust two pitch angles: that of the eyes within the head and that of the head against the trunk. As we fixate on a target, the *line of sight* (LOS) runs from the center of the retina through the midpoints of the lens and pupil, and from there to the target. Thus, inside the eyeball, the LOS is clearly established. However, a suitable reference is needed to describe the LOS direction from the eye to the visual target: the line from the ear to the eye can serve as this reference.

Looking down on the job

In the past, the so-called Frankfurt plane served as a reference, but its anatomical definition by landmarks on the bony skull

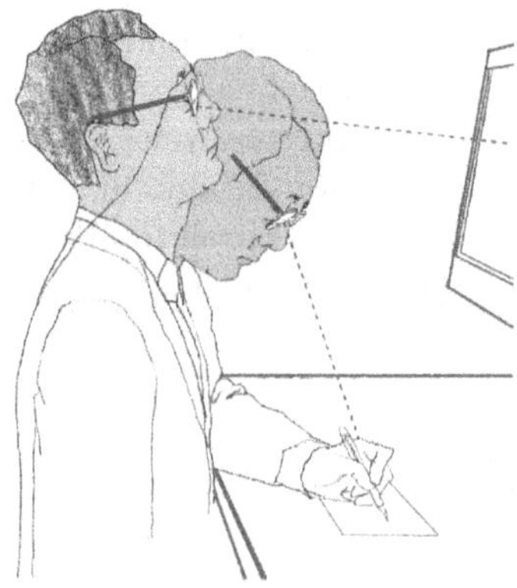

FIGURE 5.3 Excessive eye/head/neck movements can be avoided by placing visual targets close to each other. Adapted from Kroemer and Kroemer (2017).

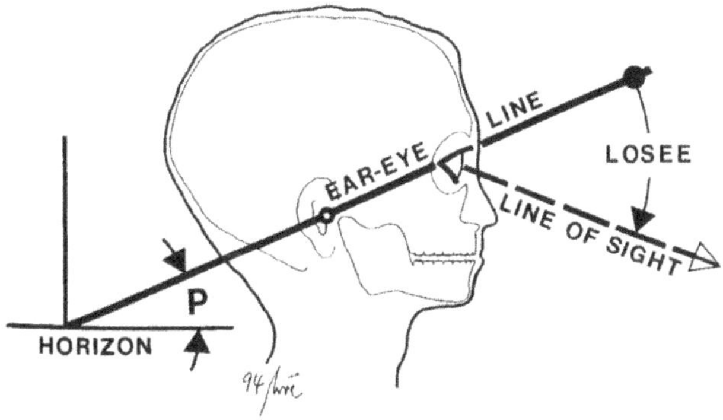

FIGURE 5.4 LOS is usually angled down from to the EE line. LOSEE is the angle between LOS and EE; P is the position angle between EE and the horizon. LOS, line of sight; EE, ear–eye.

made it difficult to use. The ear–eye (EE) line is simpler to establish: it runs through the easily discernible ear hole and the junction of the eyelids, as shown in Figure 5.4 in a side view of the head. The angle between the LOS and EE is the pitch angle of the eye, LOSEE, as it looks at the target. LOSEE* is optimal around 45° when reading text, whether on paper or a computer screen. LOSEE becomes smaller as the observed object moves farther away; it is about 15° (close to horizontal) when we look straight ahead. The position angle *P* describes how we hold the head (EE line) compared to the horizon. When the head is upright, *P* is about 15°.

Size of the visual target

If a visual target is not a point but has a noticeable length or breadth perpendicular to the LOS, the target size is typically expressed as the subtended visual angle: the angle formed at the pupil. The magnitude of this angle α depends on the size *L* of the object and the distance *D* from the eye, as shown in Figure 5.5. The subtended visual angle is usually described in degrees of arc (with 1° = 60 minutes = 60 × 60 seconds of arc):

$$\alpha(\text{degrees}) = 2 \times \arctan\left(0.5 \times L/D\right)$$

For visual angles not larger than 10°, this can be approximated by

$$\alpha(\text{degrees}) = 57.3 \times L / D$$

or, to calculate the angle in minutes,

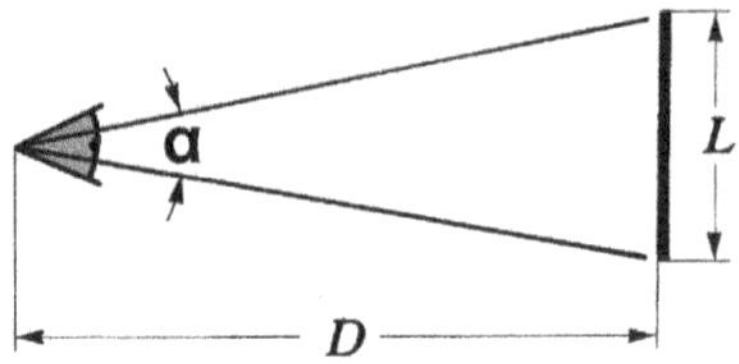

FIGURE 5.5 Subtended visual angle (α) depends on the object's size (L) and distance (D) from the eye.

Table 5.1 Visual angles of familiar objects

Object and size (L)	Distance (D)	Visual angle (α)
Instrument, 5 cm diameter	0.5 m	5.7°
The moon or the sun	> 300,000 km	About 30 minutes
Character on computer monitor	0.5 m	About 17 minutes
Individual letter on the page of a book at reading distance	0.4 m	About 13 minutes
U.S. quarter coin, 2.4 mm	At arm's length, 0.7 m	2°
	At 82 m (90 yards)	About 1 minute
	4.8 km (3 miles)	About 1 second

Source: Adapted from Table 5.2 Kroemer Elbert, Kroemer, and Kroemer Hoffman (2018).

$$\alpha\left(\text{minutes of arc}\right) = 60 \times 57.3 \times L / D = 3438 \times L / D$$

The human eye can perceive, at a minimum, a visual angle of approximately 1 minute of arc. Table 5.1 lists the visual angles of some familiar objects. For ease of use, technical products should be designed so that the angle subtends at least 15 minutes of arc, increasing to at least 21 minutes of arc under low light levels.

Diopter

The target distance D is measured in meters; its reciprocal, $1/D$, is called the *diopter*. The diopter indicates the optical refraction needed for best focus. Thus, a target at infinity has a diopter value of zero, whereas a target at 1 m distance has a diopter value of 1 (unity). Table 5.2 provides values for some typical target distances.

Table 5.2 Target distances and associated focal points

Target distance D	Focal point (diopter)
Infinity	0
4 m	0.25
2 m	0.5
1 m	1
0.67 m	1.5
0.5 m	2
0.33 m (1 foot)	3
0.25 m	4
0.2 m	5
0.1 m (4 inches)	10

Focusing

Accommodation describes the eye's ability, primarily through shaping its lens, to bring objects at varying distances into sharp focus: from infinity down to the nearest point of distinct vision. If we hold up a finger in front of the eye, we can focus sharply on it, while the background remains blurred. Alternatively, we can concentrate on features of the background, leaving the finger indistinct. We can see clearly only those objects whose images can be focused on the retina.

Incessant changes

If our gaze sweeps across various objects in the near field of vision, the lens must continuously change its curvature to adapt its focal length, ensuring sharp images appear on the retina. Even when maintaining focus on a nearby target, the muscles continuously adjust their contracting forces. For example, when we read text, the lens does not remain still but oscillates at a rate of about four times per second. Simultaneously, the iris continuously adjusts the size of its central opening (the pupil). This keeps its dilator (opening) and sphincter (closing) muscles constantly active, adjusting the diaphragm of the eye according to the light conditions in the visual field. During daylight, the pupil size is usually 3–5 mm in diameter; in low light such as at night, the pupil size exceeds 8 mm. The pupil contracts when we focus on near objects and opens when the lens relaxes. Furthermore, the pupil responds to emotional states, dilating under strong emotions such as alarm, joy, pain, or intense mental concentration, and narrowing with fatigue and sleepiness. Therefore, changes in pupil size have been used to assess a person's attention and attitude.

Overcoming ocular problems

A healthy, young eye can accommodate vision from infinity to very close distances, such as 10cm, indicating a diopter range from zero to about 10. As we age, the minimal distance increases to about 5 diopters (20cm) at around age 40 and to about 1 diopter (1 m) at age 60, on average. With increasing age, the eye's accommodation capability decreases because the lens becomes stiffer (less elastic) by losing water content. The stiffer lens is less able to focus light rays from targets at varying distances to converge precisely on the retina. If the convergence point is in front of the retina, the condition is called *myopia*; if behind, *hyperopia*. A nearsighted (myopic) person has no trouble seeing close objects but finds it difficult to focus on distant targets. This condition often improves with age, as the lens commonly remains flattened. In contrast, farsightedness (hyperopia) usually becomes more pronounced with age*, making it increasingly difficult to focus on near objects. Both problems, myopia and hyperopia, can be corrected fairly easily through the use of artificial lenses (contact lenses or eyeglasses), as shown in Figure 5.6. More invasive techniques to correct vision may include using a laser or other energy source to reshape a deformed cornea.

More light, please!

In many individuals, the pupil shrinks with age, as the tiny muscles that control it weaken. This results in less light striking the retina. To achieve sufficient visual acuity, many older people need to increase the illumination on visual objects. Another problem often encountered with age is the yellowing of the vitreous humor. The more yellow it becomes, the more energy it absorbs from the light passing through; consequently, this condition is also mitigated by increasing the illumination of the visual target. However, light rays may be refracted within the yellowed vitreous humor, leading to the perception of a light mist or veil in the visual field. If bright lights are present, the resulting "veiling glare" can significantly reduce one's vision. Artificial corrective lenses cannot rectify the problem of yellowed vitreous humor.

Floaters

So-called *floaters* appear as small flecks in front of the eye. In reality, they consist of small clumps of gel or cellular debris suspended in the vitreous humor. Often, they go unnoticed because the eye adjusts to these imperfections. Floaters are visible to the individual only when they are in the LOS, casting a shadow on the retina. They are more easily perceived when one looks at a plain background. Fortunately, occasional floaters are usually harmless.

Checking for glaucoma

Glaucoma is a leading cause of blindness in the United States and other regions, especially among older individuals. It is a disease

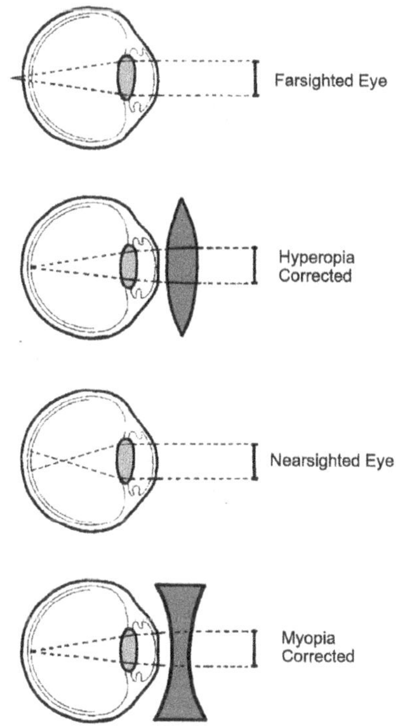

FIGURE 5.6 Convex and concave lenses can correct deficiencies in refractive (focusing) ability.

that damages the optic nerve, usually due to high pressure inside the eye. Regular eye examinations help detect the early stages of glaucoma and prevent further damage, as early glaucoma has no symptoms and damage caused by it cannot be reversed.

Help for ailing eyes

Another common issue in the aging eye is *cataracts*: cloudiness in the normally clear lens. When vision is severely impaired, an eye surgeon may remove the cloudy lens and replace it with an artificial intraocular lens implant.

Two other vision deficiencies, usually not related to aging, can often be improved through surgery or corrective lenses. *Astigmatism* occurs when the cornea or lens is not uniformly curved, causing light entering the eye to be bent unevenly. As a result, an object may not be sharply focused on the retina, depending on its position within the visual field. Often, the astigmatism is a spherical aberration, meaning light rays from an object located at the side are more strongly refracted than those from an object at the center of the field of view, or vice versa. *Chromatic aberration* occurs when the lens cannot focus all wavelengths of light onto the same point on the retina, resulting in "fringes" of color or blurriness at the edges of an object. This condition is fairly common: an eye may be hyperopic for long wavelengths (red) and myopic for short wavelengths (violet or blue).

Color perception

Even for fully functioning eyes, the perception of color requires sufficient light energy arriving at the cones to activate them. Color vision deficiencies occur in approximately 4%–8% of the male population (more common in those of European descent and less in those of Chinese descent), whereas only about 0.5% of the female population experiences this issue. Color "weakness" is more prevalent, whereby individuals are unable to distinguish as many gradations of color as those with normal color vision, particularly in differentiating between red and green. A few individuals are missing one of the cone systems and therefore cannot distinguish some basic colors from one another. Very few people can see only one color or none at all. In general, color vision also diminishes with age and at high altitudes (starting around 1500 m).

"Blind" at night

Night blindness (*nyctalopia*) is the condition in which a person has difficulty seeing in dim light—that is, when the visual target has low illumination or luminance.

5.3 Designing for dim and bright conditions

We need light to see

Without light, we cannot see. The sun, a lamp, or an electronic display all generate light. The moon, the walls of a room, or a printed page all reflect light. The most important factor for human vision is luminance, the light energy reflected from a surface (unless we stare directly into a light source). The energy that falls onto the retina stimulates rods and cones. Because the human eye adapts to lighting conditions, our eyes do not convey reliable information about the absolute lighting level, but they respond to variations in illumination, luminance, and color as these occur over time and space.

Luminance and illuminance

Luminance is the light reflected or emitted from an object. It indicates the brightness of light and can be used to quantify the brightness of a display (in units of candela per square meter). *Illuminance* or illumination is the light that falls onto an object and spreads over the surface area. Illuminance indicates how we perceive the brightness of an illuminated area (in units of lux or foot-candle).*

Seeing colors

The fully functioning human eye senses and adapts to increases and decreases in illumination of the retina over a wavelength range of about 380–720 nm, from violet to red. We can see objects if they are "bright" enough (either by generating light or reflecting it) to carry sufficient visible energy onto the retina. On the retina, the minimal intensity required to trigger light perception is 10 photons, causing an illuminance of about 0.01 lux. At such low intensity, the primary perception is of dim light, not color, because only rods are activated. When the illuminance of the retina exceeds about 0.1 lux, both rods and cones respond, and the cones report colors. We experience this at twilight and dawn, when we can see color in the brighter sky, but all dimmer objects appear only in shades of gray as reported by the rods.

Seeing in the dark

Some interesting visual phenomena can occur in dim light and darkness:

- If one stares at a single light source against a dark background, the light appears to move; this is the so-called autokinetic phenomenon.
- Night vision capabilities deteriorate with decreasing oxygen. At an altitude of about 1300 m, vision is reduced by about 5%; at 2000 m, the reduction is about 20% (and up to 40% in smokers whose blood has lost some capability to carry oxygen).
- If the horizon shows no visual cues, the lens relaxes and focuses at a distance of about 1–2 m, making it difficult for a person to notice distant objects; this is known as night myopia.

Adapting between dark and light

Between dark and bright conditions, the eye can change its sensitivity across a wide range of illumination and luminance. Figure 5.7 depicts the luminance values that we perceive.

Adjustments of the pupil, stimulation of rods and cones, and spatial summation of stimuli all contribute to visual sensitivity. The actual response of the eye during adaptation to darkness or light also depends on the luminance and duration of the previous condition to which the eye had adapted, the wavelength of the illumination, and the location of the light stimulus on the retina.

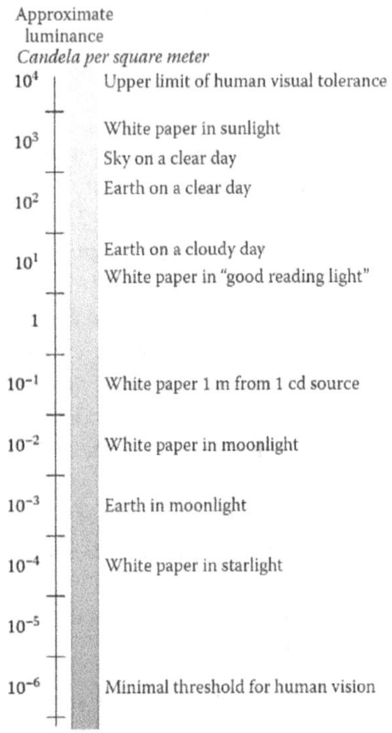

FIGURE 5.7 Luminance of objects that humans can see.

Adapting to darkness

Full adaptation from light to darkness takes up to 30 minutes; during this period, cones are initially the most sensitive, with rods following more slowly. Adaptation to darkness primarily depends on the change in threshold in the cones. Even after full adaptation, sensitivity at the fovea (which has many cones) is much less than at the periphery of the retina (which has a predominance of rods). Therefore, weak lights can be detected in the periphery of the field of view (where more rods are present), but not when looking directly at them, as they appear on the fovea (dense with cones). People who suffer from complete "dark blindness" have nonfunctional rods and can adapt only via cones. As we move from a lighted environment and adapt to darkness, our eyes adjust more quickly to red and yellow lights than to blue. Adaptation to darkness becomes slower as we age.

Adapting to light

In contrast, adaptation to light occurs quickly, usually achieved within a few minutes. During this time, the fovea perceives wavelengths in the yellow region of the visible spectrum most effectively. Therefore, it is best to illuminate instruments in vehicles driven at night with yellowish or reddish light, with yellow likely being more suitable for older adults' eyes. These colors maintain the dark adaptation of the rods, allowing the driver to observe the predominantly black, gray, and white events on the road at night. Furthermore, the driver's eyes adjust relatively quickly to red and yellow lights after being in a lit environment at night.

Visual acuity

Visual acuity can be defined in several ways, typically as the ability to detect small details and discriminate between small objects at a given distance. It depends on the shape of the object and on the wavelength, illumination, luminance, contrast, and duration of the light stimulus.

Testing acuity

Visual acuity is usually measured at viewing distances of 6 m (20 ft) and 0.4 m (1.3 ft) because the factors determining the resolution of an object differ between far and near viewing. To determine acuity, high-contrast patterns are presented to the observer. The most common patterns are Snellen letters or Landolt rings; see Figure 5.8. The smallest detail detected or identified is taken as the threshold—for example, the location of the gap in the Landolt ring or the positional offset on the Vernier scale. These measures of acuity primarily depend on the ability to see edge differences between black and white stimuli at relatively high illuminance levels. Although this measurement of static edge acuity is simple, it is neither the only nor the best measure of visual resolution capabilities. For example, individuals with perfect Snellen acuity (able to see small letters at a distance) may struggle to detect targets against a busy background or observe highway signs at specific distances.

Perceiving color

Sunlight contains all visible wavelengths of the spectrum, but objects illuminated by the sun absorb some of the radiation. Thus, the light that we see on an object is what it transmits or reflects; its energy distribution differs from the illumination the object received. However, a human observing the object does

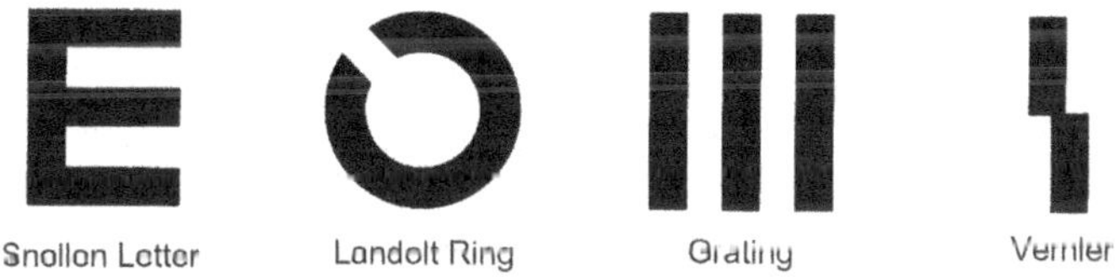

FIGURE 5.8 Examples of visual targets used in acuity testing.

not analyze the spectral composition of the light reaching their eyes. In fact, what appears to be the same color may have different spectral contents*.

Trichromatic vision

Color-matching experiments demonstrate that humans can perceive the same color stemming from various combinations of the three primary colors: red, green, and blue. Therefore, human color vision is termed trichromatic*.

Experiencing color

The brain does not measure wavelengths; it simply classifies incoming signals from different groups of wavelengths. We judge colors by comparison and we name them by habit. Some societies do not distinguish between certain colors of the spectrum, lumping blue and green together, while others have many words for shades of red. Human color perception is more of a psychological experience rather than a specific property of the electromagnetic energy we see as light.

Describing colors

The physics of color stimuli arriving at the eye can be described scientifically (although often with considerable effort); however, perception, interpretation, and reaction to colors are highly individual and variable. Thus, people find it very difficult to describe colors verbally, given the many possible combinations of individually perceived hue, lightness, and saturation values.

Reacting to colors

People may experience emotional reactions to color stimuli. For example, reds, oranges, and yellows are usually considered warm and stimulating. Violets, blues, and greens are often perceived as cool and may generate sensations of cleanliness and restfulness. However, the attraction to certain colors and their combinations varies by age, culture, and region. For example, red may evoke a sense of danger or urgency in Western countries and the Middle East, purity in India, or luck in China. Green may be linked with nature and health in Western countries, or with infidelity in China. Blue may be associated with safety and trust in Western cultures (hence, often used by banks or police uniforms) or with spirituality in Eastern cultures.

Designing illumination

The characteristics of human vision provide the basis for engineering procedures to design ergonomic environments for proper vision. Here are the most important concepts:

- Proper vision requires appropriate luminance of an object, that is, the energy reflected or emitted from it that meets the eye.

- Incident illuminance on an object, and how much of it is reflected, determine its luminance.
- Quantity and direction of illumination need to be carefully selected by the designer of the task or workplace.
- Special requirements for visibility, such as the diminished seeing abilities of the elderly, require particular care in the arrangement of proper illumination.
- The use of colors, if selected properly, can be helpful; but color vision requires sufficient light and is affected by how the user perceives the color.

Summary

The eyes provide a significant portion of the information we need in daily life and at work. To see objects and events in detail, they must be in bright light. When the light dims, objects lose their colors; in darkness, they become invisible. Even a well-lit visual target must be at a distance that allows us to distinguish specific and relevant details. If our eyesight is impaired, which is common as we age, it is particularly important to have appropriate lighting, and we may need to use artificial lenses or other interventions.

Steps to fit the human

Step 1: Ensure that the eyes function properly; if in doubt, consult an eye physician.

Step 2: Determine which specific task must be completed.

Step 3: Provide a suitable combination of visual targets and illumination for the task. Consider any accommodations that may be necessary based on the visual acuity required.

Step 4: Design, then test. Modify and iterate as necessary, adjust conditions as desired, and redesign if the user or the task changes.

Notes

The text contains markers, *, to indicate specific references and comments, which follow.

5.1 Our eyes

Optics: The branch of physics that deals with the properties and behavior of light, especially regarding vision and other instruments that use visible, ultraviolet, and infrared light.

For more on how the eye senses light (photoreception), see Molday and Moritz (2015).

5.2 Seeing the environment

The visual field: Sheedy (2006).

Computer use and eye fatigue: Symptoms related to "digital eye strain" may include blurriness, light sensitivity, dry eyes, and double vision. One common recommendation is to follow the so-called 20/20/20 rule, where the affected person takes a 20-second break every 20 minutes to look 20 feet (6 m) away; although this is an easy and low-risk intervention, the clinical evidence that this helps is weak.

LOSEE is best: Hill and Kroemer (1986), Sheedy (2006).

Farsightedness: The general term for farsightedness (decreased near-vision) is hyperopia; when it occurs in a person over about age 40 due to natural aging of the crystalline lens, it is called presbyopia.

5.3 Designing for dim and bright conditions

Luminance and illumination: Luminance is light reflected or emitted from an object, and illumination is light falling upon an object. One way to remember these terms is to note that (**L**)uminance is the light (**L**)eaving an object, whereas (**I**)lluminance refers to the (**I**)ncident light. Definition and measurement of light has been complicated in the past by the use of different procedures and units, as explained by Boyce (2014), Howarth (2015).

Spectral contents of color: For measuring and specifying colors see, for example, Howarth (2015). For ergonomic applications, see, for example, Kroemer Elbert, Kroemer, and Kroemer Hoffman (2018).

Trichromatic vision: The data of a color-matching experiment can be displayed as vectors in a three-dimensional (red, green, and blue) space, called a *chromaticity diagram*, standardized in 1931 and still in common use.

References and further reading

Boyce PR (2014). *Human Factors in Lighting*. Boca Raton, FL, CRC Press. https://doi.org/10.1201/b16707

Hill SG, Kroemer KHE (1986). Preferred declination of the line of sight. *Human Factors*, 28(2):127–134. https://doi.org/10.1177/001872088602800201

Howarth PA (2015). Assessment of the visual environment. Chapter 25 in Wilson JR, Sharples S(eds), *Evaluation of Human Work*, 4th ed. Boca Raton, FL, CRC Press. https://doi.org/10.1201/b18362

Kroemer Elbert K, Kroemer HB, Kroemer Hoffman AD (2018). *Ergonomics: How to Design for Ease and Efficiency*, 3rd ed. London, UK, Elsevier. https://doi.org/10.1016/C2016-0-02516-0

Kroemer AD, Kroemer KHE (2017). *Office Ergonomics: Ease and Efficiency at Work*, 2nd ed. Boca Raton, FL, CRC Press. https://doi.org/10.1201/9781315368603

Molday RS, Moritz OL (2015). Photoreceptors at a glance. *Journal of Cell Science*, 128(22):4039–4045. https://doi.org/10.1242/jcs.175687

Sheedy J (2006). Vision and work. Chapter 18 in Marras WS, Karwowski W (eds), *The Occupational Ergonomics Handbook: Fundamentals and Assessment Tools for Occupational Ergonomics*, 2nd ed. Boca Raton, FL, CRC Press. https://doi.org/10.1201/9781420003635

CHAPTER SIX

How we hear

Ears

Our ears sense energy from the outside world in the form of pressure waves in the air and convert these into nerve impulses, which the brain then integrates into a psychoacoustical perception of the surrounding sound environment. However, this interpretation is a subjective modification of what the ear reports. For example, the perceived sound and appeal of music differ from person to person and are likely to change as individuals grow older or form different associations with musical sounds.

People differ greatly in their interpretation of acoustical events, depending on their experiences, attitudes, and preconceived ideas. They also vary in their ability to recognize sounds and react to them, with aging often bringing significant changes. Nevertheless, there are general similarities in how human hearing functions, which allow us to develop ergonomic recommendations. We need to understand how we hear in order to fit the task to the human.

6.1 Our ears

Pathways of sound

Sound reaches the inner ear via two paths: conducted by bones or by air. For sound to be transmitted through bony structures, very high intensities are required. Normally, the sound we perceive travels through the air: from outside through the ear canal, where it excites the eardrum and then the structures behind it, as described in the following text.

A sound wave is a pressure wave generated by vibrations. The outer ear (*auricle* or *pinna*) collects airborne sound waves and funnels them into the auditory canal (*meatus*). The eardrum (*tympanic membrane*), which separates the air-filled middle ear

DOI: 10.1201/9781032674575-9

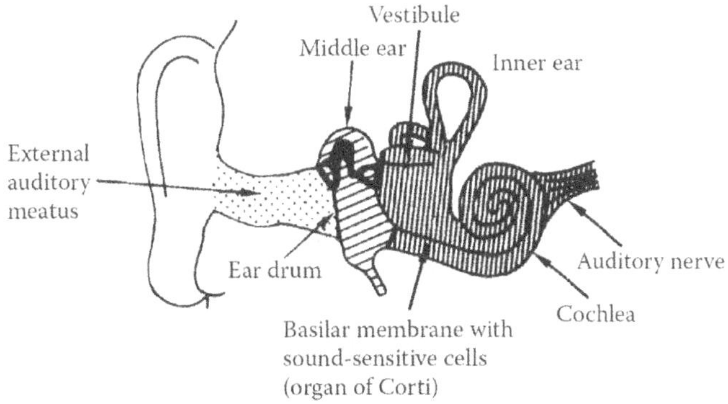

FIGURE 6.1 Main features of the anatomy of the ear: air-filled outer and middle ear sections, separated by the eardrum, and the fluid-filled inner ear with the cochlea and the vestibule.

from the environment, converts the pressure (sound) wave back into a mechanical vibration. In the middle ear, these vibrations are amplified by three tiny bones (the *ossicles*) and passed to the inner ear through two membrane-covered windows (one called *oval*, the other *round*). A watery fluid (called *endolymph* or *perilymph*) fills the inner ear and carries the vibrations to the organs that sense sound (the *organ of Corti* in the *cochlea*, connected to the *auditory nerve*) and body position (the *vestibule*, discussed in Chapter 7). The auditory nerve transmits electrical signals to the brain, which processes and interprets them as sound. Figure 6.1 shows the main features of ear anatomy.

Outer ear

As a sound wave moves through the outer ear and auditory canal, resonance amplifies the sound intensity by 10–15 decibels (dB; Section 6.2 explains the units) by the time it reaches the eardrum. After traveling through the ear canal, the sound wave arrives at the eardrum, causing it to vibrate according to the frequency of the sound.

Middle ear

The eardrum's vibrations are mechanically transmitted to three small bones (*ossicles*) in the middle ear: the hammer (*malleus*), anvil (*incus*), and stirrup (*stapes*). The ossicles further amplify the sound and transmit it to the oval window. The area of the eardrum (about 60 mm^2) is larger than the surface of the oval window (about 4 mm^2), so the sound pressure entering the inner ear at the oval window's membrane is about 15 times greater than it was at the eardrum.

Inner ear

The inner ear contains the receptors for hearing and body position (the vestibular system; see Chapter 7). Fluid shifts in the inner ear transmit sound waves from the oval window to the round window through the *cochlea*, a tube about 2 mm in diameter and 30 mm in length, wound into two and a half turns like a snail shell. The fluid's motion deforms the *basilar membrane*, which runs the length of the cochlea and houses the *organ of Corti*. These movements stimulate sensor *cilia*, which are feathery hair cells within the organ of Corti. Depending on their structure and location, the sensors respond to specific frequencies and generate impulses transmitted through the auditory nerve to the brain for interpretation.

The *Eustachian tube* connects the middle ear to the pharynx (part of the respiratory system). When the tube is open, it equalizes air pressure in the middle ear with external air pressure. However, when the tube is obstructed, such as during nasal congestion from a cold or an ear infection, pressure equalization may fail, leading to feelings of pressure or pain in the ear and impaired hearing. During airplane flights, especially during rapid descents, a clogged Eustachian tube can delay pressure equalization between the inner ear and the environment. You may try to open the tube by chewing gum or excessive yawning, but "pumping" your outer ear with your hand will not help your middle ear.

6.2 Hearing sounds

Sound

Sound is a vibration that stimulates auditory sensations. A sound contains a mixture of frequencies, whereas a pure *tone* is an oscillation with a constant frequency. The unit of measurement for frequency is hertz (Hz, oscillations per second). We often describe our personal perception of tone frequencies as "pitch."*

Frequencies we can hear

We cannot hear sounds below 16 Hz, although we may feel such infrasonic vibrations*; nor can we hear sounds above 20 kHz, though dogs and other animals can hear such ultrasonic frequencies. Infants can hear tones from about 16 to 20,000 Hz (20 kHz), a span of nearly 9 octaves. With aging, the ability to hear high frequencies diminishes significantly. Older adults can rarely hear tones above 10 kHz. Our normal hearing is most sensitive between about 1 and 5 kHz. In human speech, vowel sounds are below 1 kHz, while sibilant consonants can exceed 5 kHz; most speech occurs within the 300–700 Hz range.

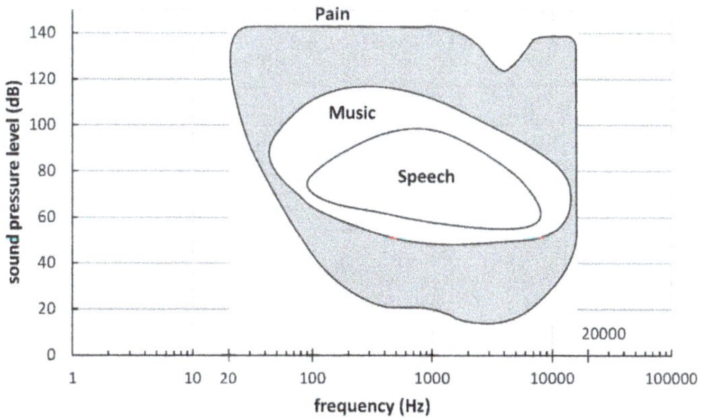

FIGURE 6.2 Typical hearing ranges of adult humans. Adapted from Kroemer Elbert, Kroemer, and Kroemer Hoffman (2018).

How loud?

We use the term "loudness" to describe our sensation of the intensity (sound pressure level (SPL), power, amplitude) of an audible vibration. The low-pressure threshold of human hearing is about 20 μPa (20×10^{-6}N/m^2; 1 Pa=1 N/m^2) in the frequency range of 1000–5000 Hz. When sound pressure exceeds 140 Pa, the ear experiences pain. Figure 6.2 shows the ranges of human hearing.

10 dB=1 bel

We perceive sound pressure in a roughly logarithmic manner. Therefore, loudness is measured in a logarithmic unit* called the decibel (dB). The threshold of hearing (20 μPa) serves as reference. The SPL is the ratio between two sound pressures: P is the current sound pressure (in root-mean-square, rms) and P_0 is the reference value at the threshold of hearing. Thus, the equation for SPL in decibels is as follows:

$$\text{SPL} = 10\log_{10}(P^2/P_0^2) = 20\log_{10}(P/P_0)$$

Thus, in decibels, the dynamic range of human hearing from reference (20 μPa $= 20 \times 10^{-6}$ Pa) to 200 Pa is as follows:

$$\text{Range of human hearing} = 20\log_{10}\left[200/\left(20\times10^{-6}\right)\right]$$
$$= 20\times7 = 140 \text{ dB.}$$

The sound intensity level (SIL) refers to the "power" of the sound and is defined (in units of dB) as follows:

$$\text{SIL} = 10\log_{10}(I^2/I_0^2)$$

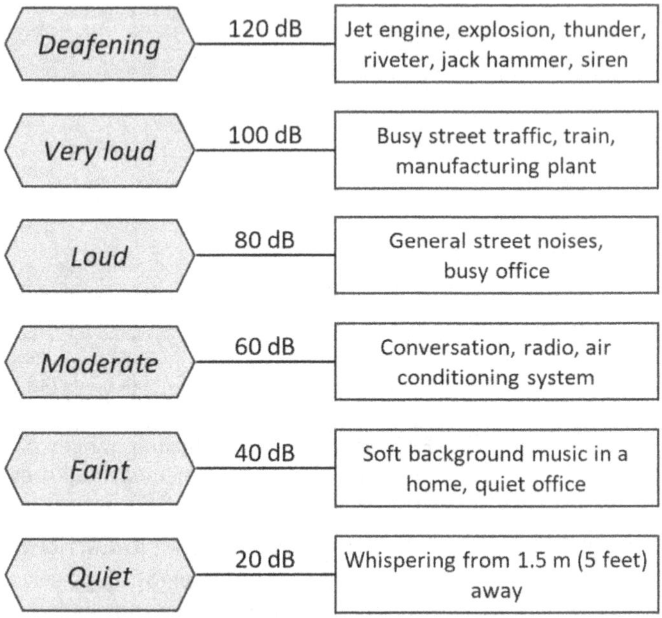

FIGURE 6.3 Typical SILs. Adapted from Kroemer Elbert, Kroemer, and Kroemer Hoffman (2018). SIL, sound intensity level.

where I is the (rms) sound intensity and I_0 is the reference sound intensity, taken to be the minimal intensity threshold of hearing (10^{-12} W/m^2). Figure 6.3 illustrates ranges of typical SILs.

Emitted sound power (such as from a loudspeaker or a running engine) is measured in watts (W). Sound power relates to sound pressure via the equation W/m$^2 = P^2$/[(air density) × (speed of sound)]. Thus, sound power is proportional to the square of sound pressure.

Sounds occurring together

Due to the logarithmic scale, doubling the sound pressure results in a 6dB increase in SPL; a 1.41-fold increase in sound pressure leads to a 3dB increase.* If two sounds occur simultaneously, one can approximate their combined SPL based on the difference between their intensities as follows:

- If the difference is 0dB, add 3dB to the louder sound.
- If the difference is 1dB, add 2.5dB to the louder sound.
- If the difference is 2dB, add 2dB to the louder sound.
- If the difference is 4dB, add 1.5dB to the louder sound.
- If the difference is 6dB, add 1dB to the louder sound.

- If the difference is 8 dB, add 0.5 dB to the louder sound.
- If the difference is 10 dB or more, consider the louder sound alone.

This illustrates that the louder sound dominates. Therefore, we can measure the sound level emitted by a specific source (for example, a horn or a rumbling machine) even when background sound is present, provided the source sound is at least 10 dB louder than the background.

Psychophysics of hearing

Physical measurements describe acoustical events, but people interpret and react to them subjectively; for example, you may find certain sounds attractive while someone else finds them annoying. The sensation of a tone or a complex sound depends not only on its intensity and frequency but also on our emotional response to it.

Loudness

Due to the specific sensitivities of human hearing, both frequency and intensity of a sound influence our perception of its loudness in a complex manner. Sounds at more than 100 dB are ear-splitting at all frequencies, but quieter sounds exhibit different patterns, particularly at frequencies below 1000 Hz. Figure 6.4 displays phon* curves, which indicate the combinations of tone frequency and sound pressure that appear equally loud as they do at 1 kHz. The highest contour represents the

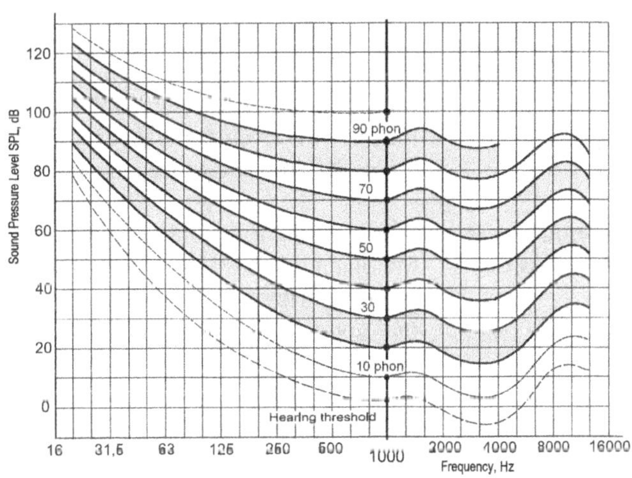

FIGURE 6.4 Phon curves represent combinations of sound pressure and frequency that we sense as equally loud. Adapted from ISO 226, most recently revised in 2023.

threshold of pain, while the lowest contour signifies the minimal threshold of hearing.

Equally loud

Following a phon contour, such as that of 50 phon, we see that at frequencies below about 500 Hz, the SPL of low-frequency tones must be increased to appear equally loud. For example, a 125 Hz tone requires an SPL of nearly 70 dB to sound as loud as a 1000 Hz tone at 50 dB. A 50 Hz tone at 45 dB is barely audible (i.e., at the hearing threshold), similar to a 200 Hz tone at 15 dB; both are near the zero phon contour. However, at higher frequencies, especially in the range of approximately 2000–5000 Hz (where healthy hearing is rather sensitive), the tone intensity can be lowered and still sound as loud as a tone at 1000 Hz. Above 8000 Hz, the SPL in decibels must be increased, above the level it had at 1000 Hz to sound equally loud.

These curves of equal loudness are valid only for pure tones because they do not reliably reflect the subjective impression of loudness when different frequencies are heard simultaneously (which is the case in most daily environments). Although the phon curves are imperfect, they illustrate the sensitivity of human hearing across different frequency ranges. For example, Figure 6.4 confirms that low-frequency (bass) tones at low intensities are difficult to hear.

Measuring like a human hears

Filters on sound-measuring equipment are used to account for human sensitivity to tones of different frequencies. These filters, usually in the form of software, adjust the physical readings to match human perception. The A-weighting filter* mimics the human hearing response at 40 dB. The notation dBA or dB(A) is used to describe A-weighted SPL values. Two sounds can have the same dB levels but different dBA levels since the human ear is less sensitive to low frequencies.

Responding to music

Music is one of the oldest forms of human expression and art. Making music has long accompanied various activities, such as singing during fieldwork or playing rhythmic tunes for marching. Yet even today, definitive research on the psychophysical effects of different types of music and rhythm on well-being and productivity is sparse, partly because these effects are subjective and depend on the specific work environment.*

"Music while you work" is intended to break up monotony, generate mild excitement, and provide an emotional impetus for activity. It is usually not continuous (as most background music is, discussed below) but programmed to occur at specific times; it commonly features varying rhythms with vocals and popular, commercial tunes. Such music played during work has been linked to improved morale and productivity, but establishing definite connections between specific components of music and arousal theory is challenging. Thus, selecting musical content, rhythm, loudness, and timing during work relies more on guesswork than on science.

Muzak

Non-specific background music, sometimes called *Muzak* (or "elevator music"), functions like acoustical wallpaper in places such as shops, hotels, and waiting rooms (including electronic waiting rooms like telephone on-hold music). Its purpose is to create a welcoming atmosphere, relax customers, reduce boredom, and mask disturbing sounds. Typically, its character is subdued, its tempo intermediate, and it lacks vocals. Although it may feel pleasant to transient customers, it can be monotonous for those who are continuously exposed to it.

Selecting music is an art

Certain types of music appeal to different individuals in varying circumstances, yet predicting the relationship between listeners and tunes, and selecting specific music for certain activities, environments, and listener populations, is an art. Providing music is popular partly because it seems to have plausible effects and partly because it serves as an effective market strategy: many people appear happier and more productive when there is music playing in the background. Moreover, even if not effective, music seems harmless when it is quiet and if the listener can control it.

Acoustic experiences

Some interesting acoustic events occur for human hearing*:

- *Directional hearing*: Humans can determine the direction of a sound based on the difference in arrival times (phase difference) or intensities (due to the inverse-square law of energy flux). However, the ability to use stereophonic cues varies among individuals and is greatly reduced when earmuffs are worn.
- *Distance hearing*: The ability to gauge the distance of a sound source is related to the fact that sound energy diminishes with the square of the distance traveled; however, human perception of energy also depends on the frequency of the sounds. A sound source is perceived as further away when it is low in intensity and frequency, and closer when it is high in intensity and frequency.
- *Doppler effect*: As the distance between the source of sound and the ear decreases, the pitch heard increases; as

the distance increases, the sound appears to have a lower frequency. The greater the relative velocity, the more pronounced the frequency shift. This Doppler effect can be used to measure the speed at which the source and receiver are moving relative to each other.

- *Common-difference tone*: When a common frequency interval of 100 Hz or more separates several tones, an additional frequency is perceived based on the common difference. This effect explains how one can hear a deep bass tone from a sound system that is physically incapable of producing such a low tone.
- *Concurrent tones*: When two tones of the same frequency are played in phase, they are heard as a single tone, with loudness equal to the sum of the two tones. When two identical tones are played exactly (180°) out of phase, they cancel each other completely and cannot be heard. This physical phenomenon, called destructive interference or phase cancellation, can be used to suppress the propagation of acoustical or mechanical vibrations and, with current technology, is especially effective at frequencies below 1000 Hz.

6.3 Addressing noise

Noise

Noise is any unwanted, disruptive, and objectionable sound, loud or quiet; many aspects of noise are psychological and subjective. For instance, the dripping of water produces brief, low-intensity sounds that can be perceived as noise, just as a neighbor's loud, complex music may be. Noise surrounds us day and night, at home, at work, and everywhere in between. There are many sources of noise, such as barking dogs, overheard conversations, automobile traffic, trains, construction, industrial equipment, spectator events, and even our own music that we enjoy through loudspeakers, headsets, and earphones.

What noise can do?

Any sound may be annoying and thus perceived as noise. The threshold for noise annoyance varies depending on individual sensitivity and mental state. Noise impacts us in many ways, including

- Creating negative emotions, such as surprise, frustration, anger, and fear.
- Delaying the onset of sleep, awakening a person from sleep, or disturbing someone's rest.

- Making it difficult to hear desirable sounds.
- Producing temporary or permanent alterations in body chemistry.
- Interfering with sensory and perceptual capabilities, thereby degrading task performance and masking important alarms.
- Temporarily or permanently altering one's hearing capability.

TTS, PTS

Changes in hearing capability due to excessive noise are physical, not merely psychological. Noise-induced hearing impacts can result from single or repeated exposure to loud sounds. *Temporary threshold shift* (TTS) is a temporary loss of hearing, usually at higher frequencies, often accompanied by a sensation of ringing (*tinnitus*). With TTS, hearing typically returns to its previous level within minutes or days. Exposure to intense sounds, such as from an explosion, can lead to a *permanent threshold shift* (PTS), which results in irreversible hearing loss. PTS may stem from damage to the bones of the middle ear or, in the inner ear, from damage to the cochlear cilia, the organ of Corti, the basilar membrane, or the nerves leading to the central nervous system. The specific injuries depend on the frequency and intensity of the incident noise. The timing and severity of the loss depend upon exposure duration, sound characteristics (intensity and frequency), and the nature of the exposure (continuous or intermittent). Damage may be immediate, such as from a firecracker, or may occur over time due to continuous noise exposure.

One of the sinister aspects of noise-induced hearing damage is that victims may not realize they are developing an injury and therefore do not take action to avoid further damage. Noise-induced threshold shift occurs first at the frequency range of the noise, typically around 4000 Hz, but then spreads to higher and lower frequencies.

Noise and task performance

Simple and repetitive tasks show little impairment from noise, but it often degrades the execution of more difficult tasks, especially those requiring attention, perception, and information processing. Unexpected and irregular noise causes greater degradation than continuous noise; sudden onset can create a startle response.

Signal-to-noise "ratio"

One of the most noticeable effects of noise is its interference with spoken communications and the perception of signals. Workers in loud environments often complain that they must shout to be heard and that they cannot understand others who attempt to communicate with them. The difference between the intensities of speech (the signal) and noise is critical in determining whether speech and signals can be detected above the noise. For example, with speech at 80 dBA in a noise environment of 70 dBA, the signal-to-noise ratio (SNR) is simply the difference, +10 dB (not the mathematical ratio, which would be 80/70). Detection of a signal is a prerequisite for discriminating it from other signals, identifying and locating its source, recognizing its intended meaning or urgency, and judging its importance.

Shouting in noise

People tend to raise their voices to speak over noise and return to normal volume when the noise subsides; this is called the *Lombard reflex*. In a quiet environment, males typically speak at about 58 dBA in a normal voice, 76 dBA in a loud voice, and 89 dBA when shouting. Women normally speak at 2 or 3 dBA less at lower volumes and 5–7 dBA less at higher volumes. People can easily increase the SNR by raising their voices (the signal) at low noise levels, but this ability diminishes as noise levels rise. Above about 70 dBA, raising one's voice becomes inefficient, and above 85 dBA, it no longer works. Furthermore, this forced effort of shouting often decreases intelligibility because articulation becomes distorted at extreme voice outputs.

Noise-induced hearing loss

High-intensity noise can cause hearing damage. In industrialized countries, noise-induced hearing loss (NIHL) typically occurs around 4000 Hz. It may spread to lower ranges but primarily extends into higher frequencies, peaking at about 8000 Hz. However, reduced hearing at (and above) 8000 Hz can also result from aging, making it difficult to distinguish between cases related to noise exposure and those related to age.

Age-related hearing loss

Losing some hearing ability with age is normal. Generally, we can expect the following reductions due to gradual age-related hearing loss (*presbycusis*):

- 10 dB at 50 years,
- 25 dB at 60 years,
- 35 dB at 70 years.

Figure 6.5 shows the representative hearing capabilities measured in the same person at 20, 70, and 80 years of age. The audiogram of the younger person indicates fairly normal hearing, with all measurement points close to the zero-loss line. However, at age 70, hearing has decreased by about 20 dB in all frequencies below 2000 Hz. Above 2000 Hz, the audiogram shows the typical age- and noise-related reduction, which is most significant near 8000 Hz. The losses are even more

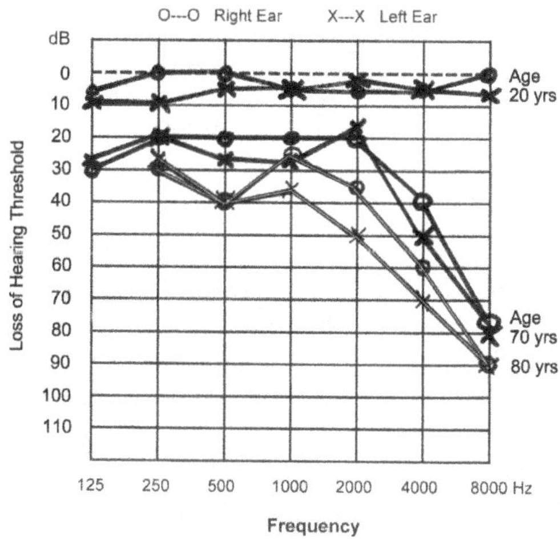

FIGURE 6.5 Audiogram taken on the same person at the ages of 20, 70, and 80 years.

pronounced at age 80; the reason for the larger decline in the left ear compared to the right side remains unexplained.

Damaging sounds Sounds that are sufficiently strong, lasting, or repetitive can damage one's hearing in the affected frequencies. Not everyone responds to sound in the same manner, but in general, sound levels of about 85 dBA or more are potentially hazardous because of the energy they contain. The magnitude of actual hearing loss directly relates to the sound level. Local laws often regulate permissible noise levels*. In the United States, current workplace regulations have a time-weighted recommended exposure limit for an 8-hour work shift of 85 dBA, 2.5 hours at 90 dBA, and 1 minute at 112 dBA. In the European Union, current workplace regulations limit daily noise exposure to 87 dBA.

Signs of hazardous sounds Simple subjective experiences can indicate exposure to hazardous sound. The presence of a dangerous sound environment is indicated by hearing sounds that are significantly louder than conversational levels, having difficulty hearing communications, perceiving ringing in the ears (tinnitus) after being in a noisy environment, and noticing that sounds seem muffled to the ears after leaving a noisy area.

Fighting NIHL Exposures to excessive sound, in either level or duration or both, generate risks of NIHL. Common sources of noise that contribute to NIHL include guns, many occupational environments such as construction and heavy industry, power tools, leaf blowers, chain saws, airplanes, vehicles, and, unfortunately, music (whether heard live or through loudspeakers or headphones). NIHL can occur regardless of whether one likes the sound. There are three major approaches to combat the damaging effects of noise.

- *Avoid generation*: The first, fundamental, and most effective strategy is to prevent the generation of excessive sound by properly designing machine parts such as gears or bearings, reducing rotational speeds, changing airflow, or replacing a noisy apparatus with a quieter one.
- *Leave the area*: Another effective strategy is to remove individuals from noisy environments altogether, or to do so at least for part of the work period to reduce exposure to loud sounds.
- *Impede transmission*: The final strategy is to impede the transmission of sound from the source to the listener. In occupational environments, this includes using mufflers on the exhaust side of a machine, encapsulating the noise source, physically increasing the distance between the source and the ear, and placing sound-absorbing surfaces in the sound's path. These sound-absorbing surfaces can include earplugs and other protective ear wear.

Planning for "no noise"

The fundamentally successful way to avoid noise is to not generate sound. Therefore, planning to avoid noise is crucial for engineers and architects. This is primarily achieved by selecting technologies and machines that do not produce unacceptable sounds. This design principle applies to offices, where machines, keyboards, and conversations should remain quiet. It also applies to assembly facilities, where no hand tools that create excessive noise, such as pneumatic cutters and mechanical riveters, should be used. This approach extends to manufacturing, where machinery and processes should be chosen for their quiet design. Noisy engines can be muffled or replaced, clinking dies eliminated, and hammering and riveting replaced by welding and gluing. Additionally, avoiding traffic noise in transportation is essential, such as placing mufflers on trucks and buses, cars and motorcycles, and implementing noise-reducing take-off and landing procedures for airplanes at airport runways.

Preventing noise propagation

Avoiding unnecessary sound by design is important, but certain machinery and jobs are inherently noisy. In such cases, the next step is to prevent noise propagation. The laws of physics indicate that sound energy decreases with the square of the distance from the source. Thus, the architect may choose to locate offices and other workplaces away from noise sources, such as airport traffic or machinery. A parking lot or structure can shield a residential neighborhood from a nearby factory. Within the factory, work areas should be as far from manufacturing noise sources as possible; intervening spaces, such as storage areas, can act as auditory buffers.

Noise barriers

An effective way to reduce noise propagation is to completely enclose a noise source. However, often we only place a partial barrier between ourselves and the source of noise. Outdoors, having trees and bushes can be beneficial, although their effectiveness as a sound barrier depends on their density: the air spaces among foliage and branches may require several rows of trees and bushes to achieve sufficient sound abatement. Bare trees in a leafless season do not block noise.

Another effective method for reducing noise propagation to individuals is to enclose them in buildings. Suitable materials for walls, doors, and windows can reduce the internal sound load by 20–30 dB. The "weak spots" in auditory protection are the openings for doors, windows, cables, and plumbing. Table 6.1 lists typical noise abatement levels that can be expected from various building materials.

Table 6.1 Noise attenuation by common building materials

Building component	Attenuation (dB)	Comment
Window, single glazing	20–25	Regular speech is clearly understandable on the other side of the window or door
Window, double glazing	25–30	
Regular single door	20–30	
Regular double doors	30–40	Loud speech is clearly understandable on the other side of the window or door
Window, soundproofed	30–35	
Heavy special door	40–45	Loud speech is audible on the other side of the door
Brick wall, 6–12 cm thick	35–40	Very loud speech may be audible on other side
Brick wall, 25–38 cm thick	50–55	No speech likely to be audible on the other side of the wall
Double brick walls, spaced apart, each 12 cm thick	60–65	

Hearing protection devices

The last resource for protecting human hearing is wearing a hearing protection device* (HPD) that reduces the harmful or annoying effects of sounds. HPDs can be worn externally (usually as sound-isolating helmets or earmuffs) or as earplugs inserted into the ear canal. These hearing protectors vary in effectiveness, depending partly on the intensity and frequency spectrum of the sound entering the ear and partly on the fit of the protective device to the wearer's ear.

Passive HPDs

Conventional HPDs are passive; they achieve attenuation by allowing sounds to pass through materials that absorb, dissipate, or otherwise impede energy flow. Despite these simple features, proper selection and correct use can make conventional HPDs highly protective by reducing ambient noise above about 80 dBA, which can strain human hearing. One disadvantage of most passive devices is that they reduce high-frequency sound more than low-frequency sound, diminishing the power of consonants and distorting speech. Individuals who have already experienced hearing loss at higher frequencies face further difficulties in understanding speech when using a conventional HPD.

Plugs and muffs

Below 500 Hz and above 2000 Hz, earplugs generally provide greater noise attenuation than earmuffs, while earmuffs are typically more effective at attenuating sound between 500 and 2000 Hz. For any type of HPD, effectiveness heavily relies on proper fitting and use. A poor initial fit, loosening of the device during activity, and not even wearing the equipment reduce effectiveness. Users typically find earmuffs easier to fit. Though earmuffs may be uncomfortable in hot and humid environments, they can be greatly appreciated by the wearer for insulating the ears in cold conditions.

Speaking and hearing with HPD

Generally, conventional HPDs cannot differentiate between the energies of speech and noise, and thus cannot selectively transmit desired signals. As a result, most devices do not directly improve the SNR. However, they can occasionally enhance clarity in loud noise by lowering the total energy of both speech and noise that reaches the ear, thereby reducing distortion due to overload in the cochlea. An HPD has little or no degrading effect on intelligibility in noise above about 80 dBA, though it can reduce hearing acuity during quiet periods (such as when worn even when protection is not needed). Some negative effects of the device may arise from the tendency to lower one's own voice because the protector amplifies bone-conducted voice feedback inside the head, primarily at low frequencies.

Consequently, we may perceive our own voice as louder in relation to the noise than it actually is, often leading to a compensatory decrease of the voice by 2–4 dB. Therefore, we should make a conscious effort to speak louder when wearing an HPD.

Active HPDs

Since conventional HPDs alter all received signals and may affect judgment of sound content, direction, and distance, newer electronic HPDs possess active attenuation qualities that adjust to prevailing noise levels, job demands, and individual users' hearing abilities. Active noise reduction HPDs use destructive interference by generating sound waves that are of the same nature as the noise but out of phase by 180°. This noise cancellation works particularly well below 1000 Hz. With high-speed signal processing, active HPDs combine noise abatement and improved communication; they typically cancel external noise and then transmit desired signals such as speech or music. The use of noise-cancelling headphones has become common in many everyday settings (including in offices or as a passenger in an airplane), but this can be risky if it causes the user to miss important warning or alarm signals (such as a bicyclist not hearing a nearby automobile).

Voice communications

Understanding verbal communications is a psychological process that requires suitable acoustical conditions. A sufficient portion of the voice message must penetrate the noise to enable understanding. Face-to-face communication provides visual cues that enhance the intelligibility of speech, even in the presence of background noise. Indirect voice communications lack these visual cues. The distance from speaker to listener, background noise level, and voice level are important considerations. The ambient air pressure and gaseous composition of the air affect the efficiency and frequency of the human voice and, consequently, speech communication.

Intelligibility

The ability to understand the meanings of words, phrases, sentences, and entire speeches is known as intelligibility. For effective communication of voice messages over noise, at least 75% intelligibility is typically required. The intensity of a speech signal relative to ambient noise is a fundamental determinant of speech intelligibility. With an SNR of 10 dB or higher, individuals with normal hearing should comprehend at least 80% of spoken words in typical broadband noise. As the SNR decreases, intelligibility drops to 70% at 5 dB, to 50% at 0 dB, and to 25% at –5 dB. Individuals with reduced hearing capability often experience even greater reductions in intelligibility.

Understanding speech

In voice communication, frequencies range from about 200 to 8000 Hz, with the range of about 1000–3000 Hz being the most important for intelligibility. Men typically use more low-frequency voice energy than women. Filtering or masking frequencies below 600 Hz or above 3000 Hz has little effect on intelligibility, but interfering with voice frequencies between 1000 and 3000 Hz significantly reduces understanding. In speech (as well as in written text), consonants are more critical for understanding words than vowels. Unfortunately, consonants have higher frequencies and generally less speech energy than vowels, making them more susceptible to masking by ambient noise. Consequently, consonants are more difficult to understand, especially for older individuals.

Components of speech communication

Speech communication has five major components:

1. The message,
2. The speaker,
3. The transmission of the message,
4. The environment, and
5. The listener.

The message itself becomes clearest when its context is expected, its wording is clear and concise, and the ensuing actions are familiar to the listener. The speaker should articulate slowly and distinctly, using common and simple vocabulary with a limited number of terms. Redundancy and the use of phonetically distinct words can be beneficial. For practical example, the International Spelling Alphabet (also known as the NATO phonetic alphabet), presented in Table 6.2, contains words intended to be easily distinguished from one another, especially over radio or telephone. The transmission of speech should utilize a high-fidelity system that produces minimal distortion in frequency, amplitude, or time. Environmental noise, of course, affects the listener's natural ability to comprehend the message.

Designing warning signals

Warning signals should be designed carefully to penetrate surrounding sounds. One way to enhance the effectiveness of auditory alarms is to use frequencies below 500 Hz. A positive side effect of low frequencies is that they diffract easily around barriers; however, they require very high SPLs to be heard (see Figure 6.4). Therefore, most warning signals fall within the range of 1000–4000 Hz, unless noise is prevalent in these frequencies. If such masking noise exists, warning signals should

Table 6.2 International spelling alphabet

A	Alfa	N	November
B	Bravo	O	Oscar
C	Charlie	P	Papa
D	Delta	Q	Québec
E	Echo	R	Romeo
F	Foxtrot	S	Sierra
G	Golf	T	Tango
H	Hotel	U	Uniform
I	India	V	Victor
J	Juliett	W	Whiskey
K	Kilo	X	Xray
L	Lima	Y	Yankee
M	Mike	Z	Zulu

not use harmonic frequencies (double, triple, etc.) of that noise. Instead, signals should contrast with the masking noise by differing in tonal qualities such as intensity and frequency, and in modulation over time, including increases and decreases. A general rule is that the intensity of a signal should be about 15 dB above any masking noise levels. It can be very effective to combine auditory warning signals with other indicators that are seen or felt, such as lights and vibrations.

Improving defective hearing

In some cases, hearing deficiencies, whether natural or caused by noise, can be addressed with situation-based assistive listening devices. These devices may include amplifiers for telephones, digital applications for smartphones, and closed-circuit systems in auditoriums with a receiver for the user. In other cases, the hearing deficiency is best addressed by hearing aids worn in or behind the ear. In contrast to older analog hearing aids, these newer digital hearing aids are smaller and less likely to malfunction (which is often heard as an undesired whine or whistle), and can actively filter out background noise, amplifying desired sounds and making them clearer.

Hearing aids

Hearing aids fitted to the ear canal are popular because they are easily hidden, but they may give the wearer a "stuffed up" feeling if they block the ear canal. That blockage may prevent the wearer from benefiting from their remaining hearing ability. Many current open-ear hearing aids sit behind the top of the ear's pinna but, in contrast to traditional behind-the-ear fittings, these new devices are small and have a barely visible thin tube that extends into the ear canal: they are fairly comfortable to wear and easy to switch on and off.

Current hearing aids often feature a surround microphone, either manual or automatic, that picks up sound from any direction. Some new devices offer various adjustments for different environments: one setting for a quiet room, one for a concert, and another for use in a restaurant, for example. These options are pre-programmed, so the user simply selects the appropriate setting, possibly with a remote control or smartphone app; some devices switch to the suitable setting automatically. Another feature is the automatic adjustment between the aids for the left and right ears, accommodating users who can hear better in one ear than the other. The technology can also automatically fine-tune the settings as the listening environment changes.

Surgical implants Several types of surgical implants can assist individuals for whom hearing aids are ineffective or unsuitable.

- Bone-anchored hearing aids conduct sound through the skull to a functioning inner ear. In cases of deafness in only one ear, a bone-anchored transmitter is attached to the skull on the deaf side to pick up sound and conduct it to the "good" side.
- Middle ear implants aid those with mild to moderate hearing loss. They attach directly to the middle ear bones and amplify sound signals. A component behind the ear houses the microphone, sound processor, and battery.
- Cochlear implants assist individuals with severe hearing loss. They convert sound into nerve impulses and send them to the brain. A transmitter is placed under the skin behind the ear, and electrodes are implanted inside the cochlea.

Some techniques can be combined, such as cochlear implants with bone-anchored hearing aids.

Summary

Our ears provide essential information for daily life and work. The acoustic information is coded in combinations of sound frequencies and intensities that change over time. We interpret that information based on our individual experiences and hearing capabilities.

Noise surrounds us. It can range from a mere annoyance to interfering with sleep and rest, reducing our well-being, affecting our performance, and even damaging our ability to hear. To avoid noise-induced damage, various noise mitigation techniques and HPDs are available, some of which actively enhance the perception of sound signals.

Steps to fit the human

Step 1: Ensure that the ears function properly; if in doubt, consult an ear physician and/or a hearing specialist.

Step 2: Determine which specific task must be performed.

Step 3: Provide a suitable acoustic environment, minimizing unnecessary noise and its effects. Consider any accommodations that may be necessary based on the hearing ability required for the task or the work environment. Evaluate whether background music is appropriate or desired.

Step 4: Wear appropriate hearing protection as needed.

Step 5: Design, then test. Modify and iterate as necessary, adjust conditions as desired, and redesign if the user or the task changes.

Notes

The text contains markers, *, to indicate specific references and comments, which follow.

6.2 Hearing sounds

Tone or pitch: In auditory physics, pure tone is a wavelength with constant frequency and amplitude. Note that musicians may refer to musical tone as being the quality or mood of a melody or voice, and pitch being the highness or lowness of the sound.

Infrasound: More on felt mechanical vibrations in Kroemer, Kroemer, and Kroemer (2018), Stanton, Hedge, Brookhuis et al. (2004).

Loudness is stated in a logarithmic unit: The measuring unit decibel, dB, is more convenient to use in acoustics than the basic unit bel, B, named after

Alexander Graham Bell. Since *deci* means one-tenth, 1 dB = 1/10 B.

Doubling the sound pressure: Since $\log_{10}(2) = 0.3$, doubling the sound pressure adds 6 dB (= 20 × 0.3), and a 1.41-fold increase in P adds 3 dB (= 20 $\log_{10}(1.41)$).

Phon curves: Developed by Fletcher and Munson in 1933, phon curves are now defined in the international standard ISO 226:2023. Sound waves of different frequencies have equal loudness levels (measured in phons) if they are perceived as equally loud by a representative young person with normal hearing.

A-weighting filter: McMin (2013).

Music at work: Responses to music at work depends on context and work tasks and may not improve productivity but may still be preferred, as discussed in Sanseverino, Caputo, Cortese et al. (2022). For background music, see also Fox (1983), Owen (2006).

Acoustic events: Adapted from Kroemer Elbert, Kroemer, and Kroemer Hoffman (2018).

6.3 Designing for noise

US workplace regulations: National Institute for Occupational Safety and Health (NIOSH) Occupational Noise Exposure publication 98-126; EC Directive 2003/10/EC – Noise 2021.

Hearing protection devices: Berger, Royster, Royster et al. (2003), Casali and Gerges (2006), Casali and Robinson (2006), Doutres, Terroir et al. (2022).

References and further reading

Berger EH, Royster LH, Royster JD, Driscoll DP, Layne M (2003). *The Noise Manual*, 5th ed. Fairfax, VA, American Industrial Hygiene Association.

Casali JG, Gerges SNY (2006). Protection and enhancement of hearing in noise. Chapter 7 in Williges RC (eds), *Reviews of Human Factors and Ergonomics*, Volume 2. Santa Monica, CA, Human Factors and Ergonomics Society. https://doi.org/10.1177/1557234X0600200108

Casali JG, Robinson GS (2006). Noise in industry. Chapter 31 in Marras WS, Karwowski W (eds), *The Occupational Ergonomics Handbook: Fundamentals and Assessment Tools for Occupational Ergonomics*, 2nd ed. Boca Raton, FL, CRC Press. https://doi.org/10.1201/9781420003635

Doutres O, Terroir J, Jolly C, Gauvin C, Martin L, Negrini A (2022). Towards a holistic model explaining hearing protection device use among workers. *International Journal of Environmental Research and Public Health*, 19(9):5578. https://doi.org/10.3390/ijerph19095578

European Commission (2021). Directive 2003/10/EC of the European Parliament and of the Council of 6 February 2003 on the Minimum Health and Safety Requirements Regarding the Exposure of Workers to the Risks Arising from Physical Agents (Noise) European Commission; Brussels, Belgium. https://eur-lex.europa.eu/legal-content/EN/TXT/PDF/?uri=CELEX:02003L0010–20190726

Fletcher H, Munson WA (1933). Loudness, its definition, measurement and calculation. *Journal of the Acoustic Society of America*, 5:82–108.

Fox JG (1983). Industrial music. In Oborne DJ, Gruneberg MM (eds), *The Physical Environment at Work*. New York, Wiley, pp. 221–226.

ISO 226 (2023). *Acoustics — Normal Equal-Loudness-Level Contours*. Geneva, CH, International Organization for Standardization.

Kroemer Elbert K, Kroemer HB, Kroemer Hoffman AD (2018). *Ergonomics: How to Design for Ease and Efficiency*, 3rd ed. London, UK, Elsevier. https://doi.org/10.1016/C2016-0-02516-0

McMin T (2013). "A-weighting": is it the metric you think it is? In *Proceedings of Acoustics 2013* – Victor Harbor, 165–168, Australian Acoustical Society, Victor Harbor, Australia.

NIOSH (National Institute for Occupational Safety and Health) (1998). *Occupational Noise Exposure publication 98-126 (1998, last reviewed 2023)*. https://stacks.cdc.gov/view/cdc/6376/cdc_6376_DS1.pdf

Owen D (2006). The soundtrack of your life: Muzak in the realm of retail theatre. *The New Yorker*, 10:66–71.

Parmanen J (2012). Some reasons to revise the international standard ISO 226 2003: acoustics—normal equal-loudness-level contours. *Open Journal of Acoustics*, 2:143–149. https://doi.org/10.4236/oja.2012.24016

Sanseverino D, Caputo A, Cortese CG, Ghislieri C (2023). "Don't stop the music," please: the relationship between music use at work, satisfaction, and performance. *Behavioural Sciences*, 13(1):15. https://doi.org/10.3390/bs13010015

Stanton N, Hedge A, Brookhuis K, Salas E, Hendrick HW (2004). *Handbook of Human Factors and Ergonomics Methods*. Boca Raton, FL, CRC Press. https://doi.org/10.1201/9780203489925

CHAPTER SEVEN

How we sense objects and energy

Body sensors

We have several types of body sensors that are usually active simultaneously, providing important and often redundant information. For example, a bicyclist feels body movement, speed, and terrain bumps through tactile and haptic sensors, sees the path ahead and nearby objects rush by, and hears warning signals from other riders or vehicles. A blind person runs their fingertips over Braille to "read" text that is not visible. A careful cook cautiously brings their hand close to an object to determine whether it is hot. A child touches an object to assess whether its surface is smooth or rough. Yet, many designers have made surprisingly little use of existing information about human sensory capabilities. For example, round doorknobs offer no indication, by feel or sight, of the direction they must be turned to open the door; emergency bars that must be pushed to open a hinged door usually provide no cues, either tactile or visual, as to which side of the door will open*. We need to understand how we sense movement, objects, energy, and even pain to fit the task to the human.

7.1 Sensing body movement

Combined signals

The central nervous system (CNS; see also Chapter 9) receives signals from various human senses simultaneously, providing us with specific information that creates a general picture of events occurring both inside and outside the body. For example, as we lift a load, muscle and skin sensors indicate the force we exert on the object. We know how one of our body segments

DOI: 10.1201/9781032674575-10

moves even if we cannot see it because Ruffini corpuscles near the joints report the position of our limbs and the angles of the joints; these cutaneous sensors offer more details since the bending of a joint stretches and relaxes the skin around different parts of the joint.

Vestibulum

The *vestibulum* provides the primary information about the body's posture and movement. It is a pea-sized non-auditory organ located in each inner ear, next to the cochlea. The vestibulum contains three semicircular canals, shown in Figure 7.1, filled with fluid (*endolymph*). The arches of the three vestibular canals are arranged at about right angles to one another, making them sensitive to different head rotations. Although the canals share a common cavity in the *utricle*, each canal operates as an independent fluid circuit. Near its junction with the utricle, each canal features a widening (called an *ampulla*) that contains a protruding ridge carrying sensory hair cells, or *cilia*, which respond to displacements of the endolymph. Cilia are also located in both the utricle and *saccule*. The hair cells generate signals that travel along the vestibular nerves to the CNS.

Sense of body balance

The vestibular system informs us about body balance by sensing the magnitude and direction of head accelerations, including gravitational pull. The system adapts to constant acceleration and may not perceive small changes in position. Placing the head in a new posture requires the brain to compare signals not only with a new spatial reference system but also with new inputs from the sensors, as the endolymph now exerts

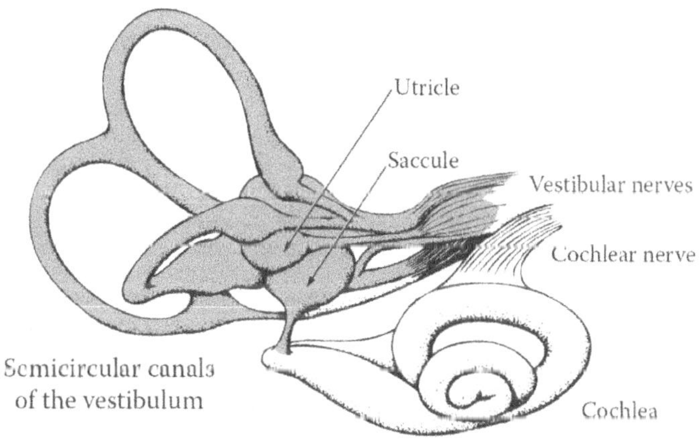

FIGURE 7.1 Three semicircular canals of the vestibulum. Adapted from a 1934 drawing by medical illustrator Max Brödel.

different loads on the cilia. With many other signals arriving simultaneously from various other sensors (such as the eyes) which also relate to the body's position and motion, the brain faces a complex task of integration and interpretation. Thus, it is not surprising that several vestibular illusions can occur. The best-known of these is motion or space sickness, likely caused by conflicting inputs from the vestibulum and other sensors. For example, if you are inside a ship that is tilting to one side, the vestibular system detects that tilt while the eyes perceive the interior of the ship as straight. Some other vestibular illusions are as follows:

- When the body is aligned with the gravitational vector, such as when a person in an airplane turns left or right, the plane's roll is not perceived.
- *Coriolis illusion*: If a person tilts their head forward or backward while in a turning airplane, two semicircular canals are triggered simultaneously, causing disorientation.
- *Tilt illusion*: A person interprets linear acceleration as body tilt.
- *Elevator illusion*: A change in gravitational force creates the visual illusion of objects rising or lowering.
- *Leans illusion*: An airplane pilot in a prolonged banked turn may perceive the plane as level.
- *Inversion illusion*: A person in zero gravity or lying prone may feel as though they are upside down.

7.2 Feeling objects, energy, and pain

Different groups of sensory capabilities rely on sensors located in the skin*:

- Mechanoreceptors, which sense taction such as contact, touch, tickling, stretching, pressure, texture, and vibration.
- Thermoreceptors, which detect warmth or cold, relative to each other and to the body's neutral temperature.
- Electroreceptors*, which respond to electrical stimulation of the skin (it is unclear if such receptors exist in humans).
- Noci(re)ceptors (from the Latin *nocere*, meaning to damage), which sense pain. Some of these sensors are deep inside the body.

Taction

The *taction* sense relates to contact and touch at the skin. A tactile sensation occurs when the stimulus acts solely at the skin, while a haptic sensation involves simultaneous information from sensors in muscles, tendons, and joints (kinesthetic proprioceptors). Most of our everyday perception is haptic.

Tactile sensors

The most common tactile sensor is a free nerve ending, the unencapsulated dendrite of a nerve fiber. Thousands of these tiny fibers extend through the skin layers, responding particularly to mechanical displacements and being most sensitive near hair follicles. In smooth, hairless skin, encapsulated receptors are also common, as shown in Figure 7.2. Near the upper skin layer, Meissner's corpuscles and Merkel's disks respond to touch (pressure) and are especially abundant in the ridges of the fingertips. The Pacinian corpuscle is an encapsulated nerve ending associated with a single, dedicated nerve fiber, sensing deep transient pressure and high-frequency vibration. These highly responsive tactile receptors are densely located in the palmar sides of the hands and fingers as well as in distal joints. They are also prevalent near blood vessels, at lymph nodes, and in joint capsules. Krause end bulbs are particularly sensitive to cold but likely respond to other stimuli as well.

In 1826, Ernst Heinrich Weber (1795–1878) demonstrated that skin sensors react to the location of a stimulus and to certain types of stimuli, particularly force or pressure, warmth, and cold. However, despite nearly 200 years of research, the sense of taction remains not fully understood.

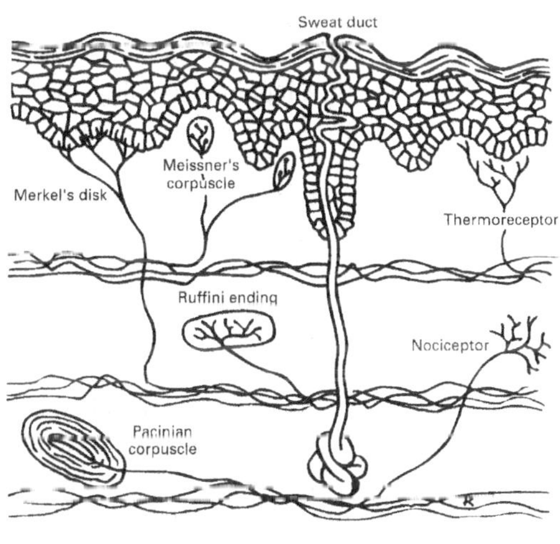

FIGURE 7.2 Common skin sensors. Adapted from Griffin (1990).

Taction stimuli are often not well defined: what is the relationship between pressure, force, and touch? Which sensors respond to each stimulus, and in what way? Do the sensors respond individually or in groups? If in groups, what patterns do they follow? Does one sensor respond to only one stimulus or to several? A single Meissner's corpuscle may connect to up to nine separate nerves, which may also branch to other corpuscles. It is unclear how such simultaneous convergence and diversion of the neural pathways can reliably generate and code neural signals.

Temperature

Our temperature sensations are relative and adaptive. Objects at skin temperature are perceived as neutral or indifferent, at "physiological zero." We classify a temperature below this level as *cool* and a temperature above it as *warm*. Slowly warming or cooling the skin near physiological zero may not elicit a change in sensation. This range of temperature neutrality varies in different body parts; for the forearm, the neutrality zone is approximately 31°C–36°C.

Feeling cold or warm

Some nerve sensors respond specifically to decreasing and cold temperatures, while others react to increasing and hot temperatures. The two scales may overlap, leading to paradoxical or contradictory information. For example, areas of the skin that consistently register cold when stimulated may also register "cold" when in contact with a warm object at about 45°C. An opposing paradoxical sensation of warmth also exists. We more easily sense changes from physiological neutrality toward warm temperatures than changes toward cold temperatures. We adapt to warm sensations quickly, except at rather high temperature levels. Adaptation to cold is slower and seems not to occur completely. Rapid cooling often causes an overshoot phenomenon; that is, for a short time, one feels colder than one physically is. Chapter 8 discusses more interactions between the body and the environment.

Sensing electricity

While we can feel electrical currents on the skin, there is still no known human receptor specialized for sensing electrical energy. Apparently, electricity can activate almost any sensory channel in the peripheral nervous system. The threshold for electrical stimulation depends heavily on the configuration and location of the electrode used, the waveform of the electric stimulus, the rate of stimulus repetition, and the individual subject. Generally, the threshold is about 0.5–2 mA for a pulse with a duration of 1 ms.

Pain

Tissue injury, hard touch or pressure, electricity, and extremes of warmth or cold can trigger unpleasant sensations such as burning, itching, or pain. Some related research results are difficult to interpret due to varying levels and categories labeled as pain. Current understanding suggests that nociceptors and other nerve cells have specialized molecules for detecting pain-causing stimuli. These cells transmit signals to nerve cells in the spinal cord, which relay the messages to the brain, where they are identified as painful. We may experience sharp or piercing pain, usually associated with events on the skin, teeth, or head, and dull or numbing pain, often originating deep within the body.

Pain can range from barely noticeable to unbearable. The threshold for pain varies greatly between individuals and in different situations, likely because pain is difficult to separate from other sensory and emotional components. Under certain circumstances and with specific stimuli, one can adapt to pain. Some people have experienced so-called second pain, which is a new wave of pain following the primary pain after about 2 seconds. Referred pain indicates a displacement of the pain's location, typically from its visceral origin to a more superficial area; for example, cardiac anginal pain may manifest in the left arm.

Not all pain is concerning. For instance, the acute pain that accompanies a minor tissue injury is protective, as it encourages us to avoid further damage. This type of pain is usually temporary and lessens over time.

7.3 Designing for tactile perception

In modern daily life, we primarily rely on our senses of vision and hearing for information input, while other senses (such as touch, smell, and taste) are used less frequently. This disparity partly relates to the ergonomic design's technical practicalities, but there is also a lack of specific and quantitative knowledge about how other human senses function. Nevertheless, new technology has led to innovative uses of these senses, such as mobile phones that produce sound and vibration simultaneously to alert us to an incoming call.

Research needs

Little is known about using our sense of smell (*olfaction*) for engineering purposes; however, we understand that our reactions to odors can vary, similar to our responses to tastes (*gustation*) in food and drink. Surprisingly, despite centuries of systematic studies on human senses, knowledge about cutaneous sensitivities remains limited. Several reasons contribute to this. First, earlier research has not always clearly defined the stimuli used. Second, sensors are distributed unevenly across the body, necessitating research that is focused on the specific locations of interest. Third, the functioning of these sensors is not fully understood: many sensors react to two or more distinct stimuli simultaneously, producing similar outputs, and it is often unclear whether specific sensors respond to a given stimulus. Fourth, the pathways of signal conduction to the CNS are complex and multifaceted, as they may involve multiple afferent paths from different body regions. Finally, the CNS interprets incoming signals in poorly understood ways. Given these uncertainties, further research is needed to provide complete, reliable, and relevant information on using cutaneous senses in human factors design*.

Using tactile signals

In the absence of reliable experimental data, the following guidelines for the design applications of the tactile sense rely primarily on extrapolations of experience and educated (though often useful) guesses.

- Mechanoreceptors can differentiate touch information by
    - The strength of the stimulus,
    - The temporal rate of change,
    - The size of the stimulated skin area (i.e., the number of receptors), and
    - The location of the stimulated skin.
- Sensitivity to tactile stimuli is highest in the facial area and fingertips, and only moderate in the forearm and lower leg. Some body areas, such as the eyes, are highly sensitive but obviously unsuitable for ergonomic design.
- Tactile sensitivity also depends on temperature; for example, if a low-intensity stimulus appears slowly on cold skin, it may go unnoticed.

Using temperature signals

Our sense of temperature is challenging to use for communication purposes for several reasons. First, it has a relatively slow response time and is poor at identifying location. Further, it adapts to a stimulus over time and may integrate multiple stimuli into a single signal across an area of skin. Finally, interactions exist between mechanical and temperature sensations; for example, a cold weight feels heavier on the skin than a warm weight.

The strength of thermal sensation depends on the location and size of the sensing body surface. Thermal sensation is enhanced by increasing

- The absolute temperature of the stimulus, and its difference from physiological zero,
- The rate of change in temperature, and
- The exposed surface; for example, we respond differently to immersion of our entire body in a bath compared to immersion of a hand or foot.

Assuming a physiologically neutral skin temperature of about 33°C, the following general rules of thumb apply for naked human skin.

- A skin temperature of 10°C is perceived as "painfully cold," 18°C feels "cold," and 30°C still feels "cool."
- The highest sensitivity to changes in coolness exists between 18°C and 30°C.
- Heat sensors respond well throughout the range of about 20°C–47°C.
- Thermal adaptation—that is, a reset to physiological zero—can be attained in the range of approximately 18°C–42°C, meaning that changes are not felt when the temperature difference is less than 2°C.
- Distinctly cold temperatures (near or below freezing) and hot temperatures above 50°C provoke sensations of pain.
- Compared to warmth, cold is sensed more quickly, particularly in the face, chest, and abdominal areas. The body's ability to feel warmth is less distinct but is strongest in hairy areas of the skin, around the kneecaps, and at the fingers and elbows.

Using smells

Olfactory information is seldom utilized by engineers due to the scarcity of research results, the varied reactions of people to olfactory stimuli, the ease with which smells can be masked, and the difficulty in arranging olfactory stimuli. Among the few industrial applications in use today is the addition of (mal) odorous methyl mercaptan to natural gas and pyridine to argon, allowing people to notice and respond to the smell of leaking gas. In a more pleasant application of the olfactory sense, some restaurants and bakeries may seek to attract customers by emitting enticing food aromas.

Using electric signals

Electricity is rarely used as an information carrier; however, it holds great potential for transmitting signals to humans. Attaching electrodes is straightforward. The transmitted energies are low, requiring only about 30 μW at the electrode–skin junction, up to a tolerable limit of about 300 mW. Coding can occur through placement, intensity, duration, and pulsation. Electrical stimulation can provide a clear, attention-demanding signal that is resistant to masking; however, responses to weak electrocutaneous signals typically have a long response latency, many misses, and false alarms. In complex environments requiring heightened vigilance, electrical signals may serve to provide signal redundancy.

Do not use pain as a signal!

Pain is not suitable for engineering applications. Ethically, one is bound not to cause pain. Moreover, damage may occur before the sensation of pain is experienced, and additional harm may result before the person can respond.

Reaction and response

Reaction time* is the period from the appearance of a stimulus (such as an electrical signal, touch, or sound) to the onset of a responding effector action (e.g., moving a hand). The additional time required to perform the action (e.g., pulling a handle) is the motion or movement time. Added together, these times yield the response time, which is the total period to sense, process, and act on information:

reaction time + motion time = response time

Simple reactions

Table 7.1 lists approximate shortest possible simple reaction times (see also the breakdown of reaction times in Chapter 9). This is a selective compilation of data from multiple experiments on reaction time conducted before the 1960s and found in engineering handbooks. Today, the origins of most underlying data, the experimental conditions under which they were measured (including stimulus intensity), the accuracy of the measurements, and the subjects of the research are unknown. Thus, this information is interesting but should not be regarded as definitive.

The compilation of reported reaction times shows little difference in reactions to electrical, tactile, and sound stimuli. The slightly longer reaction times for sight and temperature stimuli may lie within the range of both measurement accuracy and normal variations among individuals. Therefore, there may

Table 7.1 Shortest reported simple reaction times (in milliseconds)

Sensory stimulus	Approximate shortest reaction time (ms)
Electric shock	130
Touch and sound	140
Sight and temperature	180
Smell	300
Taste	500
Pain	700

be little practical difference for these. However, the reaction time following a smell stimulus is distinctly longer, and that for taste is longer still; it takes by far the longest to react to the sensation of pain. Simple reaction times are not significantly affected by age from about 15 to 60 years, but reactions are considerably slower at younger ages and slow again as one grows older.

Choice reactions

Table 7.1 presents *simple reaction* times, which apply when a person knows a particular stimulus will occur, is prepared for it, and knows how to react. When conditions become more complex—such as when there is uncertainty about the signal's appearance—reaction times increase. If a person must choose among several actions, this is known as a *choice reaction*; making a choice takes longer than a simple reaction with a single response. It takes even longer if the person has to distinguish among multiple stimuli, with only one triggering the response. Therefore, in real-life situations, we can expect reaction times to be considerably longer than those listed in Table 7.1.

Motion time

Motion time occurs after reaction time. Movements may be simple, such as lifting a finger in response to a stimulus, or complex, such as swerving a car to avoid a collision. Swerving the car involves more intricate movements than simply lifting a finger, along with larger body and vehicle masses that must be moved; all of this takes time.

Response time

Minimizing response time—the sum of the reaction and motion times—is often a goal of human factors engineering. This involves three interrelated choices, where we must select

1. The most appropriate stimulus for the situation that initiates the shortest possible reaction time;

2. The body part best suited for the task of reacting; and
3. The equipment that allows for the fastest execution of the task.

Chapter 9 expands on the discussion of actions and reactions in feedforward and feedback loops.

Summary

Much of our daily life relies on receiving information from our surroundings and interpreting it to guide our actions. This is one of the most ancient and instinctive features we use. While our senses can be tricked, the careful selection of stimuli can help elicit appropriate responses. Research conducted over the last century has explained the overall sensory processes and many of their complexities, but many details remain unknown or uncertain. The lack of technically useful information presents many opportunities and challenges for future investigations, such as using the perception of smell, temperature, and electrical signals to convey information.

Steps to fit the human

Step 1: Determine which information must be conveyed for the task to be completed.

Step 2: Select the primary stimulus to provide the critical information. Consider adding a secondary stimulus of a different type as a supplement or backup.

Step 3: Establish and fine-tune the parameters of the stimuli. Consider the impact of choice on overall response times.

Step 4: Design, then test. Modify and iterate as necessary.

Notes

The text contains markers, *, to indicate specific references and comments, which follow.

Opening doors: Cognitive engineer Donald Norman (see Norman, 2013) popularized the concepts of affordances (possible interactions between a person and an object, such as pushing or pulling a door) and signifiers (which indicate what action to take, such as a push plate or a handle). His name has been associated with the so-called Norman door: one that is difficult or confusing to use due to misalignment between affordances and signifiers. This concept is related to coding, discussed in Chapter 20.

7.2 Feeling objects, energy and pain

Skin: Sensors located in the skin are called cutaneous (from cutis, Latin for skin) or somesthetic (from soma, Greek for body); see also Jarrett (1973).

Electroreceptors: It has been disputed whether such specific sensors exist or whether the related sensations stem from the stimulation of other tactile sensors. A similar discussion concerns nociceptors: some researchers hold that there are no specific pain sensors, but that other sense organs transmit pain. More information can be found in Basbaum and Julius (2006), Boff and Lincoln (1988).

7.3 Designing for tactile perception

Research needs on human sensory capabilities for engineering applications: Gawande (2008); Chapter 2 in Kroemer (2006).

Reaction and motion times: More details can be found in Boff, Kaufman, and Thomas (1986), Boff and Lincoln (1988), Kroemer Elbert, Kroemer, and Kroemer Hoffman (2018).

References and further reading

Basbaum AI, Julius D (2006). Toward better pain control. *Scientific American*, 294(6):60–67. https://doi.org/10.1038/scientificamerican0606-60

Boff KR, Kaufman L, Thomas JP (1986). *Handbook of Perception and Human Performance*. New York, Wiley.

Boff KR, Lincoln JE (1988). *Engineering Data Compendium: Human Perception and Performance*. Wright-Patterson AFB, OH, Armstrong Aerospace Medical Research Laboratory. https://ntrs.nasa.gov/api/citations/19880019247/downloads/19880019247.pdf

Gawande A (2008). The itch. *The New Yorker*, June 30:58–65.

Griffin MJ (1990). *Handbook of Human Vibration*. San Diego, CA, Academic Press. https://doi.org/10.1016/C2009-0-02730-5

Jarrett A (1973). *The Physiology and Pathology of the Skin*. London, UK, Academic Press.

Kroemer Elbert K, Kroemer HB, Kroemer Hoffman AD (2018). *Ergonomics: How to Design for Ease and Efficiency*, 3rd ed. London, UK, Elsevier. https://doi.org/10.1016/C2016-0-02516-0

Kroemer KHE (2006). *"Extra-Ordinary" Ergonomics: How to Accommodate Small and Big Persons, the Disabled and Elderly, Expectant Mothers and Children.* Boca Raton, FL, CRC Press. https://doi.org/10.1201/9780203025246

Norman DA (2013). *The Design of Everyday Things.* Cambridge, MA, The MIT Press.

CHAPTER EIGHT

How we experience indoor and outside climates

Body and heat

The human body is always generating heat, with production fluctuating based on the intensity of physical work performed. The body must dissipate any excess heat to its surroundings: this is easy in a cold climate but difficult in a hot one. Two key factors govern heat flow within the body and the exchange of heat with the environment:

1. *Physiology*: Core body temperature must remain close to 37°C.
2. *Physics*: Heat flows from warmer to colder matter.

The physiological requirement to maintain core body temperature highlights the need for effective engineering of the environmental climate. The principle of physics that heat energy flows from warm to cold provides the tools needed to create suitable environmental conditions. We must understand how we experience and function in the climates around us to fit the task to the human.

8.1 Basics of human thermoregulation

Body temperature

We are not "cold-blooded" (a somewhat misleading term*), but "warm-blooded": we are capable of maintaining a nearly constant body temperature rather than it simply following the outside temperature. The human body has a complex control system to keep its internal body temperature near 37°C (98°F)*. A deviation in deep body temperature of just a few degrees

DOI: 10.1201/9781032674575-11

affects cellular structures, enzyme systems, and other bodily functions, impairing physical and mental work capacities. A deviation of just 6°C in core body temperature can be lethal. However, minor fluctuations in temperature occur throughout the day and from day to day, primarily due to circadian rhythms (see Chapter 15). Major impacts on the human thermal regulatory system result from the interplay between the (metabolic) heat generated within the body, especially during strenuous work, and external energy gained (in hot surroundings) or lost (in cool environments).

Producing, distributing body heat

The body produces heat in its metabolically active tissues: primarily in skeletal muscles, but also in internal organs, fat, bone, nerves, and connective tissue. Blood carries this heat energy throughout the body. The circulatory system can modulate blood flow through constriction, dilation, and shunting of its vessels, mainly in its superficial arteries and veins. This generates and controls heat dissipation from warmer to colder parts of the body. Heat exchange with the environment occurs at the respiratory surfaces of the lungs and, of course, at the skin.

Controlling body heat

The body regulates its heat content to prevent undercooling or overheating. The temperature of core tissues (such as the brain, heart, lungs, and abdominal organs) must be kept constant at about 37°C. Therefore, all management of varying temperatures occurs in the body's external tissues. As a result, there is often a significant temperature difference between core and "shell." A gradient of about 4°C between deep body and superficial skin is normal at rest in a comfortable climate, but in the cold, the temperature difference may exceed 20° or more.

Hot skin in a hot environment

Usually, the body must dissipate heat energy through its skin and lungs to avoid overheating its core. This heat dissipation is not a problem if the environment is colder than the skin and lung surfaces due to the natural flow of heat from warm to cold. However, this principle of physics makes it challenging to lose excess heat in an environment hotter than the body surfaces. Nature's solution is to increase skin temperature above the environmental temperature to evaporate water (in the form of sweat) on the skin. Evaporation of 1 cm^3 of water requires about 2430 J (580 calories) of energy; this extraction of energy cools the body. The nervous system's control of sweat gland activity is called *sudomotor* function. When the body cannot dissipate enough heat, it must reduce internal heat production, meaning we will limit physical activities as much as possible if we get too hot.

Cold skin in a cold environment

In a very cold environment, the body may struggle to conserve heat. It primarily does this by reducing blood flow to the skin, which lowers skin temperature and reduces heat loss. We can also decrease heat loss by wearing appropriate clothing to enhance insulation. When the body loses too much heat, it can increase heat production by contracting muscles, either voluntarily or through reflex shivering—both generate heat energy.

Heat exchange

The body's heat control systems of muscular, vascular, and sudomotor functions respond to and are influenced by the physical components of the environment. Heat exchanges between the body and environment occur through convection, conduction, radiation, and evaporation.

Convection or conduction

Traditionally, we refer to energy (heat) exchange with a fluid such as water or air as *convection*, while energy exchange through direct contact with a solid is called *conduction**. However, both rely on the same physical principle: energy flows from the warmer body to the colder one. Once the temperatures of the contacting surfaces equalize, energy exchange ceases. The amount of heat exchanged by convection or conduction depends on

- The area of the body surface participating in the heat exchange,
- The temperature of the body surface,
- The temperature of the medium in contact with the body, and
- The coefficient of heat conduction between the body and the medium.

The heat conduction coefficient of different materials varies. For example, cork or wood feels warm when touched because their heat conduction coefficients are lower than that of human tissue. Cool metal readily absorbs heat from our bodies and conducts it away, as metals generally have a high conduction coefficient. Heat exchange is enhanced if the medium moves quickly along the skin surface (such as a cooling breeze on a hot day), as this helps maintain a temperature differential. As long as there is a temperature gradient between the skin and the medium, there will be natural movement of air or fluid, known as free convection. Forced action (from fanning the air or from swimming in water instead of floating motionless) can produce more movement, referred to as induced convection, shown in Figure 8.1.

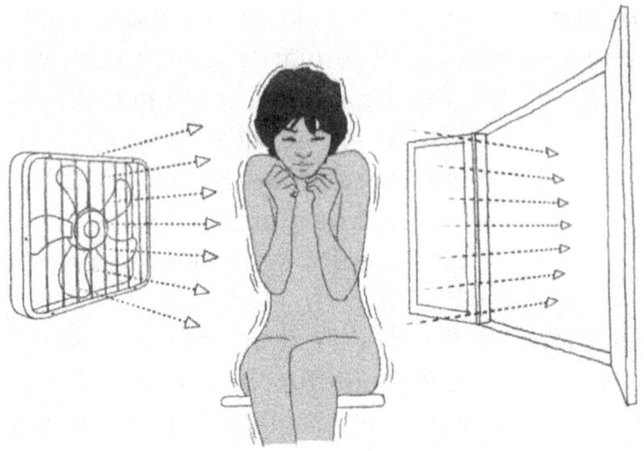

FIGURE 8.1 Air flow helps heat exchange by convection. Adapted from Kroemer and Kroemer (2017).

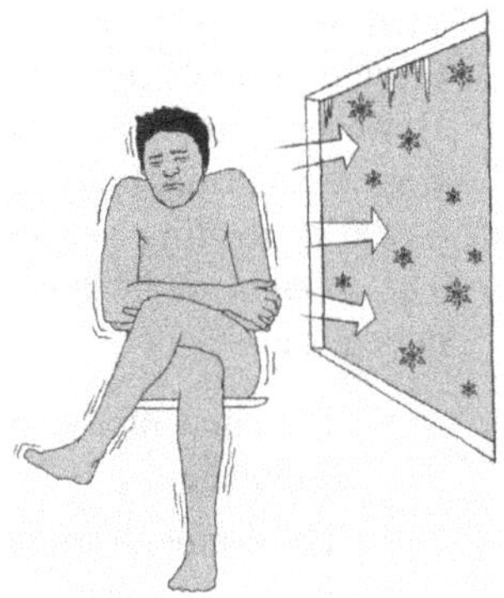

FIGURE 8.2 Heat radiated from the body to a cold windowpane. Adapted from Kroemer and Kroemer (2017).

Radiation

Heat exchange through *radiation** primarily depends on the temperature difference between two opposing surfaces, such as between a windowpane and a person's skin. Heat always radiates from the warmer to the colder surface, such as from the body to the cold window in winter or from a sun-heated pane to the body in summer (see Figures 8.2 and 8.3). This radiative heat exchange is not affected by the temperature of the air between the two surfaces but depends instead on

- The area of the body surface involved in the heat exchange,
- The temperature of the emitting surface,
- The emission coefficient of the emitting surface,
- The absorption coefficient of the receiving surface, and
- The temperature of the receiving surface.

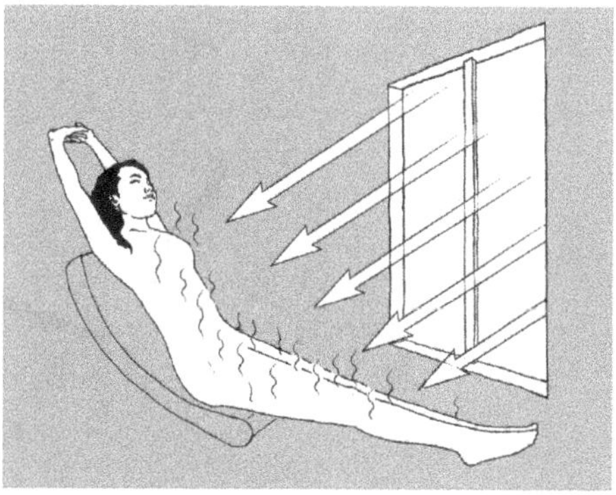

FIGURE 8.3 Heat radiated by the sun to the body. Adapted from Kroemer and Kroemer (2017).

The wavelengths of radiation from the human body are in the infrared range. Hence, it radiates like a so-called black body, having an emission coefficient close to 1 (unity), independent of the color of the radiating human skin. However, its absorption coefficient depends on skin color.

Evaporation

Heat exchange by *evaporation* occurs only in one direction: the human body loses heat through evaporation. There is no condensation of water on the skin, so there is no mechanism which would add heat. Evaporation of water (sweat) from the skin requires an energy of about 2430 J (580 cal) for every cubic centimeter of water. The heat lost through evaporation from the skin depends on

- The volume of sweat evaporated (partly a function of the wet body surface area),
- The vapor pressure at the skin,
- The relative humidity of the surrounding air, and
- The vapor pressure in the surrounding air.

Of course, heat loss by evaporation requires that the surrounding air is less humid than the air directly at the skin. Additionally, the movement of the air layer at the skin enhances convection and increases heat loss through evaporation because drier air replaces humidified air, as shown in Figure 8.4. Evaporative heat loss occurs even in a cold environment since there is always some evaporation of water in the warm lungs and continued secretion of sweat at the skin surface during physical activity.

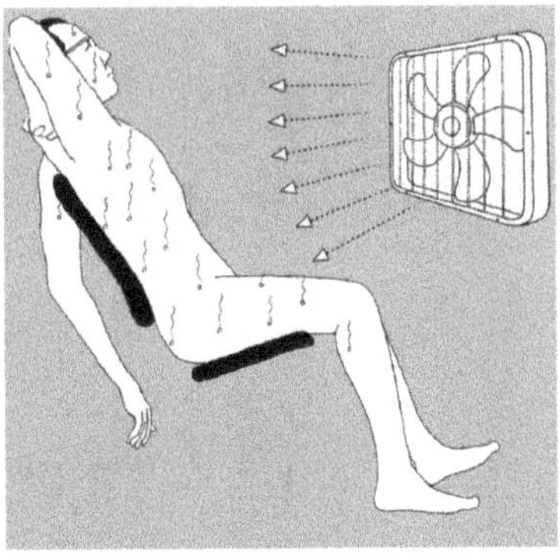

FIGURE 8.4 Air flow helps evaporation. Adapted from Kroemer and Kroemer (2017).

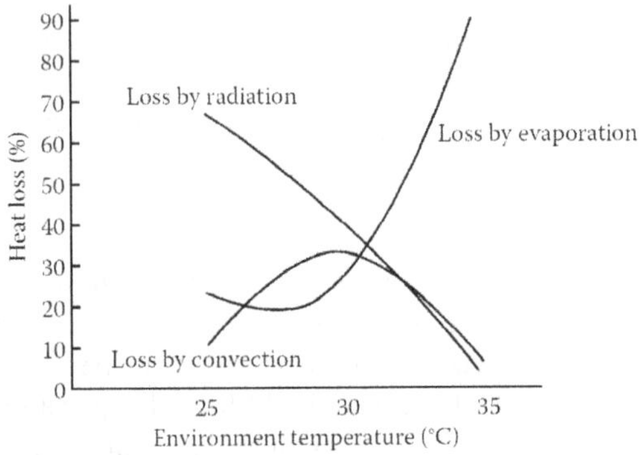

FIGURE 8.5 Cooling the body in a warm environment. Adapted from Kroemer Elbert, Kroemer, and Kroemer (2018).

The best way to regulate body heat depends on the surrounding climate temperature. At temperatures between 25°C and 30°C, we primarily lose heat by radiation. When the environmental temperature exceeds 30°C, evaporation becomes far more effective for cooling the body than convection and radiation, as illustrated in Figure 8.5.

8.2 Climate factors: temperatures, humidity, and drafts

Four physical factors define the thermal environment:

1. Air (or water) temperature,
2. Air humidity,

3. Air (or water) movement, and
4. Temperature of surfaces.

The combination of these four factors determines the physical conditions of the climate, how it affects us, and how we perceive it.

Measuring temperature

Temperature measurement is commonly performed with thermometers (now usually filled with alcohol instead of mercury), thermistors, or thermocouples. Regardless of the technique used, it is essential to ensure that ambient air temperature is not affected by humidity and air movement. To measure the so-called dry temperature of air, the sensor must remain dry while being shielded from radiated energy: these are the main properties of the dry-bulb (DB) thermometer.

Measuring humidity

Air humidity can be measured with a hygrometer, which originally used human or horse hair that changed length with wetness, or a bar of charcoal that gained water weight on a humid day; now, it is an instrument whose electrical conductivity (resistance) changes with ambient humidity. Another instrument is the psychrometer, which uses one DB and one wet-bulb (WB) thermometer; the WB is cooled further by evaporation in proportion to the humidity of the air. A psychrometer is called *natural* if there is no artificial air movement and *forced* if there is.

Vapor in the air

Air humidity can be expressed either in absolute or relative terms. The highest absolute vapor content in the air exists when any further increase would lead to the development of water droplets, which would then fall out of the gas. This so-called dew point depends on air temperature and barometric pressure: warmer air at lower pressure can hold more water content than cooler air at higher pressure. We usually refer to relative humidity, which indicates the actual vapor content in relation to the maximum possible content at the given air temperature and pressure.

Measuring air or water flow

The movement of air (both speed and direction) is measured with various types of anemometers. Historically, these were basically windmills that spun as air flowed past, such as due to wind or ventilation. Now, various electric or electronic techniques have largely replaced the traditional measuring mechanisms. Water flow is also measured using various mechanical or electronic means, depending on the specific application (e.g., pipe, stream, or ocean).

Measuring radiant temperature

Radiant temperature refers to the heat radiating from a warm object, such as the sun, a furnace, or a hot cup of tea. It can be determined by measuring the temperature with a thermometer placed inside a black globe, which absorbs radiated energy. The globe temperature (GT) measured by the black-globe thermometer requires a correction for the existing air temperature.

Interactions among climate factors

Air temperature is the climate factor that most obviously affects body temperature; in fact, we habitually use the labels "hot" or "cold" to describe our overall perception of a climate. *Air humidity* determines our ability to sweat, which is the body's last resort to avoid overheating. The *movement of air* around our skin affects our ability to exchange energy by convection and to lose heat by evaporation. The energy exchange by *radiation* depends primarily on the temperatures of the emitting and receiving surfaces.

These examples demonstrate that most climate factors can influence the measurements of other components. Furthermore, the various combinations of these factors, which we collectively refer to as *climate*, can be perceived differently by individuals and lead to significantly different effects on them. Various techniques exist to express the combined effects of environmental factors in one model, or index—such as the Effective Temperature chart, used widely in the late 20th century, which relied on measurements of air temperature, air velocity, and relative humidity.

Wet-bulb globe temperature

One convenient way to account for all four physical factors of the thermal environment is to use a combined assessment called the *wet-bulb globe temperature* (WBGT). A WBGT combines three temperature measurements into one index to assess how the heat environment affects humans. WBGT applies primarily to warm environments and weighs the effects of several climate parameters:

- WB is the (wet-bulb) temperature of a sensor with a wet wick exposed to natural air currents.
- GT is the globe temperature at the center of a black sphere of 150 mm diameter.
- DB is the (dry-bulb) air temperature.

For outdoors, the WBGT is calculated using all three temperatures including the solar load:

$$\text{WBGT(outdoors)} = 0.7\ \text{WB} + 0.2\ \text{GT} + 0.1\ \text{DB}$$

For indoors (or for outdoors without the sun), the WBGT is calculated without the DB:

$$\text{WBGT(indoors)} = 0.7\ \text{WB} + 0.3\ \text{GT}$$

The effect of clothing can also be considered as an adjustment, resulting in an "effective WBGT" value.

8.3 Our personal climate

Microclimate

What matters most to the individual is usually not the general, overall climate (macroclimate), but the localized climatic conditions (microclimate) that one interacts with directly. Each person prefers a microclimate that feels comfortable based on their work situation, adaptation, and clothing. The ideal microclimate is highly individual and variable. It can depend somewhat on age, as muscle tonus tends to decrease with increasing years: older individuals are generally less active, have weaker muscles, consume fewer calories, and begin to sweat at higher skin temperatures. A person's perception of a suitable microclimate also depends on their surface-to-volume ratio, which (for example) is much larger in children than in adults, and on their fat-to-lean body mass ratio.

The type and intensity of work performed are major determinants of individual thermal comfort. Physical work in cold conditions may lead to increased heat production, resulting in reduced sensitivity to the cold environment, while strenuous physical work in heat can make it challenging to achieve an energy balance. The effects of microclimate on mental work are less clear, but it is commonly understood that extreme climates hinder mental performance.

Clothing

Clothing can significantly affect the body's thermal balance and overall comfort. It has three main characteristics: insulation, permeability, and ventilation.

Clothing insulation

Insulation is a measure of the resistance to heat exchange through convection, conduction, and radiation. In hot or cold environments, it inhibits energy exchange through convection and radiation either cooling or heating. Light-colored clothing minimizes heat gain from solar radiation on sunny days, while dark-colored clothing absorbs the sun's radiated heat. In cold conditions, clothing helps reduce body heat loss through convection and radiation. It also lowers the risk of thermal injury by conduction when in contact with very hot or cold objects.

The insulating value of clothing is expressed in units of clo, which is the reciprocal of thermal conductivity: 1 clo = 0.115 m^2 K/W. One clo represents the insulation needed for a person at rest to maintain thermal equilibrium in an environment at 21°C (70°F) in a normally ventilated room (0.1 m/s air movement). Table 8.1 lists the insulation values for typical clothing. Note that air bubbles in the clothing material or between clothing layers provide additional insulation in both hot and cold environments.

Clothing permeability

Permeability is a measure of how clothing allows water vapor to move through the fabric. This is vital in hot environments, where good permeability facilitates evaporation from the skin, thereby cooling the body. In cold climates, allowing water vapor to escape is important to prevent the "clammy" feeling of trapped moisture in and under clothes. Permeability can also help protect the body against chemical exposure.

Clothing ventilation

Ventilation is a measure of the ability of ambient air to flow through clothing. Such airflow is beneficial in a hot environment as it enhances evaporative and convective cooling; however, it is often undesirable in a cold environment.

Table 8.1 Typical insulation values from clothing

Type of clothing	Insulation (Clo)
None	0 (zero)
Light informal summer clothing (short-sleeve shirt, short pants, or skirt)	About 0.3
Light work outfit (long-sleeve shirt, trousers, or skirt)	About 0.6
Light business suit (long-sleeve shirt, trousers or skirt, or light suit jacket)	About 1
Heavy business suit (long sleeve shirt, heavy trousers, heavy suit jacket)	About 1.5
Snow suit	About 2
Arctic quilted thermo parka outfit	About 2.5

Source: Adapted from Havenith (2005).
Note: 1 clo = 0.155 m^2 (kelvin)/W.

Exposed skin

Clothing determines the surface area of exposed skin. A larger exposed skin surface area promotes better heat dissipation in hot environments but can lead to excessive cooling in cold conditions. The head and neck, being close to the warm body core, have warm surfaces, which is beneficial in heat but not in cold. Fingers and toes require special protection in cold conditions because they are far from the warm body core.

Wind chill

Clothing significantly affects the individual microclimate by minimizing wind chill, a specific type of convective heat loss. When air moves swiftly along exposed skin surfaces, cooling effects are intensified. Figure 8.6 shows perceived temperature, expressed as equivalent wind chill temperatures, based on wind speed and actual DB air temperature (but not considering humidity) on the naked human face, as determined in 2001 by the Canadian and US National Weather Services*. As wind speed increases and air temperature decreases, the cooling of exposed skin (and hence the risk of frostbite) intensifies, as illustrated in both Table 8.2 and Figure 8.6. Clearly, simply covering the skin, such as with a face mask, enhances insulation against heat loss due to wind chill.

Acclimatization

Acclimatization refers to the adjustment of a person's body (and mind) to changed environmental conditions, while the process

	Actual air temperature (degrees C)											
Wind speed (km/h) **0**	**5**	**0**	**-5**	**-10**	**-15**	**-20**	**-25**	**-30**	**-35**	**-40**	**-45**	**-50**
5	4	-2	-7	-13	-19	-24	-30	-36	-41	-47	-53	-58
10	3	-3	-9	-15	-21	-27	-33	-39	-45	-51	-57	-63
15	2	-4	-11	-17	-23	-29	-35	-41	-48	-54	-60	-66
20	1	-5	-12	-18	-24	-30	-37	-43	-49	-56	-62	-68
25	1	-6	-12	-19	-25	-32	-38	-44	-51	-57	-64	-70
30	0	-6	-13	-20	-26	-33	-39	-46	-52	-59	-65	-72
35	0	-7	-14	-20	-27	-33	-40	-47	-53	-60	-66	-73
40	-1	-7	-14	-21	-27	-34	-41	-48	-54	-61	-68	-74
45	-1	-8	-15	-21	-28	-35	-42	-48	-55	-62	-69	-75
50	-1	-8	-15	-22	-28	-35	-42	-49	-56	-63	-69	-76
55	-2	-8	-15	-22	-29	-36	-43	-50	-57	-63	-70	-77
60	-2	-9	-16	-23	-29	-36	-43	-50	-57	-64	-71	-78
65	-2	-9	-16	-23	-30	-37	-44	-51	-58	-65	-72	-79
70	-2	-9	-16	-23	-30	-37	-44	-51	-58	-65	-72	-80
75	-3	-10	-17	-24	-31	-38	-45	-52	-59	-66	-73	-80
80	-3	-10	-17	-24	-31	-38	-45	-52	-60	-67	-74	-81
	Equivalent windchill temperature (degrees C)											

FIGURE 8.6 Equivalent temperature considering wind chill, as a function of wind speed and air temperature. Shading indicates temperatures with high risk of frostbite. Adapted from the Canadian and US National Weather Services, Osczevski and Bluestein (2005).

Table 8.2 Wind chill index as perceived by a person dressed in arctic clothing

Wind chill index	Effect on person dressed in arctic clothing
50	Hot
200	Pleasant
400	Cool
1000	Very cold
1200	Bitterly cold
1400	Exposed flesh freezes
2500	Intolerable

Source: Adapted from Parsons (2003).

Note: $WCI = 1.16\ (10v^{1/2} + 10.45 - v)\ (33 - DB)$, with v denoting the air velocity in m/s and DB denoting the dry-bulb air temperature in °C.

of adjustment is called *acclimation.* When acclimating to a hot environment, a person improves the control of blood flow to the skin, facilitates sweating, and increases stroke volume of the heart without elevating heart rate. A healthy, fit individual can achieve heat acclimatization within 1 or 2 weeks, but can lose it just as quickly. In contrast, physiological acclimation of the whole body in moderately cold environments is typically minimal. Most adjustments to cold are related to proper clothing, beneath which the body remains at its usual microclimate. However, blood flow to exposed surfaces of the face and neck, as well as to the hands and feet, can show some acclimation to cold conditions.

A climate that initially feels uncomfortable and restricts physical work may become quite agreeable after a few days or weeks of acclimation. Similarly, seasonal changes in climate, type of work, clothing, and personal attitude significantly influence what feels acceptable. In regions with distinct summer and winter seasons, most people are willing to accept summer conditions that are warmer, more humid, and draftier than what they would tolerate in winter.

8.4 Working in hot environments

The human body has a few natural defenses against hot environments. We can adjust our behavior, such as wearing suitable clothing, seeking shade, or moving to cooler locations. The body has three primary ways to regulate its temperature in the heat: re-distributing blood flow, sweating, and decreasing metabolic rates.

Blood distribution

The human body produces heat, much of which it must release to the surroundings to prevent overheating. As discussed, the physical means to disperse heat are convection, conduction, radiation, and evaporation. These heat dissipation methods work best when the skin temperature is well above the temperature of the immediate environment. To achieve this, the body's circulatory system re-distributes blood flow to facilitate heat transport to the skin: skin vessels dilate, and superficial veins are fully open (opposite actions are taken in the cold). Blood flow may become four times larger than at rest, increasing tissue conductance and creating reddened skin.

Sweating

When heat transfer to the outside by blood re-distribution is insufficient, the body activates its sweat glands: evaporation of sweat cools the skin. Sweating reduces the body's water content, which must be replenished by water in drinks and food. The recruitment of sweat glands varies across different areas of the body, and there are significant differences in the ability to sweat among individuals: most adults have 2–4 million sweat glands, while some have fewer. During acclimation, the body learns to control sweating. The amount of sweat produced and evaporated depends on the number of glands, clothing, environment, work requirements, and the individual's acclimatization.

Reducing physical effort

If heat transfer by blood distribution and sweat evaporation is insufficient, the body reduces muscular activity: this lowers the amount of energy generated through metabolic processes. In fact, this is the final and necessary action of the body to prevent core temperature from exceeding a tolerable limit. If forced to choose between excessive overheating and continuing physical work, the body will prioritize maintaining core temperature, leading to a reduction or cessation of work activities.

Signs of heat strain

If the body is overheating, there are several clear signs of severe *heat strain*. One sign is increased circulatory activity: cardiac output rises, primarily due to a higher heart rate. This may be associated with a reduction in systolic blood pressure. Another sign of excessive heat strain is an elevated and superfluous rate of sweating. Even at rest, the body maintains a baseline of insensible perspiration of about 50 cm^3/hour. Sweat production increases from this baseline depending on the amount of heat that must be dissipated. During normal work, about 1 L of sweat is produced in 1 hour (approximately 20 times the insensible perspiration). Maximum sweat rates are 2–4 L an hour and

10–14 L/day during extremely strenuous efforts in hot climates. Sweat begins to drip off the skin when the sweat generation reaches about one-third of the maximum evaporative capacity. The sweat that runs down the skin (and is not evaporated) contributes very little to heat transfer.

Drinking water

Normally, we replenish our body's water needs from food (which is mostly water by volume) and by taking sips of fluids. However, sweating due to strenuous efforts, especially in a hot environment, can dehydrate the body. Dehydration of only 1% or 2% of body weight can critically affect the body's ability to function (1% of 100 kg body mass converts to 1 L of water). Generally, sufficient fluid levels can be easily maintained by frequently drinking small amounts of water. Sweating extracts not only water from the blood plasma but also carries some salt from the blood onto the skin. Generally, it is not necessary to add salt to drinking water, since the salt in the food of most diets is more than sufficient to replace salt lost through sweating.

Heat distress

Among the first reactions to heavy exercise or physical work in excessive heat are sensations of discomfort and possibly skin eruptions ("prickly heat" rash) associated with clogged sweat ducts. As a result of sweating, *heat cramps* may develop; these are muscle spasms related to local salt deficiency that may occur after quickly drinking large amounts of fluid. *Heat exhaustion* results from a combination of dehydration and circulatory system overload. This is associated with fatigue, headache, nausea, and dizziness, and often is accompanied by giddy behavior. *Heat syncope* indicates a failure of the circulatory system, demonstrated by fainting. *Heat stroke* indicates an overload of both the circulatory and sweating systems and is associated with hot dry skin, increased core temperature, and mental confusion.

Working hard in heat

The ability to perform short bursts of maximum muscle strength is unaffected by heat or the body's water loss. However, the ability to perform high-intensity and endurance-type physical work is significantly reduced before full acclimatization to heat. Even after acclimatization, the body must balance the demands on the cardiovascular system for heat dissipation with the need for sufficient blood supply to the muscles. When forced to choose, the body favors heat dissipation, resulting in a proportional reduction in a person's capability to perform physical work.

8.5 Working in cold environments

The human body has few natural defenses against cold environments. We can adjust our behavior, such as putting on suitably heavy clothing to cover exposed skin, seeking shelter, or using external sources of warmth. The body has only two major ways to regulate its temperature in the cold: re-distributing blood flow and increasing metabolic rates.

Distributing blood

To conserve heat, the body lowers the temperature of the skin to reduce the temperature difference with the outside environment. This is achieved by constricting blood vessels near the body surface, displacing circulating blood away from the skin and towards the core. The effects of this re-distribution can be dramatic; for example, blood flow in the fingers may drop to only 1% of what exists in a moderate climate. Vasoconstriction results in the appearance of pale skin in the cold.

The *hunting reflex* is a cold-induced, automatic body response: after initial vasoconstriction, blood vessels suddenly dilate to allow warm blood to return to the skin (such as the hands), re-warming that section of the body before vasoconstriction resumes. This sequence may repeat several times.

Wearing gloves, caps

Displacing blood away from the skin and toward central circulation is an effective way to keep the body core warm. However, the temperature in peripheral tissues may approach that of the environment, leading to very cold extremities such as fingers, toes, and noses. This presents the danger of potential tissue damage (frostnip or frostbite) if temperatures near freezing occur. Wearing insulated gloves, boots, and face mask can protect sensitive extremities.

The blood vessels in the head do not undergo as much vasoconstriction as those in the skin of the extremities, allowing the head to stay relatively warm even in cold environments, with less risk to the tissues. However, the resulting large temperature difference between the head and the environment leads to significant heat loss. This can easily be prevented by wearing a cap and a collar or scarf to create an insulating layer.

Increasing metabolism

The other major reaction of the body to a cold environment is to increase metabolic heat generation. This may occur involuntarily: shivering usually begins in the neck, apparently to warm the most important blood flow, which is to the brain. Shivering is caused by muscle units firing at different repetition frequencies and out of phase with each other. As this muscle use produces no mechanical work externally, the entire activity is transformed into heat production. This results in an increase in the metabolic rate, which may reach four times the resting rate. If the body still does not warm, shivering can become rather violent when motor unit firings synchronize, causing large muscle units to contract. Such shivering can further increase the metabolic rate to five or more times the resting level but can only be sustained for a short time.

Voluntary muscular activities can also enhance heat production, either by increasing dynamic muscular work performed (such as working more vigorously) or by moving additional body segments (such as flexing the fingers and stomping the feet). Since the body's energy efficiency is low, dynamic muscular work can easily boost metabolic heat generation tenfold compared to rest.

"Goose bumps"

Sometimes, skin exposed to cold develops so-called goose bumps. This puckering of skin around hair follicles helps retain a layer of stationary air close to the skin, which is relatively warm and acts as an insulating envelope, reducing heat loss at the skin. The hair raised by these bumps would be more effective insulators if we had more hair (or fur).

Feeling the cold

A person's decision to stay in the cold or seek shelter depends on their subjective assessment of how cold body surfaces or the body core are. Dangerous situations can arise either when a person fails to perceive and react to the body's signals that it is becoming dangerously cold or if the body temperature drops so low that further cooling is below the threshold of perception. The perception of the body getting cold relies on signals received from thermal receptors at the surface and from sensors in the body core. As skin temperatures decrease below 35.5°C, the intensity of the cold sensation increases; near 20°C, the sensation of cold is strongest; at even lower temperatures, sensitivity to cold decreases. It is often difficult to separate feelings of cold from feelings of discomfort or pain.

The conditions of cold exposure can significantly influence the perception of coldness. It makes quite a difference whether we are wearing suitable protective clothing and what we are doing when exposed to cold air or water. Experiments have shown that subjects find it difficult to accurately assess their level of cold: neither core nor surface temperatures reliably correlate with sensations of cold. When the outside temperature plunges, the decrease can generate an overshoot sensation in cold receptors, which react not only to the difference in temperature but also to the rate of change in temperature. Exposure to very cold water amplifies the overshoot phenomenon observed in cold air, likely because the thermal conductivity of water is about a thousand times greater than that of cold air at the same temperature.

Signs of cold strains

The subjective sensation of cold is an unreliable and potentially dangerous indicator of core and surface body temperature. Measuring ambient temperature, humidity, air movement, and exposure time, then acting on these objective physical measures, is a better strategy than relying on personal, subjective assessments.

If vasoconstriction and metabolic rate regulation cannot prevent serious energy loss through body surfaces, the body will experience cold stress. As mentioned earlier, the skin is the first part of the body to suffer cold damage while the body core is protected for as long as possible. If skin temperature approaches freezing, ice crystals form in cells and lead to cell damage, known as *frostbite*.

In the hands, joint temperatures below 24°C and nerve temperatures below 20°C hinder our ability to perform fine motor tasks. Manual dexterity declines as finger skin temperatures fall below 15°C. Tactile sensitivity diminishes below 10°C and severely decreases as the skin temperature falls below 8°C. At about 5°C, skin receptors for pressure and touch cease functioning, resulting in numbness.

When nerve temperature drops to 8°C, peripheral motor nerve velocity may decrease to near zero, causing a *nervous block*: this explains the rapid onset of physical impairment due to local cooling. Deep cooling renders people incapable of performing even basic activities that might save them (e.g., "cannot light a match").

Reduction of core temperature can have serious consequences. Vigilance begins to drop at core temperatures below 36°C. At core temperatures of 35°C and below, central nervous system coordination suffers, making even simple activities difficult or impossible. As the core temperature drops further, apathy sets in ("let me sleep"), and confusion ensues. Loss of consciousness occurs around 32°C, while heart failure may happen at about 26°C. At around 20°C, vital signs disappear, but the oxygen supply to the brain may still be sufficient to allow revival from severe hypothermia.

Working hard in cold

With appropriate protective clothing, a cold environment does not restrict our ability to perform medium or heavy work. However, bulky clothing may hinder the execution of certain tasks, and fine manual manipulations can be difficult with insulated gloves. Limited dexterity resulting from protective clothing should be considered during the ergonomic design of tasks and controls.

8.6 Climate effects on mental tasks

Evaluating the effects of moderate heat or cold on mental and intellectual performance is challenging due to significant subjective variations and a lack of practical, objective testing methods. However, in general, mental performance deteriorates with rising temperatures, beginning at about 25°C for unacclimatized individuals; that threshold increases to 30°C or even 35°C for those accustomed to heat. Brain functions are particularly vulnerable to heat; keeping the head cool enhances tolerance to elevated body temperatures. A high level of motivation may also mitigate some of the negative effects of heat. In laboratory tests of perceptual motor tasks, performance decline can occur in the low 30°C WBGT range, while very simple mental tasks are often not significantly affected by heat as high as 40°C WBGT.

8.7 Designing comfortable climates

Comfortable climates

In confined spaces, such as buildings or vehicle cabins, we have various technical methods to create an environment* that suits individuals performing light physical work while dressed appropriately. For most Western populations, comfortable climate conditions range from about 21°C–27°C in warm climates or during summer, and from 18°C to 24°C in cool climates or during winter. Preferred relative humidity levels are between 30% and 70%. In rooms, the difference between air temperature at the floor level and head level should be less than 6°C, ideally about 3°, to keep the head cool and the feet warm. Temperature differences between body surfaces and windows, walls, or other surfaces should not exceed approximately 10°C. The velocity of cool air, especially if irregular, should not exceed 1.5 m/s.

The technical means to influence the climate should be designed while considering the work to be performed, the

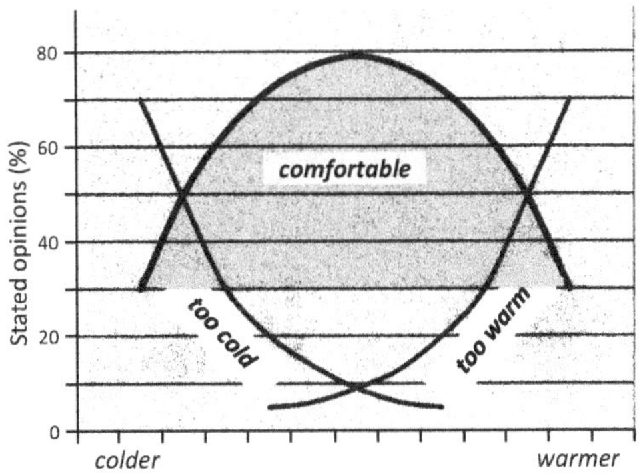

FIGURE 8.7 Different opinions about climates.

acclimatization conditions of the individuals, their clothing, and their psychological inclination to either accept given conditions or consider them uncomfortable. The preferred personal microclimate varies from person to person and can also depend on specific circumstances. For example, selecting clothing and adjusting the intensity of physical effort can significantly impact the perceived suitability of the physical climate conditions (temperatures, humidity, and air movement). Various combinations of these climate factors can subjectively appear similar. However, with so many variables, we should not be surprised that a combination of climate factors that many people find agreeable may still feel too warm to some and too cold to others, as illustrated* in Figure 8.7. Selecting appropriate clothing is often a simple and effective way to create a comfortable personal microclimate, by adding layers of clothing when one feels cold and changing into lighter clothes when one feels hot.

Summary

The body must maintain a nearly constant core temperature near 37°C, even in the presence of significant changes in internally developed energy (heat and physical work), in heat energy received from a hot environment, or in heat energy lost to a cold environment.

Heat energy may be gained from or lost to the environment through convection, conduction, and radiation, and may be lost from the body to the environment through evaporation.

In a hot environment, the body tries to keep the skin hot to prevent heat gain and to achieve heat loss. Sweating is the ultimate means to cool the body surface. In a cold environment, the body tries to keep the skin cool to minimize heat loss. Proper clothing significantly affects energy exchanges in both hot and cold environments.

The thermal environment is determined by combinations of

- Air temperature (affecting convection and evaporation),
- Air humidity (mostly affecting evaporation),
- Air movement (affecting convection and evaporation), and
- Temperature of surfaces in contact with the body (affecting conduction), and adjacent to the body (affecting radiation).

The combined effects of all or some of these physical climate factors can be integrated into a climate index. Several indices are in use, with the wet-bulb globe temperature (WBGT) being the most common. It identifies certain ranges of air and surface temperatures, as well as air velocity and humidity, which are physically suitable and perceived as "comfortable" for specific tasks and clothing.

Steps to fit the human

Step 1: Determine what work is to be done, how, and where.

Step 2: If indoors, establish a comfortable climate. If outdoors, set up a suitable work regimen and provide appropriate clothing and shelter.

Step 3: Encourage individual adjustments, especially in clothing. Allow for breaks for both rest and nutrition.

Step 4: Design, then test. Modify and iterate as necessary.

Notes

The text contains markers, *, to indicate specific references and comments, which follow.

8.1 Basics of human thermoregulation

Warm-blooded and cold-blooded: These terms are imprecise ways to categorize animals; it is more accurate to say *endotherm* (body temperature controlled by internal metabolic processes), *ectotherm* (body temperature primarily influenced by environmental heat sources), and *mesotherm* (body temperature regulated by both internal and external means). Humans are ectotherms who are also homeothermic, meaning they can maintain a stable internal body temperature.

Temperature scales: The physicist Daniel Gabriel Fahrenheit (1686–1736) set the temperature of ice at 32° and boiling water at 212°: the Fahrenheit scale has 180°F between freezing and boiling water. In 1742, the astronomer Anders Celsius (1701–1744) proposed a metric scale: water freezes at 0°C and boils at 100°C. This centigrade scale, known as Celsius, is in common use everywhere except in the United States. The physicist and engineer William Thomson, Baron Kelvin (1824–1907), established the lowest limit of temperature (absolute zero) as −273.15°C, equal to 0 K (kelvin). One kelvin has the same magnitude as 1°C.

The conversion of temperature values from one scale to another takes into account different freezing and boiling points, as well as the number of degrees between them (thus, the ratio 100/180 becomes 5/9):

Fahrenheit to Celsius: (°F – 32) × (5/9) = °C

Celsius to Fahrenheit: (°C) × (9/5) + 32 = °F

Celsius to Kelvin: (°C) – 273.15 = K

Heat exchange through convection or conduction follows Newton's law of cooling.

Heat exchange through radiation follows the Stefan–Boltzmann law of radiative heat transfer. For equations related to heat exchange via convection, conduction, radiation, and evaporation, see Bernard (2002), Bernard and Dukes-Dobos (2002), Havenith (2005), Youle (2005).

8.3 Our personal climate

Wind chill: The latest version of the wind chill index is available at https://www.weather.gov/ama/WindChill. This table was developed in the United States jointly with the Canadian weather service and is used in both countries without regionalization. Wind chill was determined experimentally by measuring heat flow from the exposed faces of 12 volunteers who were in a chilled wind tunnel (see Osczevski and Bluestein (2005)). The equivalent wind chill temperature

(WCT) is calculated as WCT = 13.12 + (0.6215 T) − (11.37 $v^{0.16}$) + (0.3965 T $v^{0.16}$) where T is the DB air temperature in °C and v is the air speed in m/s. Other countries have their own versions of wind chill charts, available from local meteorological agencies or the World Meteorological Organization.

8.4 Working in hot environments

Designing comfortable climates: Further information for the built environment, such as offices in Europe and North America, is contained in the ANSI-ASHRAE Standard 55 and in ISO standards, as listed by Parsons (2003, 2014) and referenced below.

Sketched curve: This up–down "mound" (or "inverted U") curve (and its mirror image, the down–up "valley" or "U") is often employed to describe human physical and physiological behavior; for another example, see Figure 11.3 (task complexity and performance) and Figure 11.4 (frequency and observation of signals).

References and further reading

ANSI/AAMI HE75 (2009). R2018, Human Factors Engineering - Design of Medical Devices. ANSI/ASHRAE Standard 55, Thermal Environmental Conditions for Human Occupancy, latest edition 2023.

Bernard TE (2002). Thermal stress. Chapter 12 in Plog BA (ed), *Fundamentals of Industrial Hygiene*, 5th ed. Itasca IL, National Safety Council.

Bernard TE, Dukes-Dobos FN (2002). Heat stress. In Anna D (ed), *Chemical Protective Clothing*, 2nd ed. Fairfax, VA, AIHA Press.

Havenith G (2005). Thermal conditions measurement. Chapter 60 in Stanton N, Hedge A, Brookhuis K, Salas E, Hendrick HW (eds), *Handbook of Human Factors and Ergonomics Methods.* Boca Raton, FL, CRC Press. https://doi.org/10.1201/9780203489925

ISO 7243 (2017). *Ergonomics of the Thermal Environment—Assessment of Heat Stress Using the WBGT (Wet Bulb Globe Temperature) Index.* Geneva, CH, International Organization for Standardization.

ISO 7933 (2023). *Ergonomics of the Thermal Environment—Analytical Determination and Interpretation of Heat Stress Using Calculation of the Predicted Heat Strain.* Geneva, CH, International Organization for Standardization.

ISO 8996 (2021). *Ergonomics of the Thermal Environment — Determination of Metabolic Rate.* Geneva, CH, International Organization for Standardization.

ISO 9920 (2007). *Ergonomics of the Thermal Environment—Estimation of Thermal Insulation and Water Vapour Resistance of a Clothing Ensemble*. Geneva, CH, International Organization for Standardization.

ISO/CIE 8995-3 (2018). *Lighting of Work Places, Part 3: Lighting Requirements for Safety and Security of Outdoor Work Places*. Geneva, CH, International Organization for Standardization.

ISO/CIE DIS 8995-1 (2024). *Lighting of Work Places, Part 1: Indoor (Expected 2024)*. Geneva, CH, International Organization for Standardization.

ISO/DIS 7730 (2005). *Ergonomics of the Thermal Environment — Analytical Determination and Interpretation of Thermal Comfort Using Calculation of the PMV and PPD Indices and Local Thermal Comfort Criteria*. Geneva, CH, International Organization for Standardization.

Kroemer AD, Kroemer KHE (2017). *Office Ergonomics: Ease and Efficiency at Work*, 2nd ed. Boca Raton, FL, CRC Press. https://doi.org/10.1201/9781315368603

Kroemer Elbert K, Kroemer HB, Kroemer Hoffman AD (2018). *Ergonomics: How to Design for Ease and Efficiency*, 3rd ed. London, UK, Elsevier. https://doi.org/10.1016/C2016-0-02516-0

Osczevski R, Bluestein M (2005). The new wind chill equivalent temperature chart. Bulletin American Meteorological Society, 86:1453–1458. https://doi.org/10.1175/BAMS-86-10-1453

Parsons KC (2003). *Human Thermal Environments: The Effects of Hot, Moderate, and Cold Environments on Human Health, Comfort and Performance*. London, UK, Taylor & Francis. https://doi.org/10.1201/9781420025248

Parsons KC (2005). Ergonomic assessments of thermal environments. Chapter 23 in Wilson JR, Corlett N (eds), *Evaluation of Human Work*, 3rd ed. London, UK, Taylor & Francis.

Parsons KC (2014). *Human Thermal Environments*, 3rd ed. Boca Raton, FL, CRC Press. https://doi.org/10.1201/b16750

Youle A (2005). The thermal environment. Chapter 20 in Gardiner K, Harrington M (eds), *Occupational Hygiene*, 3rd ed. Oxford, UK, Wiley-Blackwell. https://doi.org/10.1002/9780470755075

CHAPTER NINE

How we think

Two internal systems control human functioning: the endocrine system and the nervous system. The endocrine system essentially consists of glands of internal secretion, a group of organs that produce hormones and secrete them into the bloodstream. Hormones are substances that affect activities in body cells elsewhere. While the biochemistry of hormones and their effects is understood, the endocrine system does not lend itself to ergonomic exploitation and is therefore not a topic in this book.

Nervous system

The other controller of the human body is the nervous system, governed by the brain. Its anatomy and functions are well studied, but its actual functioning is still only generally understood. Our current knowledge provides ample opportunities to appreciate the interactions between body and brain actions, indicating to the human factors engineer how to design work and equipment to leverage the intricate capabilities of the nervous control system. We need to understand how we think and respond in order to fit the task to the human.

9.1 Our brain–nerve network

The brain

The brain, also known as the *encephalon*, is the control center of the body. It contains nearly 100 billion interconnected neurons and creates the human "mind," which generates emotions, moods, beliefs, memories, behaviors, and thoughts. The brain receives sensory signals to relate to body position and movement, touch, smells, sounds, and sights. It coordinates and interprets these signals, developing reactions and subsequent

DOI: 10.1201/9781032674575-12

actions, often within milliseconds. For this, the brain requires a continuous supply of oxygen, so the blood flow to the brain accounts for about 20% of the heart's total output.

Observing brain functions

In the 1920s, psychiatrist Hans Berger (1873–1941) attached metal electrodes to the heads of patients to record "brain waves." These waves represent the brain's electrical neural activity through electroencephalograms (EEGs). Around the same time, physiologist Walter Hess (1881–1973) demonstrated that electrical impulses in cat brains affect their emotions and behaviors. Several decades later, neurophysiologist José Rodríguez Delgado (1915–2011) showed similar results in humans. Since the 1990s, a non-invasive technique called functional magnetic resonance imaging (fMRI) has been used to detect changes in blood flow in the brain, with stimulated regions appearing on a screen as blotches of color. Since brain regions only "light up" on fMRI when large numbers of neurons become active, interpreting these images can be challenging when separate neuronal events blur together. Today's fMRIs are not precise enough to reveal what individual brain cells are doing; however, they can be used to map which parts of the brain handle specific functions. Newer techniques are continually being developed. One recent example is optogenetics*, a technology that has been used to study live, freely moving animals. It makes individual brain cells photosensitive and activates them with flashes of light delivered through fiber optic wires, allowing observation of distinct circuitry and providing a means to study behavior changes induced by stimulation of identified cells.

Understanding brain function

While the anatomy of the human brain is well mapped, neuroscientists still do not clearly understand how this organ, with its billions of neurons and their various connections, actually functions as the mind. The brain controls our movements and bodily functions; what we consider the mind also includes reasoning, emotion, and understanding. We can generate, in milliseconds, emotions, thoughts, behaviors, and actions. Lacking sophisticated knowledge and useful technology, we still rely on rather broad perspectives to describe the brain's functions somewhat superficially. For example, sleep features are recorded and classified by EEGs and electrooculograms (which describe eye movements, particularly their rapidity); however, why the brain needs sleep is still unclear (more on this in Chapter 15).

Neuroplasticity

Over the past few decades, our understanding of the brain's blueprint and its workings has undergone a major conceptual change. Well into the 20th century, it was commonly accepted that, after growing and developing for about the first 20 years of life, the anatomy and functions of the human brain became "fixed"—as if in a wondrous computer with hardwired circuits set to perform certain unchangeable functions. If not injured, the adult brain was thought to change only during its decline with age.

By the 1960s, researchers recognized that even the adult brain is not immutable. It can improve its functioning with every activity performed. At times, the brain can re-organize itself, altering its structure and developing new ways to function; for example, when young brain cells fail to develop properly or when injury damages cells. This ability of the brain to modify its morphology and pathways is called *neuroplasticity**. Neuroplasticity can be structural (changes in connections between neurons, akin to rewiring computer circuits) and functional (changes in the functional properties of neurons). The brain's ability to adapt to damage can be seen in the rehabilitation of patients who are trained to re-use parts of their bodies, such as hands, which had been rendered nearly useless by brain damage after a stroke.

Brain injury

Traumatic brain injury is one of the leading causes of death and disability for young adults. Head trauma can damage brain cells in several ways. The most direct damage is the so-called coup/contrecoup injury. The initial coup injury occurs at the moment of impact between the head and an object (and thus, between the brain and skull). The next contrecoup occurs when the brain bounces back within the skull and hits the opposite side. The coup/contrecoup can also damage the axons, preventing them from carrying information from one cell to the next. The brain may also swell inside the skull, compressing blood vessels and causing further cell damage.

Another source of brain injury is stroke, which is a leading cause of death and disability in adults worldwide, estimated to affect approximately one-quarter of the global population. Stroke occurs when blood supply to the brain is disrupted, either due to a blockage or excessive bleeding (hemorrhage). Either cause results in damage to brain cells within minutes.

Parts of the brain

The three major anatomical components of the brain, shown in Figure 9.1, are the cerebrum, cerebellum, and brain stem.

Cerebrum

The *cerebrum* consists of dense, convoluted masses of tissue divided into two halves, the left and right cerebral hemispheres,

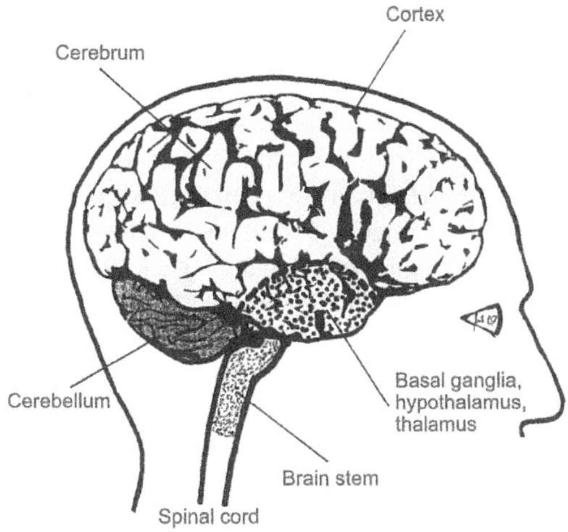

FIGURE 9.1 Major components of the brain.

which are connected by nerve fibers. It contains the frontal, occipital, parietal, and temporal lobes. The frontal lobes control the so-called executive functions, including skilled motor behavior, speech, mood, thought, and planning. The occipital lobes are responsible for vision, while the parietal lobes interpret sensory information from the body and control movement. The temporal lobes generate memory and emotion, process and retrieve long-term memories, and initiate communication or action. The cerebral hemispheres have an inner core of white matter and an outer layer (the cortex) of gray matter. The cortex is associated with voluntary movement, thought, perception, memory, language, and individuality.

Cerebellum

The *cerebellum*, located beneath the cerebrum and just above the brain stem, coordinates the body's voluntary movements (including eye movements) and maintains balance and posture. A special collection of nerve cells located at the base of the cerebrum includes the basal ganglia, hypothalamus, and thalamus. The basal ganglia help smooth out movements. The hypothalamus coordinates automatic body functions (such as sleep and wakefulness), maintains body temperature, and regulates the body's water balance. The thalamus organizes sensory messages to and from the highest level of the brain, the cerebral cortex.

Brain stem

The *brain stem* connects the brain to the spinal cord. It automatically regulates body functions such as metabolism, posture, breathing, swallowing, heartbeat, sleep, and consciousness, increasing alertness when needed. Severe damage to the brain stem can destroy these functions and result in death.

Spinal cord

The *spinal cord* is an extension of the brain: it carries commands from the brain to the body, transmits sensory information from the body to the brain, and coordinates reflexes. This long (approximately 45 cm) and fragile structure begins below the brainstem and runs down the back side of the vertebral column. The vertebral bodies protect the spinal cord within the bony "rings" encircling the vertebral foramen, as shown in Figure 2.10 of Chapter 2. Figures 9.2 and 9.3 provide details of a representative vertebral segment with the spinal cord and its nerve extensions. Motor nerves at the front (anterior) of the spinal cord transmit information from the brain to the muscles. Sensory nerves at the rear

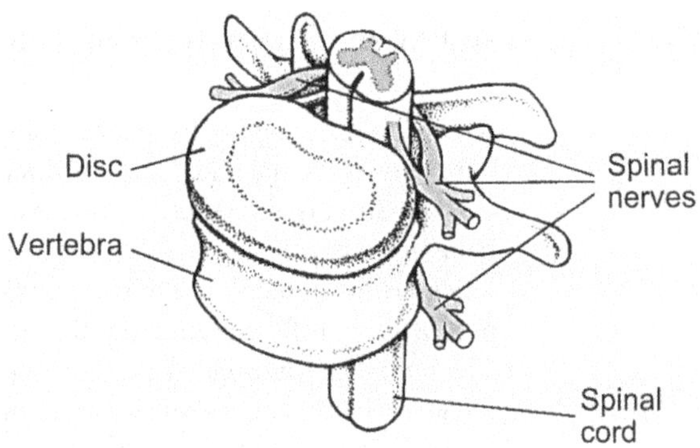

FIGURE 9.2 Typical segment of the vertebral column showing the spinal cord and emanating spinal nerve extensions, as well as vertebra and intervertebral disc.

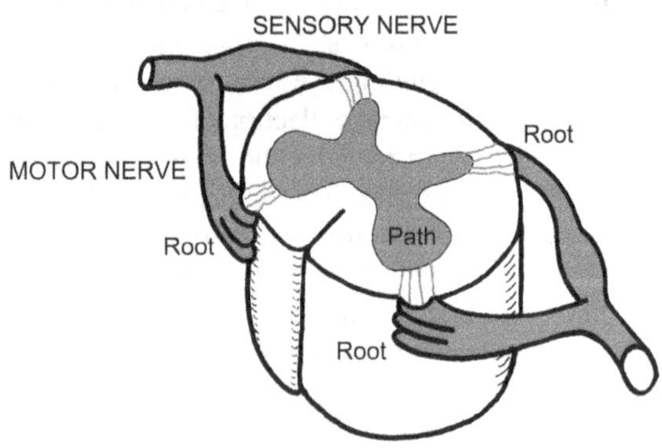

FIGURE 9.3 Typical segment of the spinal cord, with motor and sensory nerves and nerve roots.

(posterior) carry information from sensory receptors in the body to the brain. The spinal cord can be damaged by displacements of the vertebral bones, such as in an automobile accident or a fall, leading to deficits in sensation and movement.

Spinal nerves

The spinal cord provides the main pathways for nervous signals, relaying both outgoing (*efferent*) motor commands and incoming (*afferent*) sensory messages. There are 31 pairs of spinal nerves that branch off the spinal cord, emerging (looking like roots) between pairs of vertebrae, as illustrated in Figure 9.3.

Nerve impingements

Displaced vertebrae and material from a ruptured disc can impinge on the spinal nerve extensions, with dire consequences for both efferent and afferent signals. Injury to a motor nerve may cause weakness in or even paralysis of the muscle it serves. Damage to a sensory nerve can lead to feedback sensations of pain and numbness that, in some cases, may mislead the brain. For example, pain may seem to originate from body sections such as the buttocks and thighs, which are innervated by the sciatic nerve, while the actual issue might be at the point of impingement of the sciatic nerve at the spinal column.

Neurons

The basic functional unit of the human nervous transmission system is the nerve cell, or *neuron*. The nervous system contains about one billion of these cells that extend like strings throughout the body. A neuron has a cell body, or *soma*, that is a few thousandths of a millimeter thick, and short extensions called *dendrites*, where connections from other neurons arrive across synapses. Figure 9.4 sketches a motor nerve cell with its single long extension, called an *axon*, which can be more than a meter

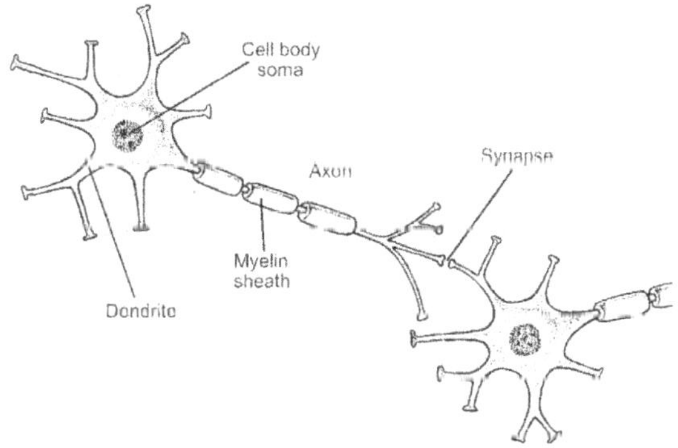

FIGURE 9.4 Schematic view of a typical motor neuron, with soma, dendrites, axon, and synapse. Adapted from Kroemer, Kroemer, and Kroemer-Elbert (2020).

long. Nerves transmit their messages electrically in one direction: from the axon of one neuron to the dendrites of the next neuron(s). At the synapses, the contact points between neurons, the axons secrete chemical neurotransmitters that trigger receptors on neighboring neurons to generate a new electrical current.

Signal transmission

Synapses are not just simple transmitters; they also serve as filters or switches by inhibiting the transmission of incoming signals that are infrequent and weak. Neurons vary in type and size, with some able to transmit up to a thousand impulses per second, while others may convey fewer than 25 signals per second. The velocity of impulse travel is constant for each nerve fiber, ranging from about 0.5 to 150 m/s. Conduction speed is slowest in thin fibers, such as those transmitting pain, and highest in thick axon fibers with myelin sheaths that serve skeletal muscles.

Action potential

Incoming signals, arriving via synapses, may be too weak to stimulate the nerve cell. However, if they are strong enough, they can prompt the cell to generate an electrical spike, shown in Figure 9.5. The peak is approximately 100 mV above the resting potential; this impulse, called *action potential*, travels to the next neuron. In a motor neuron, the impulse passes from the axon to the motor end plates in the muscle, triggering a contraction.

Motor unit

At the muscle, a motor nerve divides into several fibers, each controlling multiple muscle fibers. A *motor unit* consists of all the muscle fibers innervated by one motor neuron. In muscles that perform precise movements, a motor unit contains only

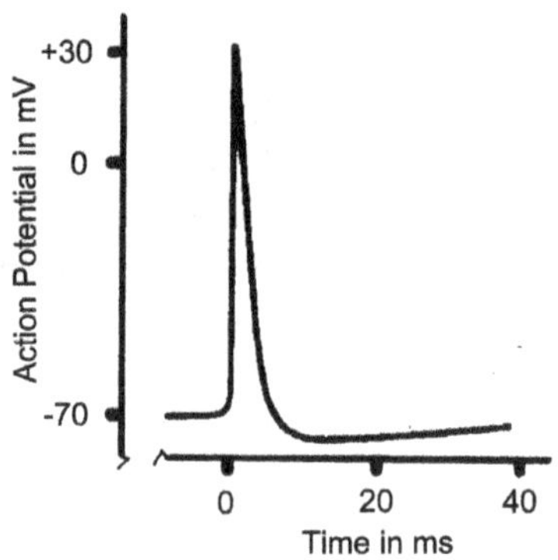

FIGURE 9.5 Action spike of about 100 mV above the resting potential of −70 mV.

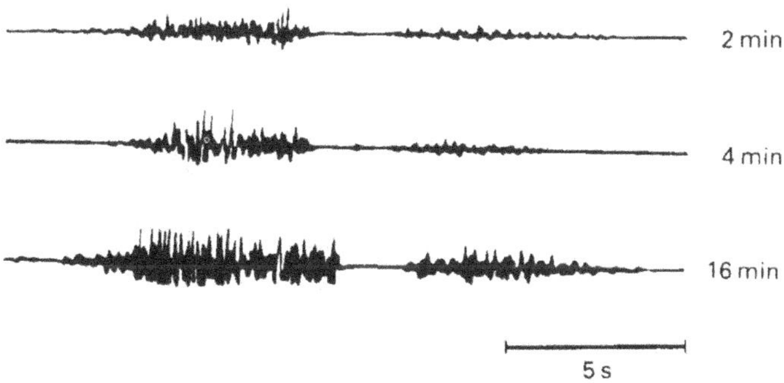

FIGURE 9.6 Sample EMGs taken after 2, 4, and 16 minutes of contractions of equal strength.

three to six muscle fibers, while muscles designed for heavy work may have 90 or more muscle fibers controlled by a single neuron.

EMGs

Electromyograms (EMGs) allow us to observe the strength and frequency of activation commands sent from the central nervous system (CNS). These electrical signals, detected by surface or indwelling electrodes, provide information on both the frequency of muscle unit activations and their fatigue status. Figure 9.6 shows representative EMGs of the extensor muscle of the upper arm. In this example, the muscle executed a series of contractions of equal strength, while EMGs were recorded after 2 minutes, 4 minutes, and 16 minutes of contractions. Comparing the EMG traces after 2 minutes and 4 minutes, we see increased amplitude in the EMG related to the same contraction, indicating muscle fatigue; after 16 minutes, the even larger signal amplitude indicates extensive fatigue.

Feedforward/ feedback loop

The innervation of a muscle by its motor nerve and the sensory feedback loop to the spinal cord and brain are shown in Figure 9.7. Feedforward information is represented by the incoming action impulses that transmit contraction commands to the muscle at the motor end plates. Feedback information about the ensuing events travels in sensory nerves toward the spinal cord. Important signals originate at Golgi organs (G), which respond to tension in tendons. Signals also arise from muscle spindles (S), which sense muscle stretch. Additionally, signals can include messages of pain (P). This feedback partially reaches the spinal cord for reflexive automatic adjustments of muscular effort, while other signals go to the brain for continuous evaluation and control of task performance.

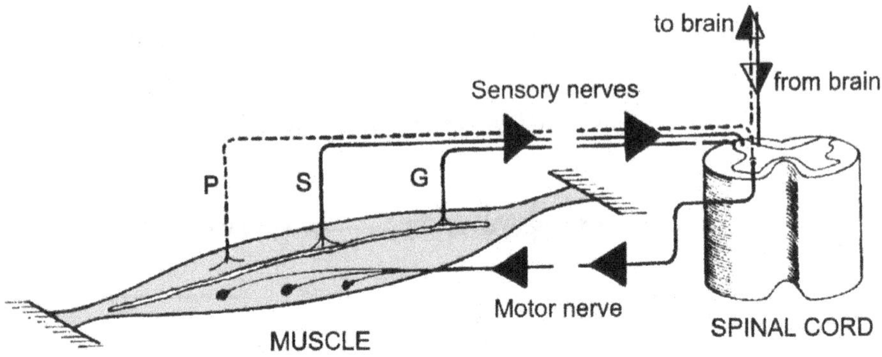

FIGURE 9.7 Muscle innervation and feedback control: pain signal (P), signal from muscle spindle (S), and signal from the Golgi organ (G).

Motor nerves carry feedforward action orders from the brain to muscles, causing muscle contractions, while sensory nerves transmit feedback reports on muscle conditions. Integrating the sequence of feedforward and feedback information provides the means to control the muscular activities.

Designing for simple movements

The body can perform some motions without direct involvement of the brain; however, most muscular activities require fine regulation. This involves varying degrees of engagement from higher brain centers such as the cerebral cortex, basal ganglia, and cerebellum. The transmission of information along lengthy paths through numerous neurons and decision processes is labor-intensive and time-consuming. Thus, learning to execute complex movements can be challenging and slow, as we know from our own experiences of acquiring new and complicated skills and particularly as we age. This suggests that job activities should be performed as simply as possible, minimizing decision-making and maximizing efficiency of information transmission while employing the least amount of body mass. This principle is illustrated by a counter-example: controlling the speed of an automobile through complicated and forceful movements of the feet on pedals not visible to the driver is not a good example of modern human factors engineering.

The brain–nerve network

The brain and nerves form a complex communication system, capable of sending and receiving vast amounts of information simultaneously. The brain acts as the control center, communicating with the body through the nerves that run up and down the spinal cord. Sensory nerves carry information to the brain about pressure, pain, heat, cold, vibration, and the feel and shape of objects. They also transmit messages from internal body sensors, for example regarding the positions and motions of body parts and the related tension in muscles and tendons. This information about events external and internal to the body enables the brain to make decisions on how to respond; the information loop is completed when the brain then receives feedback about the results of the actions performed.

Reflexes

Although to a much lesser extent than the brain, the spinal cord also coordinates certain actions, such as reflexes. A reflex typically starts with the stimulation of a sensory receptor, for example one located in the patella tendon just below the front of the kneecap. This sends a signal to the spinal cord, which produces an immediate response command that rapidly travels to the appropriate muscle, for example the quadriceps muscle in the thigh to produce the knee–jerk reflex. In this way, a muscle reaction can occur within a few milliseconds of the stimulus since no time-consuming higher brain functions are involved.

CNS and PNS

The human nervous system is commonly divided into components. One approach is to divide by anatomy: central or peripheral. The *central nervous system* (CNS) consists of the brain and spinal cord. It controls the body by gathering information, making decisions, and initiating actions, serving as the body's control center. The *peripheral nervous system* (PNS) consists of the nerves outside the brain and spinal cord. The PNS connects sensors located throughout the body to the CNS, and from there to organs, particularly muscles and glands. Peripheral nerves are bundles of single nerve fibers; some are very thin, less than half a millimeter in diameter, while others are thicker than 5 mm. The PNS does not control but instead delivers signals*.

Somatic and autonomic

Another way to further divide the PNS is by function: somatic or autonomic. The *somatic nervous system* is under voluntary control and comprises the sensory and motor nerves of the PNS, all of which are necessary for conscious actions and mental activities. The somatic system links the body with the outside world through perception, awareness, and actions, especially through muscle activation. The *autonomic nervous system* represents other functional components; it regulates involuntary processes, governing the internal organs essential to life, such as blood circulation, breathing, and digestion. This system (also called *visceral*) is responsible for the body's automatic functioning, emergency responses, and emotions.

Sensory receptors

The nervous system monitors all sensations from inside the body and its surfaces, as well as through the eyes and ears. If the signals are strong enough, they are transmitted to the CNS, where the information is received and processed, leading to a judgment about whether an action is needed. If necessary, the CNS activates muscles and monitors the results of the actions taken.

External receptors

We perceive the world around us through sight, hearing, and touch. The senses of gustation and olfaction allow us to taste and smell, while we also can detect temperature, electricity, pressure, and pain (as discussed in Chapter 7). Different types of surface sensors are embedded in the skin's layers in varying concentrations. These sensors can be mapped to spinal nerves: Figure 9.8 shows divisions of skin surface areas (*dermatomes*) that convey information to specific sensory nerve extensions of the spinal cord. Certain parts of our bodies, such as our lips and fingertips, have a high concentration of sensors, while other areas, such as our backs, are relatively numb.

9.2 Taking up and processing information

Cognitive concepts of information processing

Cognition is the act or process of knowing in the broadest sense. In the domain of human factors, the intellectual processes by which knowledge is gained from perception are of particular interest. In cognitive psychology, the common concept is that information must pass through three stages of mental processing: sensory input, short-term memory, and long-term memory. "Stage" and "parallel distributed" processing models are prominent among current theories of how humans process information.

Mental work

Much of our "mental work" consists of sensing information, processing it, and finally acting on it. The response is then generally monitored. Of course, most of these activities intermingle with other mental tasks occurring simultaneously. Nevertheless, it is convenient to consider these activities as occurring in stages that follow each other in series. Figure 9.9 illustrates the concept of the human as a linear processor of signals. First, a sensor detects information and sends related impulses along the sensory pathways of the PNS to the central processor in the CNS. It interprets the information (also using short-term and long-term memory) and chooses an action (which may include the option of not taking any action). Then the CNS generates

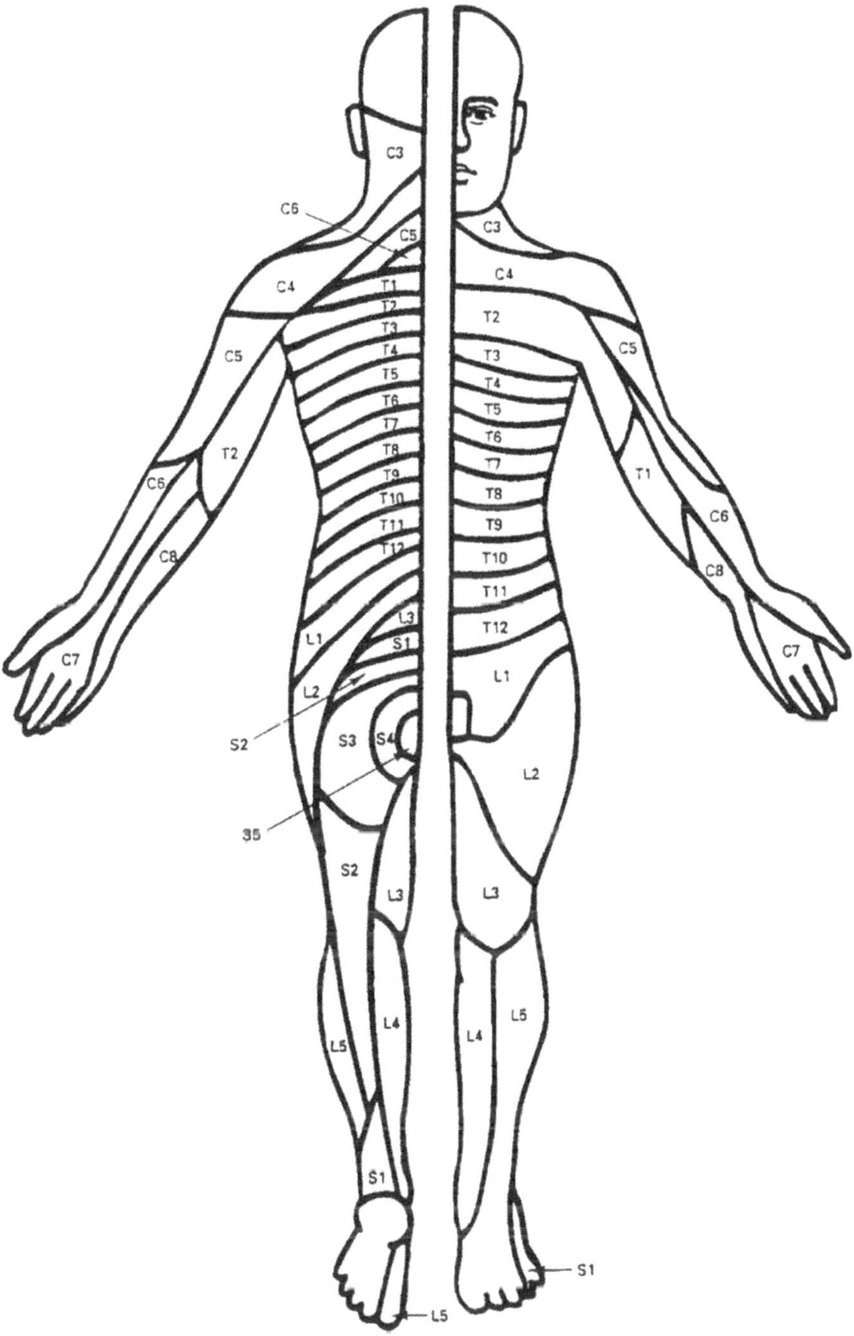

FIGURE 9.8 Sensory dermatomes identified by their nerve roots (C, cervical; T, thoracic; L, lumbar, S, sacral) at the spinal column. Modified from Sinclair (1973).

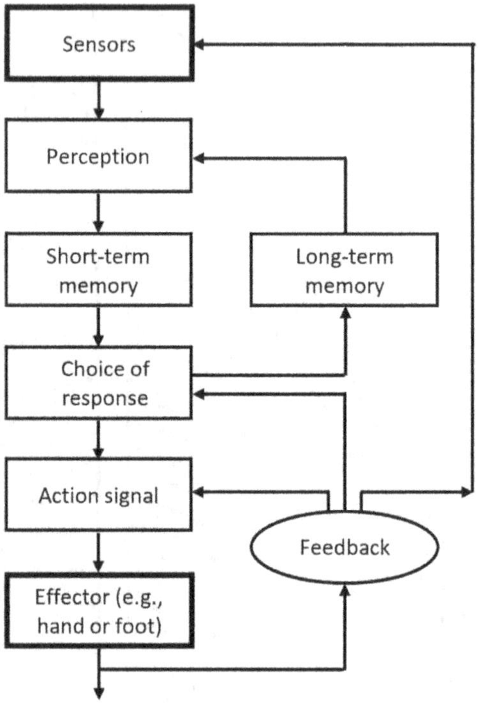

FIGURE 9.9 Model of human information processing. Adapted from Kroemer (2006).

appropriate feedforward impulses and sends these along the motor pathways of the PNS to the effectors, usually voice or the muscles in the hands or feet. A sensory feedback loop compares the output of the system with the desired performance. If the comparison between actual and intended performance shows a difference, that disparity leads to revised action.

Sensors inside the body

Interoceptors are sensors inside the body, including the Golgi organs and spindles associated with muscles. Ruffini corpuscles are receptors in the body joints that provide information on their angulation. Other sensors are located in the semicircular canals of the ears; these vestibular receptors detect and report the position of the head in space, as discussed in Chapter 7. Additional interoceptors respond to events within the internal organs, such as in the abdomen, chest, and head. They relay sensations such as pain and pressure.

Sensors near the surface

Exteroceptors provide information about the external environment. They are involved in sight, sound, taste, smell, and touch. Some senses are particularly important for controlling body activities: the sensations of touch, pressure, and pain provide feedback regarding the direction and intensity of muscular activities transmitted to an external object. The receptors are located throughout the skin, though in varying densities; see the dermatomes shown in Figure 9.8. The sensors mainly consist of free nerve endings, Meissner's, and Pacinian corpuscles. Since the PNS nerve pathways leading to the CNS interconnect extensively within the PNS, the reported sensations are not always specific to a given input; for example, very hot or very cold sensations can be associated with pain, which itself can also result from hard pressure on the skin.

Adaptation and speed

Almost all sensors respond vigorously to a change in stimulus but then report less and less as the load remains constant. This adaptation allows us to live with continuously present but unimportant stimuli, such as the pressure of clothing. The speed of adaptation varies among different sensors. Additionally, the velocities for transmitting sensations to the CNS differ for various sensory pathways: light and sound, for example, are transmitted quickly, while pain is reported more slowly.

Modifying input signals

Humans can experience a wide range of sensations, yet some signals are imperceptible to us. For instance, we cannot hear ultrasound or infrasound, frequencies of great importance to many animals such as dogs and elephants. Another example is X-rays, for which our body has no sensory organs. One major task of the ergonomist is to modify the physical properties of external signals, particularly those we cannot sense but are important, so that we become aware of them. If the signal is entirely outside the human sensory "bandwidth," the engineer must alter its nature of the signal into one that can be sensed. One example of such a technical solution is generating a buzzing sound and/or flashing red light to indicate the presence of ionizing radiant energy.

Providing proper input signals

In other cases, we can adjust the signal or environment so that the signal penetrates the surrounding sensory clutter. Examples include control buttons that blink, lights on emergency vehicles that rotate, flash, and change color, vehicle sirens that modulate their sound, and loudspeaker qualities that adjust (such as lowering background music volume and playing an alert tone before making an emergency announcement). Computer offices provide further examples: here, overall illumination levels are often kept low to avoid glare or washed-out images on a computer screen, but in certain areas, we increase illumination to highlight exit stairs or use separate "task lights" to make it easier to read paper documents, especially for older individuals (even if this may create glare). Providing proper input signals can be a daunting ergonomic task, but it is essential. Figure 9.10 presents a schematic showing how we can transform external (distal) signals for human perception.

Detecting signals

Signals need to be detected and interpreted correctly to be useful. For the ergonomist, signal detection theory (SDT) addresses how the constantly active brain distinguishes important signals from extraneous noise. If a signal is present and we recognize it, it is classified as a *hit*; if we fail to recognize it, this is a *miss*. If there is no signal but we believe there is, it is a *false alarm*. For example, when we notice a machine's oil pressure warning light coming on and there is an actual mechanical issue that needs fixing, that is a successful stimulus–response. However, if we

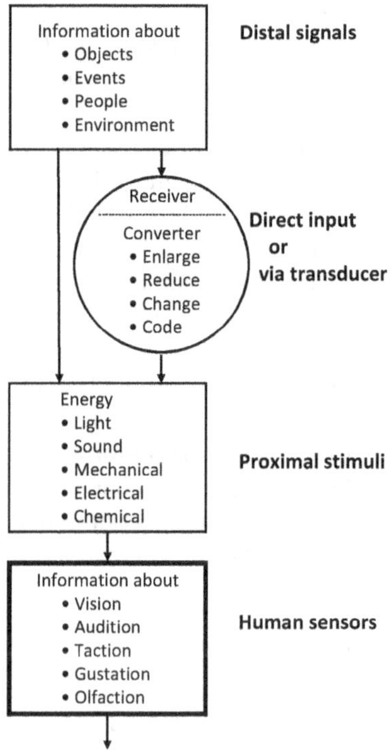

FIGURE 9.10 Transforming distal signals into proximal stimuli that can serve as inputs to the human processor. Adapted from Kroemer (2006).

fail to notice the warning light, then it is a miss. If the warning light is too dim, or if there are too many warning lights, causing the signal to get lost in the noise, we should consider additional ergonomic interventions. The intensity of these interventions should reflect the risk and severity of the consequences of a miss or a false alarm.

9.3 Making decisions

Models of IP

Traditional models of information processing (IP) depict a sequence of stages (similar to flowcharts used in industrial and systems engineering) through which information passes. The simplest model of "mental work" consists of two distinct phases: first, the evaluation of occurring events, and then, the execution of actions to bring about changes. The resulting new environment is then evaluated again, initiating a new sequence of phases.

Parallel IP

A refinement of this linear model acknowledges that multiple sources of information are present simultaneously, which we evaluate using various processing facilities. At each stage, our mind filters and transforms information from previous stages, integrating and comparing it with information from memory* before it moves to the next stage. According to this model, a person's overall cognitive performance depends on both the number and type of stages required and the efficiency of operations at each stage.

Channel capacity

Some models of IP emphasize limits on the amount of information (a type of channel capacity) that can be processed. For human factors engineers, awareness of channel capacity often leads to the recommendation that systems should be designed to minimize the amount of input information and require the lowest possible processing demands.

Perceiving sensory signals

In the first stage of IP, we select from the overwhelming onslaught of sensory signals only the most important information for further processing. The brain accomplishes this selection by comparing new inputs with information stored in our memory, searching for and recognizing familiar features. This means that our past experiences generate expectations which guide our perception and selection of information. New or surprising incoming sensory information may be rejected or, conversely, may easily penetrate our expectation-based barriers—though we may struggle to interpret the unfamiliar inputs.

Short-term memory

Human memory can be classified into short-term and long-term memory, depending on duration and capacity. Short-term (or "working") memory retains information for brief durations, such as a few seconds, allowing us to maintain a representation of the rapidly changing world around us for comparison and filtering. An example of short-term memory is looking up a phone number and keeping it active in memory until dialing it. This short-term working memory appears to have very limited capacity, both in duration and channel capacity. One theory suggests that the boundaries of short-term memory may be the limiting capacities in IP.

Long-term memory

In contrast, long-term memory has a much larger capacity and no obvious duration constraints. A fundamental distinction in human long-term memory is that between memories involving general knowledge, called semantic memory, or specific events, called episodic memory. Often, these are linked, as by an event from our past, such as an accident. The ability to remember (encode, store, and retrieve) essential information from our long-term memory is important for many tasks. The forgetting of information may result from not successfully encoding information in the memory "bank" or from having the information present but inaccessible: we may struggle to retrieve it due to a lack of memory cues or because other but irrelevant information clutters our memories.

Information stored in chunks

To interpret information, such as shown on a display, and then select an appropriate response requires us to access stored, task-relevant information. Activation of related knowledge in long-term memory may be triggered by recalling past events or actions taken under similar conditions. Two major factors influence the availability of information and its reactivation from long-term memory. One factor is the strength of the information trace, determined by its initial importance, the number of times it has been activated, and how recently that occurred. The other factor is its association with related items and events, such as a loud bang or the smell of smoke when an accident occurred. These associations allow us to store much information in chunks rather than in isolated pieces.

Making decisions

Our brain must integrate incoming sensory stimuli with other relevant information and then process that knowledge to select a suitable response. This aspect of our "mental work" is often called central processing*, consciousness, or thinking. It requires keeping the overall goal of the task in mind to select the appropriate response from all possible options. Thus, decision-making and response selection include the essential necessity of both understanding the outcomes of several possible responses and recognizing which actions are needed to execute the proper response.

Limiting IP requirements

The thinking process must integrate many pieces of information, some newly acquired and most already stored in long-term memory. Many of our daily tasks involve this type of central processing. Reducing the IP requirements, such as by limiting inputs and integration (processing) demands, simplifies the task. One example is driving an automobile in dense traffic, where we decide to either follow a preceding car or change lanes to overtake it. Following the preceding car by accelerating and braking is usually easier and less dangerous than passing it (and many newer cars have sensors that assist in following a car more safely). Overtaking the preceding car requires checking whether a neighboring lane is clear, turning the steering wheel in both directions, and carefully regulating our car's speed relative to the one being overtaken. Here, an important ergonomic modification for the driver may involve reducing unnecessary distractions, such as from the radio, mobile phone, or conversation with a passenger.

Fitting the human to the job

A traditional model of human IP is illustrated in Figure 9.9. It shows that, in the CNS, signals transmitted from sensors are perceived, decoded, interpreted, and processed through sequential stages. Little is known about how this complicated procedure actually develops; it appears to involve comparing current information with previous experiences by linking it to related knowledge stored in long-term memory. Such processing comes naturally with experience and aging, and intentionally through teaching and training. The necessity for teaching and training novices in decision-making is an example of fitting the human to the job, as opposed to fitting the job to the human.

New models of IP

The predominant premise in cognitive psychology (originally developed in the latter half of the last century) is that the brain can be modeled as a complex computing system. The computer metaphor continues to underpin most current concepts of human mental work, relying on the assumption that our brains operate similarly to computers in processing information (though, naturally, comparing a living organism to a technical product is always imperfect). For example, computers have specialized input and output devices, somewhat akin to the specialized sensory and response modalities of humans. However, computers and humans have different types of memory stores that can be accessed or searched in varying ways; searches, evaluations, and decisions follow hierarchies of importance, and capacities are often limited in both quantity and function over time.

Advanced concepts in cognitive psychology* include models based on new findings in neurophysiology, such as analogies to multiple and parallel feedforward and feedback paths combined with distributed processing in neural networks. These models can overcome the limitations of serial-stage models with discrete processing steps; they can offer a more detailed and realistic description of information storage and processing. Nevertheless, for now, the traditional stage-based model still provides a useful conceptual framework and guidance for the human factors engineer.

9.4 Designing for actions and reactions

Transducers

After the CNS has chosen a response to the received signals, it generates action signals to effectors such as the hand or foot, activating muscles to (for example) fight or run. In the workplace, this typically involves applying force to a handle, pedal, or other input control element of machinery, whether it be a simple hand tool, computer input keys, or the thrust lever of an airplane. If the effector tool is a hammer or screwdriver, the task is straightforward. However, with more complex machinery, such as an airplane or ship, so-called transducers are involved, modifying the body's actions, as Figure 9.11 illustrates. For example, in a bicycle or small boat, turning the handlebar or steering wheel immediately converts human movement into vehicle motion using mechanical gears, linkages, and hydraulic transmissions. In complex human-operated systems, like a large ship or spacecraft, the vehicle's response to the action of the human operator often takes longer to materialize. In this case, a computerized system may be useful for predicting the future results of current actions.

Designing transducers and their feedback signals presents a challenge for human factors engineers. Assessing the effect of swinging a hammer requires no special machinery: the result of the first blow is immediately clear and helps guide the next stroke to drive a nail deeply into a wooden beam while avoiding one's other hand. However, for complex human–machine

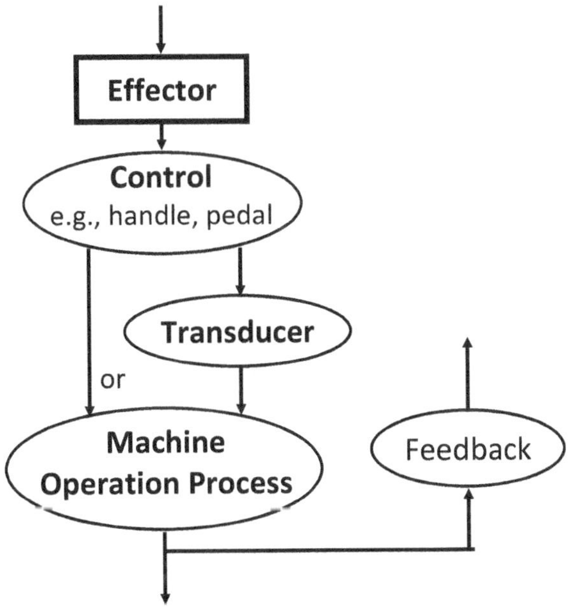

FIGURE 9.11 Transforming effector output for machine operation. Adapted from Kroemer (2006).

interactions, it is important to distinguish among specific aspects of human actions and reactions.

Response time

As discussed in Chapter 7, *reaction time* is the interval from the appearance of a stimulus (such as a light) to the initiation of a responding effector action (such as moving a hand). The additional time to perform the action (such as pulling a handle) is referred to as *motion* or movement *time*. Added together, these constitute the *response time**, the total time to sense, process, and act on information:

reaction time + motion time = response time

Reaction time

Time elapses between the appearance of a signal on the sensory input side of the nervous system and the emergence of a (re) action on the output side. Time delays occur at the sensor, during afferent signal transmission along the PNS, during processing by the CNS, during efferent signal transmission in the PNS, and during muscle activation. Estimates of these time delays (in milliseconds) are as follows:

- At the receptor: 1–38 ms
- Along the afferent path: 2–90 ms

- In CNS processing: 70–90 ms
- Along the efferent path: 9–20 ms
- Muscle latency and contraction: 30–70 ms

Simply adding the shortest time delays leads to the theoretically shortest possible reaction times. In reality, there is little reason to expect all conditions to be ideal; hence, actual reaction times should not be expected to be at the shortest end of the range.

If a person knows that a particular stimulus will occur, is prepared for it, and knows how to react, the resulting time is called *simple reaction time*. Its duration depends on the type and intensity of the stimulus. Many different tables of reaction times have appeared in engineering handbooks, but the original data—how they were measured, under what conditions, at what stimulus intensity, with what accuracy, and from which test subjects—are no longer known. Although the following information is dubious and should be re-evaluated experimentally, the shortest possible reaction times that are generally quoted are (as listed in Table 7.1):

- Electric shock, touch and sound: at least 130 ms
- Sight and temperature: at least 180 ms
- Smell: at least 300 ms
- Taste: at least 500 ms
- Pain: at least 700 ms

This listing shows no practical time difference in reactions to electrical, tactile, and sound stimuli. The slightly longer reaction times for sight and temperature stimuli (as in Table 7.1) likely fall within the range of measuring accuracy. However, the time following a smell stimulus appears distinctly longer, and reaction to taste is longer still; it takes by far the longest to react to the sensation of pain. Simple reaction times are not significantly affected by age from about 15 to 60 years, but reactions are substantially slower at younger ages and slow again as one grows older.

Choosing between reactions

If a person has to choose among several feasible actions, *choice reaction time* is involved. Choice reactions take longer than simple reactions. It takes even longer if it is difficult to distinguish among several similar stimuli, of which only one should trigger the response. The length of choice reaction time can be modeled as a logarithmic function of the number of alternative stimuli and responses, known as Hick's law (after psychologist William Edmund Hick). The mathematical formula is as follows:

$$\text{choice reaction time} = a + b \log_2 (N + 1)$$

where a and b are empirical constants and N is the number of choices. The use of a logarithm reflects the observation that increases in choice time become smaller as more possibilities are available. If the available choices have unequal probabilities, then $(N+1)$ may be replaced by the probability p of any particular alternative; this means $p = 1/(N+1)$, and the preceding equation becomes as follows:

$$\text{choice reaction time} = a + b \log_2 \left(1/p\right).$$

Motion time

Motion time (sometimes called movement time) occurs after reaction time. Movements may be simple, such as lifting a finger in response to a stimulus, or complex, such as swinging a tennis racket. Swinging the racket involves more intricate movements than lifting a finger and engages larger body and object masses, requiring more time.

Motion time depends on the distance of the movement and its required precision. This logarithmic relationship, known as Fitts's law (after psychologist Paul Morris Fitts), is as follows:

$$\text{motion time} - a + b \log_2 (2D / W)$$

where D is the distance covered by the movement (e.g., the distance to the center of the target) and W is the width of the target. The constants a and b depend on the particulars of the situation (such as the body parts involved, the masses moved, and the tools or equipment used), the number of repetitive movements, and skill (training and experience). The parameter $\log_2 (2D/W)$ is called the index of difficulty. Motion time may decrease with practice, increase with age, vary between dominant and non-dominant hands, and generally is not affected by gender.

Response time

Minimizing response time—the sum of reaction and motion lags—is often a goal of human factors engineering. Obvious ergonomic choices include optimizing the stimulus and selecting the body part best suited to the task. This usually requires both careful selection of the task (as well as equipment and procedures) and assessment of the relevant capabilities of the prospective user. This process may be iterative until the best match becomes evident.

Task performance The goal of improving performance may involve reducing response time and/or increasing the accuracy of the response. Ergonomic recommendations depend on the type of task and the information channels (such as signal lights or audible alarms) used.

When the task involves selective attention (focusing on a specific task while blocking out background stimuli), consider the following:

- Use fewer channels of information.
- Train the operator on which channels are most important and how to monitor them.
- Ensure channels do not interfere with each other (i.e., color of signal lights or volume of audible alarms).
- If intense focus is needed, make the primary channel more distinct (brighter, louder, more central).

If the task involves divided attention (multitasking or alternating focus between specific tasks), also consider:

- Further train the operator on the relative importance of channels.
- Make the tasks as easy as possible (such as using small, low-effort motions)
- Make the tasks as dissimilar as possible (such as using different channels for different tasks).
- Ensure the operator has practiced the responses.

If the task involves sustained attention (maintaining focus for long time intervals), also consider:

- Adjust the operator's work–rest schedule and allow for task variation.
- Further enhance the signal's intensity (such as blinking colors or warbling sounds).
- Make the signal as predictable as possible.
- Provide feedback to the operator on detection and task performance.
- Ensure the operator is motivated and understands the importance of task performance.

Summary

The brain is the center of the human nervous system. The CNS can perceive and interpret afferent signals sent from body sensors, process that information by integrating it with knowledge from memory, and ultimately make decisions about actions to take. Signals are then sent along the efferent pathways of the PNS to body organs, especially muscles, to initiate action. Body sensors pick up the results of these actions, along with new independent information, and corresponding signals are sent along the afferent pathways of the PNS to the brain for further decision-making.

Current concepts of brain function assume parallels between how the human brain and computers operate. Thus, these models depict IP in sequential stages. We can anticipate the use of more realistic models and their practical applications, however for now, we rely on serial stage models. By considering the total response time as the sum of reaction and motion times, we can apply ergonomic recommendations to improve performance and accuracy. These provide guidance to ergonomists in designing human-controlled systems.

Steps to fit the human

Step 1: Determine the work to be done and the input needed for the task.

Step 2: Select external signals that human sensors can easily and quickly accept, and that are distinct from background stimuli and from each other.

Step 3: Plan for decisions that depend on sensory inputs, ensuring they are familiar, secure, and quick to make. Consider whether the operator's attention is selective, divided, or sustained.

Step 4: Design the task to require minimal decision-making and employ the smallest possible body mass to execute an action.

Step 5: Design for appropriate quick reactions. Test the system, and modify and iterate as necessary.

Notes

The text contains markers, *, to indicate specific references and comments, which follow.

9.1 Our brain–nerve network

Optogenetics: See, for example, Colapinto (2015).

Neuroplasticity: A healthy or injured brain can reroute signals through existing pathways and grow new connections between brain cells. See, for example, Doidge (2007).

The PNS…delivers signals: The PNS was known to be "plastic" (and able to regenerate or adapt) when the CNS was still thought to be "hardwired."

9.2 Taking up and processing information

Memory: Arswell and Stephens (2001), Stanton, Hedge, Brookhuis et al. (2005).

Central processing: Bailey (1996), Wickens, Lee, Liu et al. (2004).

Advanced concepts: Arswell and Stephens (2001), Muchinsky and Culberson (2016), Vicente (2002).

9.4 Designing for actions and reactions

Response times and simple reaction times: More information can be found in Boff and Lincoln (1988), Boff, Kaufman, and Thomas (1986), Kroemer (2006), Kroemer Elbert, Kroemer, and Kroemer (2018), Swink (1966), Wargo (1967).

References and further reading

Arswell CM, Stephens EC (2001). Information processing (pp. 256–259). In Karwowski W, *International Encyclopedia of Ergonomics and Human Factors*. London, UK, Taylor & Francis.

Bailey RW (1996). *Human Performance Engineering: Designing High Quality, Professional User Interfaces for Computer Products, Applications, and Systems*, 3rd ed. Upper Saddle River, NJ, Prentice Hall.

Boff KR, Kaufman L, Thomas JP (1986). *Handbook of Perception and Human Performance*. New York, Wiley.

Boff KR, Lincoln JE (1988). *Engineering Data Compendium: Human Perception and Performance.* Wright-Patterson AFB, OH, Armstrong Aerospace Medical Research Laboratory. https://ntrs.nasa.gov/api/citations/19880019247/downloads/19880019247.pdf

Colapinto J (2015). Lighting the brain. *The New Yorker*, 18 May issue, 74–83.

Doidge N (2007). *The Brain that Changes Itself: Stories of Personal Triumph from the Frontiers of Brain Science.* New York, Penguin.

Kroemer KHE (2006). *"Extra-Ordinary" Ergonomics: How to Accommodate Small and Big Persons, the Disabled and Elderly, Expectant Mothers and Children.* Boca Raton, FL, CRC Press. https://doi.org/10.1201/9780203025246

Kroemer Elbert K, Kroemer HB, Kroemer Hoffman AD (2018). *Ergonomics: How to Design for Ease and Efficiency*, 3rd ed. London, UK, Elsevier. https://doi.org/10.1016/C2016-0-02516-0

Kroemer KHE, Kroemer HJ, Kroemer-Elbert KE (2020). *Engineering Physiology: Bases of Human Factors Engineering/Ergonomics*, 5th ed. Heidelberg, Springer. https://doi.org/10.1007/978-3-030-40627-1

McMillan GR, Calhoun GL (2001). Gesture-based control. In Karwowski W (ed), *International Encyclopedia of Ergonomics and Human Factors.* London, UK, Taylor & Francis, pp. 237–239.

Muchinsky PM, Culberson SS (2016). *Psychology Applied to Work*, 11th ed. Summerfield, NC, Hypergraphic Press.

Sinclair DC (1973). Mapping of spinal nerve roots. In Jarrett A (ed), *The Physiology and Pathophysiology of the Skin*, Volume 2. London, UK, Academic Press.

Stanton N, Hedge A, Brookhuis K, Salas E, Hendrick HW (2004). *Handbook of Human Factors and Ergonomics Methods.* Boca Raton, FL, CRC Press. https://doi.org/10.1201/9780203489925

Swink JR (1966). Intersensory comparisons of reaction time using an electro-pulse tactile stimulus. *Human Factors*, 8(2):143–145. https://doi.org/10.1177/001872086600800205

Vicente KJ (2002). Ecological interface design: progress and challenges. *Human Factors*, 44(1):62–78. https://doi.org/10.1518/0018720024494829

Wargo MJ (1967). Human operator response speed, frequency and flexibility: a review and analysis. *Human Factors*, 9(3):221–238. https://doi.org/10.1177/001872086700900303

Wickens CD, Lee J, Liu Y & Gordon-Becker S (2004). *An Introduction to Human Factors Engineering*, 2nd ed. Upper Saddle River, Prentice-Hall/Pearson education.

SECTION III

Body and mind working together

As we seek to understand humans, we often consider certain characteristics separately, as if they function independently. However, in most cases there are strong interactions between these characteristics.

The first section of this book addresses specific physical aspects of the human body, while the second section focuses on how we perceive our environment through our senses. In this third section, physiological and sensory aspects are merged with psychological facets to better represent the human experience, where, indeed, the body and mind work together.

We recognize and address differences in how people respond so that we can design tasks, gear, equipment, and procedures to fit their specific needs. Therefore, the following chapters provide relevant information for ergonomic design:

- Chapter 10, Hard Physical Work
- Chapter 11, Light and Moderate Work
- Chapter 12, Task Load and Stress

DOI: 10.1201/9781032674575-13

CHAPTER TEN

Hard physical work

In many areas, mechanization has reduced the demand for human labor. Nevertheless, heavy physical work remains common in agriculture and forestry; demanding labor is frequent in mining, deep-sea fishing, building construction, and other industries. It also exists in newer jobs, such as baggage handling by airline personnel. Intervals of heavy physical demands may alternate with periods of lighter duty. We must understand how we perform and respond to hard physical work to better fit the task to the human.

10.1 Basics of work physiology

Hard labor is activity that intensely employs skeletal muscles. These muscles convert chemical energy into work (physical energy) that moves body segments against internal and external resistances. From a resting position, a muscle can increase its energy generation by up to 50-fold. This significant variation in metabolic rate not only requires the ability to quickly adapt the supply of fuel to the muscle, but also produces large amounts of internal waste products that need to be removed. The bloodstream, powered by the heart, provides the means for transport of fuel supply and waste product removal.

The body's ability to maintain internal equilibrium determines how much demanding work it can perform. The capacity for work largely depends on the proper functioning of the respiratory and circulatory systems that serve the involved muscles. Their services essentially consist of supplying energy carriers (nutrients) and oxygen, as well as removing waste products and heat. The control of body temperature, given the heat generated, is particularly important in hot and humid environments (see Chapter 8).

DOI: 10.1201/9781032674575-14

Assessing labor demands and worker capacities

Heavy work requires significant physical exertion with high energy consumption. Thus, it places high demands on the worker's metabolic functions, with consequent strains on the circulatory and respiratory systems. Usually, the performance capability of an individual is limited by their ability to accommodate energy consumption and cardiac effort. Therefore, measuring the demands on metabolic and heart functions often serves to assess the severity of the physical task. Physiological measures of the worker's capacities* in metabolic, cardiovascular, and respiratory functions, along with assessments of strength and mobility, help evaluate their ability to perform heavy physical work.

Assessing physical fitness

To evaluate a person's aerobic physical fitness, we can measure the maximal amount of oxygen their body can use during a specified period (such as 1 minute) of intense exercise. This measure, VO_2max*, is also used to predict endurance capacity during prolonged, sub-maximal exercise. The higher the percentage of VO_2max that an activity requires, the shorter the time period the person can (or should) safely perform that activity.

10.2 Energy consumption

Skeletal muscles enable the body to work by moving body segments against both internal and external resistance. Muscles require energy for contraction. This energy is provided by mitochondria within muscle cells, which convert chemical energy into physical energy (see Chapter 3). Operating this "human energy machine" involves a complex metabolic process akin to the combustion of fuel in a mechanical engine:

- Fuel combustion requires both fuel and oxygen to proceed.
- Fuel combustion yields energy to move parts.
- Fuel combustion generates heat and other byproducts.

Comparing combustion with human "energy machine"

In the cylinder of a combustion engine, explosive combustion of a fuel–air mixture transforms chemically stored energy into physical kinetic energy and heat. This energy moves the engine's pistons, which rotate the crankshaft, which transfers the motion to the vehicle's wheels through gears and linkages. The excess heat generated must be removed by cooling the engine to prevent overheating, and waste byproducts from the fuel combustion must be eliminated by venting.

In the human "machine," muscle fibers act as both cylinders and pistons, while bones and joints serve as gears and linkages. The fuel, primarily derived from carbohydrates and fats in nutrients, requires oxygen to produce energy through slow combustion. When muscles work, they generate metabolic byproducts, including heat, that must be removed. Figures 10.1 and 10.2 illustrate the sequence of events in the human body associated with the conversion of nutrients into mechanical energy and heat.

Metabolism

Metabolism is a fundamental biological process in which the body ingests food and drink that contain chemically stored energy and converts them into mechanical energy. As Figures 10.1 and 10.2 show, food travels from the mouth to the stomach, where it is liquefied; alcohol, when present, is absorbed in the stomach and enters the bloodstream. Digestion breaks down food and drink into usable forms: first mechanically and then chemically. The digestion of the primary energy carriers in

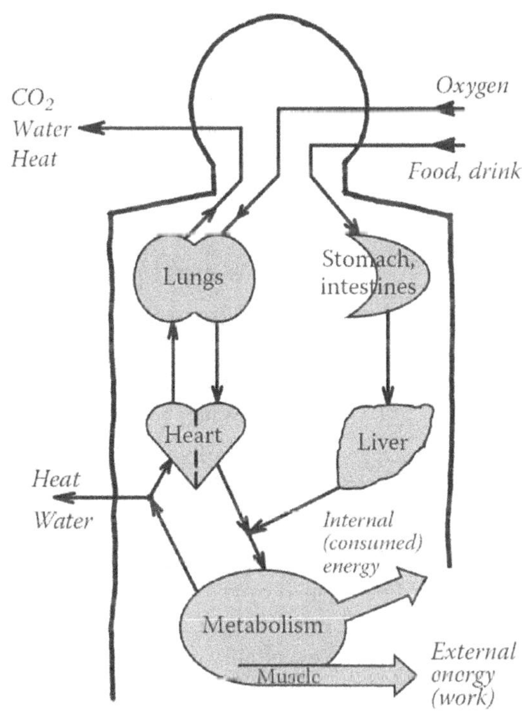

FIGURE 10.1 Diagram of the energy flow in and out of the human body. Adapted from Kroemer, Kroemer, and Kroemer-Elbert (2020).

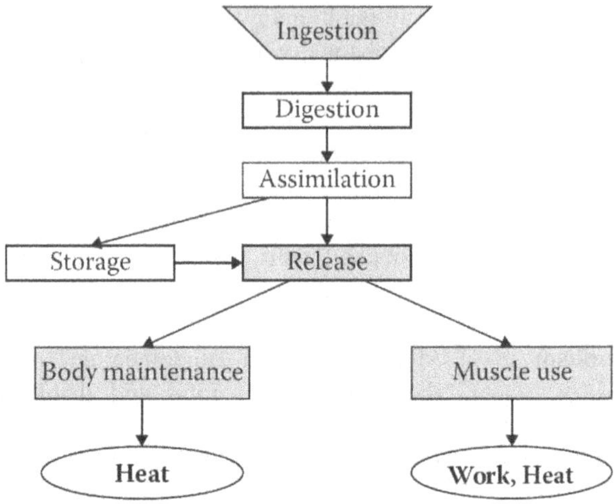

FIGURE 10.2 Diagram of the energy conversions from ingestion to heat and work. Adapted from Kroemer, Kroemer, and Kroemer-Elbert (2020).

food and drink—carbohydrates, fats, and proteins—occurs in the intestines. Digestion involves chemically converting large, complex molecules into smaller ones that can pass through the membranes of intestinal cells and be absorbed into the blood and lymph. The liver largely regulates what happens to the absorbed energy carriers: they are assimilated (re-assembled) into new molecules that can be

- Stored as energy reserves (mostly as fats*),
- Used for body growth and repair with the rest converted into heat, and
- Degraded to use their energy content (glucose and glycogen). When energy is needed for work, glucose is used first, followed by glycogen. Stored fat provides the body's largest energy resource but is the last to be used.

Metabolic byproducts

Human metabolism is a complicated process. Only a portion of the chemically stored energy is converted into mechanical work performed by muscles; most of the chemically stored energy is used to build and maintain the structures of the human body and is eventually converted to heat. Since the core temperature of the human body must remain at 37°C (98°F), the excess heat must be dissipated into the environment. Heat is transported within the body through the bloodstream: some is sent to the lungs where it is transferred to the air to be exhaled, and some is sent to the skin where it is released to the outside air, often with the help of evaporating sweat. Dissipating excess heat can be challenging in hot and/or humid climates (see Chapter 8).

Another byproduct of cellular metabolism is water, which the blood also transports to the lungs and skin for elimination, similar to excess heat. A third byproduct of metabolism is carbon dioxide (CO_2), which the blood carries to the lungs for expulsion. Proper management of these metabolic byproducts is essential for the body to maintain energy generation and continue strenuous physical labor.

Energy units

It is important to understand the units of energy (work) and power (work per unit of time), as different conventions may be used across scientific disciplines. The units for energy (work) are joule (J) or calorie (cal) with 4.18 J=1 cal; alternatively, 1 J=1 Nm=0.239 cal=10^7 ergs=0.948×10^{-3} BTU=0.738 ft-lb. The correct alternative expression for kcal (kilocalorie) is Cal. The units for power are watts, with 1 W=1 J/s, or 1 kcal/h=1.162 W.

Energy content of food and drink

The kilojoule (1 kJ=1000 J) and the kilocalorie (1 kcal=1 Cal=1000 cal) commonly serve as measures of the energy content of food and drink. The average nutritionally usable energy contents per gram for various types of food are given as follows:

- Fat: 37 kJ (9 Cal)
- Alcohol (ethanol): 29 kJ (7 Cal)
- Protein: 17 kJ (4 Cal)
- Carbohydrates (glucose, sucrose, starch): 16 kJ (4 Cal)
- Fiber: 8 kJ (2 Cal)

Prepackaged food and drinks in many countries carry labels that list the number of kilojoules or kilocalories (often misprinted as "CAL" or falsely printed as "cal") per "serving size" (which may be in rather fanciful amounts). The label usually also provides a breakdown of the energy content in terms of fats, proteins, and carbohydrates.

Basal metabolism

A minimal amount of energy is necessary to keep the body functioning even without any activities: this represents the *basal metabolic rate* (BMR). Under strict conditions (complete physical rest in a neutral ambient temperature, after fasting for 12 hours, and free from emotional stress), one can measure the minimum metabolism needed to sustain the body. The results show that basal metabolic values depend primarily on age, gender, height, and weight—the last two variables are occasionally expressed as body surface area. Among healthy adults, there is little variation in BMR; hence a commonly used value is 1 kcal (4.2 kJ) per kg per hour, or 4.9 kJ/min for a person weighing 70 kg (154 lb).

Resting metabolism

Often, it is impractical to achieve the highly controlled conditions needed to measure basal metabolism. Instead, we often measure metabolism before the working day, with the subject as relaxed as possible. Depending on the given conditions, *resting metabolic rate* is around 10%–15% higher than the BMR.

Work metabolism

The increase in metabolism from resting to working is called *work metabolism*. This increase above resting levels represents the energy needed to perform the work. Often, to describe the demands of work, we measure the total energy used by the body: this total includes resting or basal levels as well.

Measuring heaviness of work

One way to assess the heaviness (intensity) of work is simply to ask the individual to describe how hard the effort feels. To better standardize this subjective assessment, it is advantageous to employ an established rating procedure, such as the Borg scale described in Chapter 12. However, in many cases, objective measurements are desired: for this purpose, three different procedures are commonly used*. One procedure is to observe the energy supplied to the body over a given time. The second procedure measures the volume of oxygen consumed during work. The third procedure assesses heart rate (HR) during work, which is discussed in the next section.

Energy supply to the body

One of the oldest techniques to assess the heaviness of work is to account for energy input by observing what a person eats and drinks and weighs. The underlying assumption is that, after deducting what is needed to maintain the body, all surplus energy is used for performing work. Naturally, this approach requires long observation periods—days or weeks—during which the observed person performs various types of physical activities interspersed with rest periods, along with careful measurement of intake. This method is notoriously inaccurate unless performed under strictly controlled conditions.

Oxygen consumption at work

Another technique to quantify the heaviness of work is to measure oxygen consumption. When the body performs work, oxygen uptake (along with carbon dioxide release) indicates the associated metabolic energy production. Various measurement techniques exist, all relying on the principle that differences in oxygen content between inhaled and exhaled air indicate the oxygen absorbed in the lungs. Assuming an average energy value of 5 kcal (21 kJ) per liter of oxygen, the volume of oxygen (VO_2) absorbed allows the calculation of the energy the body expends during an activity in the observation period.

RQ

An enhancement of the technique of monitoring oxygen is to also consider the carbon dioxide produced. The *respiratory quotient* (RQ) provides a more detailed assessment of the nutrients metabolized. The RQ compares the volume of carbon dioxide (VCO_2) exhaled to the volume of oxygen (VO_2) consumed. Metabolizing one gram of carbohydrate requires 0.83 L of oxygen and releases the same volume of carbon dioxide. Therefore, for carbohydrate metabolism (such as glucose), the RQ is 1.0 (unitless since it is a ratio of volumes). The energy released is 18 kJ/g of O_2, or 21 kJ/L of O_2. The RQ for protein conversion is 0.8, while for fat and alcohol conversion, it is about 0.7. Measuring the volumes of CO_2 and O_2 during work allows us to determine which energy source is being metabolized.

10.3 HR and work demands

Heart rate during work

Counting HR during work exertions is a time-honored and widely used method, both formally by doctors and engineers and informally by amateur athletes. It is based on the understanding that metabolizing muscles require a continuous supply of oxygen and removal of metabolic byproducts to function properly. The higher the energy requirements, the greater the blood flow needed. To achieve this, the heart must produce higher outputs primarily by increasing the number of beats per minute. Consequently, the HR (pulse) varies according to work demands.

HR and oxygen uptake

During dynamic work, there is a close relationship between metabolic processes and their circulatory support systems, as just mentioned. As a result, HR (an indicator of circulatory function) and oxygen consumption (an indicator of metabolic activity) correlate in their responses to physical effort. (Note that these specific correlations vary among individuals and may change with physical training or de-conditioning.) Therefore, counting HR can often serve as a substitute for measuring oxygen intake. This shortcut is appealing because HR responds more quickly to changes in work demands than oxygen consumption, and counting the pulse is simpler than measuring oxygen uptake.

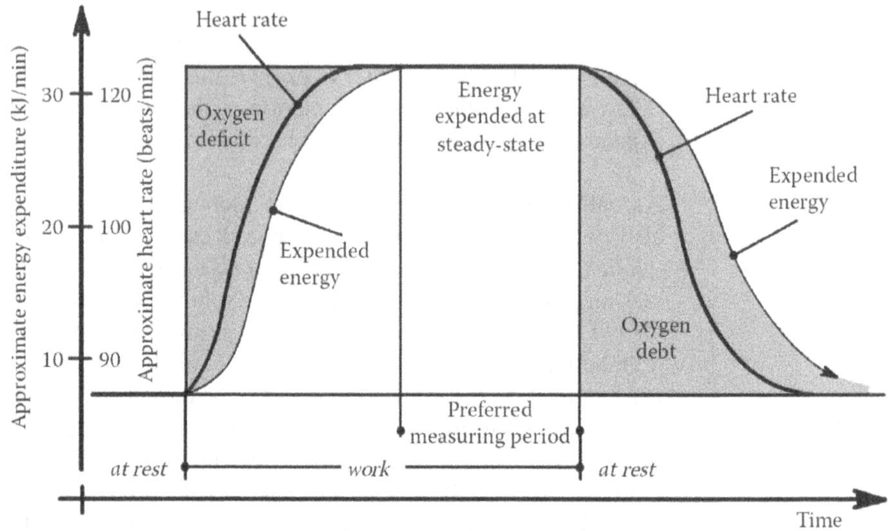

FIGURE 10.3 Representation of energy expenditure and HR before, during, and after steady-state work. Adapted from Kroemer, Kroemer, and Kroemer-Elbert (2020). HR, heart rate.

Oxygen intake and work

As Figure 10.3 illustrates, at the start of vigorous physical effort, there is an immediate need for oxygen, but actual oxygen uptake lags behind this demand. After a slow onset, oxygen intake rises rapidly and then levels off to meet the requirements of steady work. Thus, during the onset of labor, the body incurs an "oxygen deficit" due to the delayed increase in oxygen supply. After work ends, the body's oxygen intake begins to decline until it returns to resting levels; this return takes about twice as long as the initial rise in oxygen uptake at the start of work.

Oxygen deficit and debt

The first burst of energy comes primarily from degrading the energy carrier adenosine triphosphate (ATP) to adenosine diphosphate (ADP) and creatine phosphate (CP) stored in the body. During recovery, the body resynthesizes ATP and CP and replenishes oxygen stores in hemoglobin in the blood and myoglobin in muscles. This process requires elevated biochemical, cardiac, and respiratory functions, resulting in a heightened oxygen intake: typically, the oxygen "debt" repaid is approximately twice as large as the oxygen deficit initially incurred.

Breathing hard

Our body's ability to develop an oxygen deficit explains why we can perform short-term work at high intensity. Work that requires tremendous amounts of energy for a short time is also work that we cannot sustain for long time periods. Our body's need to repay the oxygen debt also explains why we continue to breathe hard after strenuous physical effort.

HR and work

Given the close interaction between the circulatory and metabolic systems, both our HR and oxygen update react to bouts of work similarly. However, HR increases faster at the start of work than does oxygen uptake; during recovery, HR returns to resting levels more quickly than oxygen uptake (as shown in Figure 10.3).

Steady-state work

If a work effort remains below a person's maximal capacity, then metabolic processes, oxygen supply, blood flow, and respiration can achieve and maintain their required levels. This condition of stabilized functions during work is called steady state (as shown in Figure 10.3). Measurements of work demands on the body are reliably obtained during this state. As expected, a physically fit person can attain this equilibrium between demand and supply at a relatively high workload, whereas an untrained or less fit person can achieve a steady state of work only at a lower demand level.

Classifying work demands

Energy expenditure and HR during extended work are objective indicators of the "heaviness" of the demands of work on a person. When asked for descriptive words, a worker might call some demands light or easy, while others may be labeled heavy or hard. However, such descriptions can vary with circumstances and experiences: some grandparents, for example, still accustomed to physical labor, might call a certain level of work demands moderate, while their grandchildren might find it heavy. Table 10.1 contains classifications of work demands. The energy values listed include basal and resting metabolism; however, all values are relative and do not reflect fitness or gender (so some stronger or fitter women may find the physical work easier, while some less strong or fit men may consider it harder).

Table 10.1 Relative descriptions of work demands

Classification description	By energy expenditure		By HR (beats/min)
	kJ/min	kcal/min	
Light, easy	10	2.5	90 or fewer
Medium, moderate	20	5	100
Heavy, hard	30	7.5	120
Very heavy	40	10	140
Extremely heavy	50	12.5	160 or more

HR, heart rate.

Table 10.2 Typical changes in physiological functions from rest to maximal effort

Function	Change from rest to maximal effort	
Energy consumption	From 1 to 20 kcal/min	×20
Oxygen uptake	From 0.2 to 4 L/min	×20
Cardiac activity	Heart rate: from 60 to 180 beats/min	×3
	Stroke volume: from 50 to 150 mL/beat	×3
	Cardiac output = heart rate ×minute stroke volume: from 5 to 35 L/min	×7
	Blood pressure, systolic: from 90 to 270 mmHg	×3
Respiration	Breathing rate: from 10 to 50 breaths/min	×5
	Minute volume = breathing rate × tidal volume: from 5 to 100 L/min	×20

Source: Adapted from Kroemer, Kroemer, Kroemer-Elbert KE (2020).

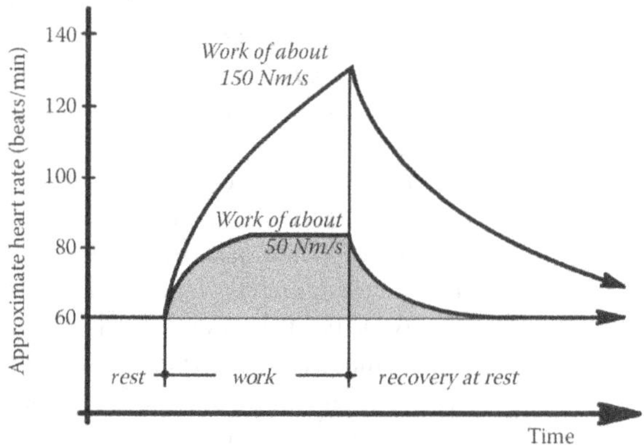

FIGURE 10.4 Response of the HR to exhausting work, light work with steady state, and return during recovery. HR, heart rate.

10.4 Limits of human labor capacity

A maximal effort significantly raises many bodily functions, as Table 10.2 shows. As long as the body can meet work demands by remaining in a "steady state," the work can continue, as Figures 10.3 and 10.4 depict. However, if work demands exceed the body's capabilities, HR and other supply functions cannot achieve a steady state and will increase their activities until they reach their limits. These physiological limits force a stoppage of work; then HR and other functions slowly return to their resting levels, as Figure 10.4 illustrates. The level of demand an individual can handle depends on their physical fitness and trained skills.

Measuring fitness to do heavy work Most medical and physiological assessments of human energy capabilities primarily rely on measuring oxygen consumption, with HR as a secondary indicator. Standardized tests allow for comparisons of capacities among individuals. These tests use generally accepted and idealized forms of external work, such as bicycle ergometers, treadmills, or steps. These forms of physical exertion primarily stress the leg muscles. Since leg mass and musculature are significant components of the body, their extensive exercise during a bicycle test also strains the pulmonary, circulatory, and metabolic functions, though not all of them. The treadmill primarily strains lower body capabilities as well, but unlike bicycling, the legs must support and propel the body's entire weight. Thus, the treadmill test challenges the body more comprehensively than bicycling; however, both omit trunk and arm capabilities from consideration. These examples demonstrate that the selection of test equipment and procedures can lead to different evaluations of people's physical fitness. For instance, a well-trained bicyclist may perform less well on a treadmill, while a well-trained long-distance runner may perform less well on a bicycle, even if both athletes are quite "fit." Therefore, a major problem with these tests is that they do not resemble actual work conditions. Consequently, test results on laboratory equipment may have limited value in predicting individuals' abilities to perform physically demanding work tasks.

Selecting fitness for heavy work Health and fitness tests are important for ensuring that only those fit enough to perform heavy physical work are employed to do so. However, from an ergonomic perspective, it is better to design work to fit the human rather than to select the human to meet the work demands: tasks and equipment should be designed to impose relatively small demands on human physical capabilities. This approach not only prevents people from being overtaxed by work but also allows more individuals, including those with less-than-ideal physical capabilities, to perform the job with a reduced risk of injury.

"Static work" Maintaining a motionless posture requires that the muscles generating that posture maintain continuous contraction. If such static (isometric) contraction* exceeds about 15% of muscular strength, the muscle compresses its own arteries and veins, reducing its blood flow. Further contraction causes more compression, which further reduces the blood flow and may even cut it off completely, despite the heart's attempts to increase blood pressure to overcome the obstruction. This leads to muscle fatigue, as reduced or obstructed blood flow interferes with fuel and oxygen delivery and waste removal.

Fatigue ultimately forces us to abandon the tiring posture to relax and recover. Thus, the ability to endure a static contraction depends on the magnitude of contraction, as Figure 10.5 illustrates.

Static or dynamic muscle effort

In contrast to fatiguing static work, suitable dynamic work facilitates blood flow through muscle tissues due to rhythmic alternations between tension and relaxation —this is known as the "muscle pump," illustrated in Figure 10.6. Dynamic work is shown at the top: blood is expelled from the muscle during contraction and enters during relaxation. Static effort is shown at the bottom: the muscle's contraction during static work increases the pulse rate as the heart strives to raise blood pressure to overcome flow resistance in the compressed blood

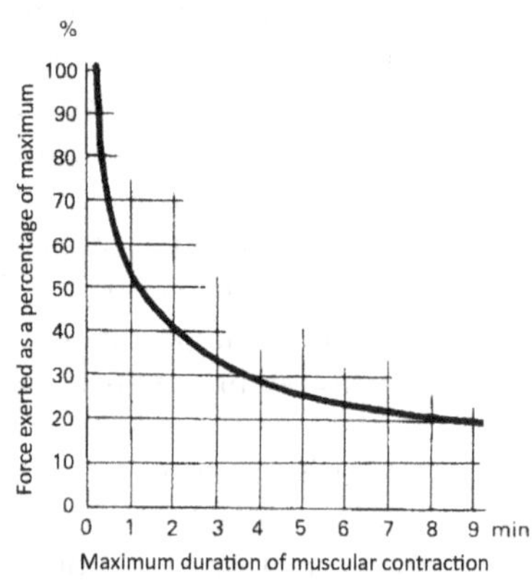

FIGURE 10.5 Maximal duration of static muscle contraction.

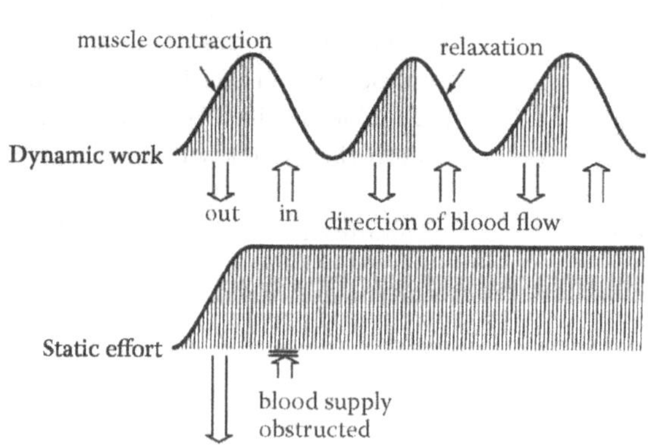

FIGURE 10.6 Blood flow through a muscle doing dynamic and static exertions.

vessels within the straining muscle. An increased HR associated with sustained static muscular effort is indeed an indicator of physical strain; however, since blood flow is restricted, metabolism is also reduced. Therefore, in the case of static effort, the direct relation between HR and energy consumption, mentioned above, no longer holds.

10.5 Designing heavy human work

Human energy efficiency at work

Engineers and economists express work efficiency as the ratio of output to effort. If we assume that the body's energy storage remains constant (meaning the observed person does not gain or lose weight) and that the body neither absorbs heat from the environment nor loses any, we can state a simple energy balance equation as follows:

$$I = H + W$$

where energy input I is from food and drink (which the body transforms into metabolic energy), heat H is produced (as a byproduct of metabolism), and work W is performed.

Human energy efficiency (work efficiency) e is then the ratio of work performed to energy input:

$$e\left(\text{in percent}\right) = 100\ W/I$$

In everyday activities, the human body is very inefficient in terms of energy usage. Only about 5% or less of the energy input converts into "work" output, which is the energy usefully transmitted to external objects; even highly trained athletes may achieve, under favorable conditions, only as much as 25%. The remainder (that is, most) of the energy input is converted into heat, typically at the end of a long chain of internal metabolic processes. Since humans are so inefficient at converting energy to work, our capabilities are better utilized for thinking and managing, controlling machinery and processes, rather than serving as physical movers.

Designing work to fit the human

The engineer or designer of the task determines the required "work" and how it is to be done. To find a suitable match between demands and capabilities, the engineer needs to adjust the work to be performed (and the work environment) to the body's energetic capabilities. These human capabilities are influenced by the individual's capacity for energy output (physique, training, health), by neuromuscular function characteristics (such as coordination of motion, muscle strength, etc.), and by psychological factors (such as motivation).

Avoiding exhausting work

Coal miners and lumberjacks, using traditional hand tools, are among the workers with the highest energy consumption: about 19,000 kJ (4500 kcal) per day, as measured in male laborers in the 1960s. Such extreme efforts are becoming less common in many countries as mechanized tools and modern machinery make work less physically demanding. The daily energy consumption for moderately demanding work is around 12,000–15,000 kJ for men and 10,000–12,000 kJ for women.

Providing rest breaks

Breaks in physical work provide recovery and rest. Providing breaks is essential when performing hard physical labor and is desirable even during lighter work for physiological and psychological reasons. Taking many short breaks is more beneficial than allowing fewer, longer interruptions, since recovery is most significant at the beginning of a break, as illustrated by the curves of energy and HR in Figures 10.3 and 10.4.

No "static work"

As mentioned, the typical understanding of physical work is that it consists of dynamic actions. Under these conditions, HR and energy consumption are generally closely related. However, many actual work tasks include static efforts, where parts of the body must be kept in a "frozen position" for a while. Although static effort does not constitute "work" in physics terms (defined as applying a force over a distance), it still strains the body, especially the cardiovascular system: HR increases while energy consumption does not. Static muscle efforts can be tiring but not productive. Therefore, they should be "designed out" from tasks by changing the process or replacing human effort with mechanical solutions.

Summary

Figure 10.7 provides an overview of the human traits and working conditions that interact to determine how much hard work a person can perform. The ergonomist has no control over individual characteristics (in gray in the figure). However, the ergonomist has both the authority and responsibility for the layout of work tasks and schedules, as well as for the equipment and tools used.

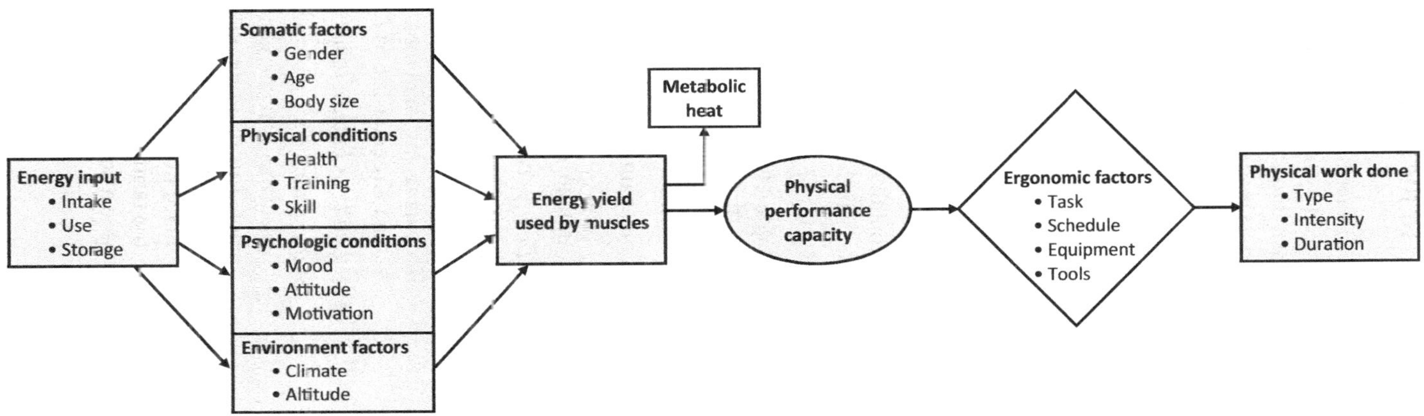

FIGURE 10.7 Traits of the human operator and the work conditions (in gray) interact with the design of work to determine the person's ability to perform heavy physical work. Adapted from Kroemer Elbert, Kroemer, and Kroemer Hoffman (2018).

Steps to fit the human

Step 1: Determine whether hard physical labor must be performed by humans. Can a machine do it? Can it be broken down into smaller, less physically demanding tasks?

Step 2: Alleviate the workload, provide supportive equipment and tools, ensure a suitable climate, minimize highly repetitive tasks, and reduce static work positions.

Step 3: Provide frequent rest breaks and encourage taking time off work as needed.

Step 4: Design the tasks and tools, then test. Modify and iterate as necessary.

Notes

The text contains markers, *, to indicate specific references and comments, which follow.

10.1 Basics of work physiology

Physiological measures of the worker's capacities: Astrand, Rodahl, and Stromme (2004), Hall (2015), Kroemer, Kroemer, and Kroemer-Elbert (2020), Rodahl (1989), Kenny, Wilmore, and Costill (2015).

VO_2max ($maxVO_2$) is also called maximal oxygen consumption, maximal oxygen uptake, peak oxygen uptake, or maximal aerobic capacity.

10.2. Energy consumption

Fats: Note that people can "get fat" without ever eating any, see Flier and Maratos-Flier (2007).

Measuring the heaviness of work: Kroemer, Kroemer, and Kroemer-Elbert (2020).

10.4. Limits of human labor capacity

Static (isometric) contraction: Because there is no displacement in an isometric (static) muscle contraction, this muscular effort does not produce "work" as defined in physics: work is a force acting over a distance (the integral of force and displacement). Hence, the term "static work" is a misnomer but is often used (see also Chapters 3 and 11).

References and further reading

Astrand PO, Rodahl K, Dahl HA, Stromme SB (2004). *Textbook of Work Physiology: Physiological Bases of Exercise*, 4th ed. Champaign, IL, Human Kinetics.

Flier JF, Maratos-Flier E (2007). What fuels fat. *Scientific American*, 297(3):72–81. https://doi.org/10.1038/scientificamerican0907-72

Hall, JE (2015). *Guyton and Hall Textbook of Medical Physiology*, 13th ed. Amsterdam, NL, Elsevier.

Kenny WL, Wilmore J, Costill D (2015). *Physiology of Sport and Exercise*, 6th ed. Champaign, IL, Human Kinetics.

Kroemer Elbert K, Kroemer HB, Kroemer Hoffman AD (2018). *Ergonomics: How to Design for Ease and Efficiency*, 3rd ed. London, UK, Elsevier. https://doi.org/10.1016/C2016-0-02516-0

Kroemer KHE, Kroemer HJ, Kroemer-Elbert KE (2020). *Engineering Physiology: Bases of Human Factors Engineering/Ergonomics*, 5th ed. Heidelberg, Springer. https://doi.org/10.1007/978-3-030-40627-1

Rodahl K (1989). *The Physiology of Work.* London, UK, CRC Press. https://doi.org/10.1201/b12599

CHAPTER ELEVEN

Light and moderate work

When we are at rest or when our work involves only light physical effort, our bodies' basic metabolic functions operate similarly to those during hard labor (see Chapter 10), but at a lower level: energy stored in chemically in food and drink is digested and then assimilated into simpler molecules. These molecules serve to build and maintain our organs and provide the energy that muscles need to function. The lungs collect oxygen and eliminate metabolic byproducts (carbon dioxide, water, heat), while the skin dissipates heat. This places demands on the circulatory and respiratory systems, though naturally less so than during heavy physical work.

Table 10.1 describes light work as having energy demands below 10 kJ (2.5 kcal) per minute and resulting in fewer than 90 heartbeats per minute, while moderate work demands up to 20 kJ/min and 100 beats/min. Therefore, in contrast to harder labor, our performance during light and moderate work is not limited by either energy use or blood flow. The main demands during light to moderate work are our willingness and ability to work diligently with endurance and attention to detail.

The work environment can significantly influence our perception of the task as being easy or demanding. In physics terms, the work environment primarily depends on air temperature and humidity, surrounding sound and noise, and lighting, as discussed in Section II of this book. Organizational conditions at work are crucial: the psychosocial environment determines, for example, how we get along with our coworkers; feeling comfortable at work also relies on the arrangement of our working hours, including work shifts—all discussed in Section IV. Understanding how we perform and respond to light and moderate physical work is essential to fitting the task to the human.

DOI: 10.1201/9781032674575-15

11.1 Basics of work physiology and psychology

Most jobs in modern offices, or while driving vehicles, for example, are sedentary and require little physical effort, as listed in Table 11.1. Therefore, many individuals with desk jobs try to increase their daily energy expenditure by engaging in demanding leisure activities, such as "working out" and playing sports. Such exercise activities activate otherwise idle body functions and help to compensate for the intake of excess energy from food and drink, which could otherwise lead to weight gain. One simple way to exercise is to walk up and down stairs instead of taking the elevator in multi-story buildings. Stair climbing requires significant muscle use to lift the body and therefore helps to maintain the metabolic and circulatory systems. Figure 11.1 shows energy expenditures for various activities.

Table 11.1 Energy expenditures at sample activities

Activity	Energy expenditure
Sedentary office work	About 2 kJ/min (0.5 kcal/min)
General house work	520 kJ/min (1.2–4.8 kcal/min)
Walking, on level ground at 4 km/h	About 14 kJ/min (3.3 kcal/min)
Walking with 30 kg load on the back, on level ground at 4 km/h	About 23 kJ/min (5.5 kcal/min)
Running, on level ground at 10 km/h	About 45 kJ/min (11 kcal/min)
Climbing stairs, 30° incline, 100 steps/min gaining a height of 17 m/min	About 60 kJ/min (14 kcal/min)

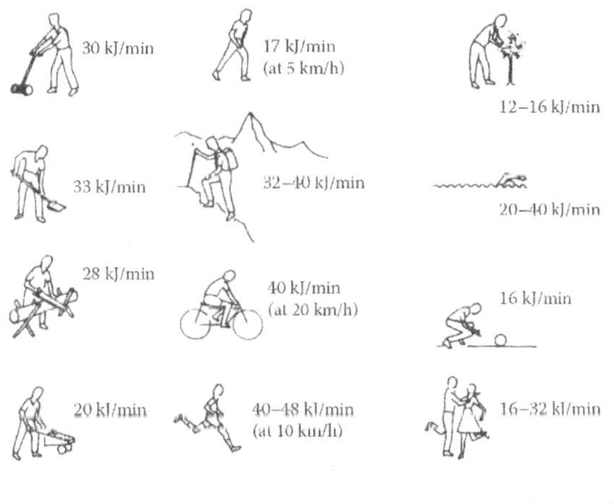

FIGURE 11.1 Energy consumption (in kJ/min) during leisure activities.

Measuring workload effects

When the energy requirements of work are low, measuring the worker's oxygen consumption provides little useful information. However, heart rate (HR)—which is easy to obtain—may be a useful indicator because it also responds to static muscle loading, such as when maintaining the body or its segments in a given position. The sensitivity of HR may be advantageous over energy assessment, but its responsiveness also extends to mental reactions, such as excitement or anger, meaning HR reflects not only physical work but also emotional states. Other specific approaches to measure workload effects* rely on electrical events in the body, such as related to muscle activation (electromyograms), brain activity (electroencephalograms), and eyelid closure frequencies. Simply observing general work output, especially as it fluctuates with changing working conditions, is another realistic and nonintrusive, yet nonspecific measure.

Scaled judgments

Another way to assess satisfaction or discontent with working conditions is to ask people for their opinions and judgments. This can be done informally or through a more structured and standardized approach by using prepared questionnaires. For example, a questionnaire might use the following pairs of descriptors:

Fresh	——	Weary
Sleepy	——	Wide awake
Vigorous	——	Exhausted
Weak	——	Strong
Energetic	——	Apathetic
Dull, indifferent	——	Ready for action
Interested	——	Bored
Attentive	——	Absent-minded

The person may be instructed to choose between one of the paired descriptors; this is called a *forced choice*. Alternatively, the individual may place a mark on a line between two opposing statements to indicate their level of interest or boredom, for example. Various other, often complex questionnaires may also be valuable for assessing comfort or discomfort at work, depending on the intended use of the results.

Three kinds of scales

Responses are often collected using one of three commonly used scales. *Ordinal scales* arrange judgments in rank order, allowing only the statements of "higher" or "lower." *Interval scales* require judgments set at equal distances from each other. *Ratio scales* are interval scales anchored by an absolute zero;

thus, numbered responses in ratio scales enable mathematical calculations. (Borg's RPE Scale in Chapter 12, Table 12.2, is an interval scale, whereas the CR-10 Scale is a ratio scale.)

Subjective or objective appraisals

Due to the inherent difficulties in obtaining objective numbers, many studies utilize subjective judgments (made by the worker or by another expert) to assess the existing work situation. While procedures vary widely among researchers, most rely on templates that solicit discomfort or pain statements. The method of administering the survey may influence the outcome—positively or negatively—since even just participating in such a study can heighten awareness of workplace problems*. From an ergonomic perspective, any positive outcome should be in everyone's interest.

Nordic Questionnaire

A commonly used, well-standardized inquiry tool is the Nordic Questionnaire, which employs either forced binary or multiple-choice answers. The tool consists of two sections: the first asks for general information, while the second focuses specifically on regions of the body. This second section features a sketch of the body divided into nine regions, as shown in Figure 11.2. The respondent indicates whether they experience any musculoskeletal problems in these areas. Although various modifications to the Nordic Questionnaire have been proposed,

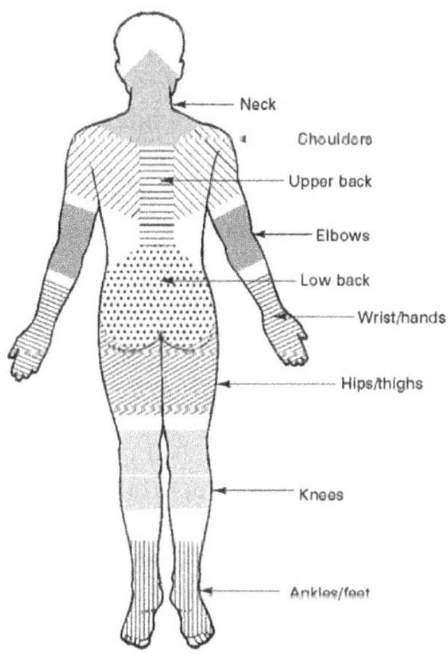

FIGURE 11.2 Body regions for use in the Nordic Questionnaire.

it has been widely administered for over 30 years and therefore provides long-term and internationally standardized information for comparisons.

11.2 Addressing tiredness, boredom, and alertness at work

Tiredness and boredom

Feelings of tiredness and boredom are familiar in everyday life. By strict definition, *tiredness* results from fatigue: the physiological effect of having expended energy, overstraining our muscles, and requiring our bodies to rest to recuperate from the preceding exertion (as discussed in Chapter 3). In contrast, *boredom* is a psychological or emotional condition characterized by a lack of events that lulls us into a state of indolence. However, in everyday language, we say we are "tired" due to either (physical) fatigue or (psychological, emotional) boredom. Thus, "I am tired of …" and "I am bored by …" have come to mean almost the same thing: we take little interest in our current task and feel disinclined, or even unable, to continue what we were doing.

Diversity and monotony

Some individuals enjoy jobs with diverse tasks that change and variously challenge their mental and physical capabilities. These workers maintain interest due to the variety of demands, finding satisfaction in successful solutions. However, other people prefer jobs that predictably present consistent tasks, or at least similar ones. They find fulfillment in skillful repetition, such as on an assembly line, while thinking, daydreaming, or conversing. Thus, monotony (the lack of unusual stimuli) is something to loathe or to like, depending on one's inclinations. For most people, however, performance—and job satisfaction—is highest with a job that is neither overly complex nor too simple, as depicted in Figure 11.3. An individual's preferred spot on that mound-shaped curve* may change based on skill, health, and mood.

Vigilance and event frequency

Many jobs require alert watchfulness (vigilance) over long periods. For example, the ability to sustain concentration was especially important for radar operators during World War II. It was observed that the frequency with which radar operators reported submarines on their screens diminished with the length of their watch. Half of all sightings were reported during the first 30 minutes on duty; then the reports of sightings declined, reaching just 10% after 2 hours. Here, alertness* deteriorated with time on watch.

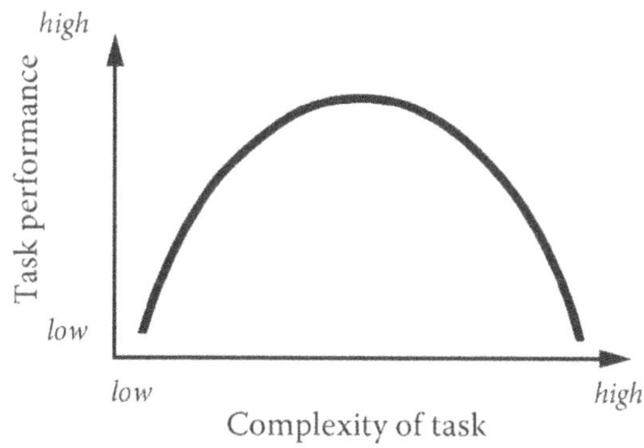

FIGURE 11.3 Relative relationship between task complexity and performance.

Signal frequency

Other experiences have shown that the frequency of signals greatly influences the number of reported signals. If so few signals occur that the observer becomes tired of waiting for them and pays little attention, a signal may not even be recognized when it finally happens. Conversely, if signals occur at very high frequencies, the observer may struggle to follow and report them. Between these extremes of too few and too many signals lies a broad range of signal frequencies in which signal appearances are well reported. This finding supports the hypothesis that, with very few events, the observer is underloaded and so the mind drowsily wanders; however, with too many events, the observer is overloaded and hence fails to recognize and report all of them. Figure 11.4 depicts these relationships: it suggests that the observer's performance follows a mound-shaped curve, also called an "inverted U" (note the similarity to Figure 11.3).

Operator performance

These observations opened an important field of research that generated extensive knowledge and many theories about vigilance and signal detection. The key conclusion is that performance decreases with time. It also worsens if the intervals between signals vary significantly, if the observer is fatigued or under mental stress, or if the observer must perform in unpleasant environments, such as those characterized by heat, noise, or vibration. On the other hand, strong signals at suitable frequencies that are meaningful to the task improve performance. (See also signal detection theory, defined in Chapter 9.)

Monotonous jobs

In the 1920s, industrial managers thought it appropriate to divide jobs, when possible, into a small number of identical tasks for the operator to repeat over and over. This approach, often traced

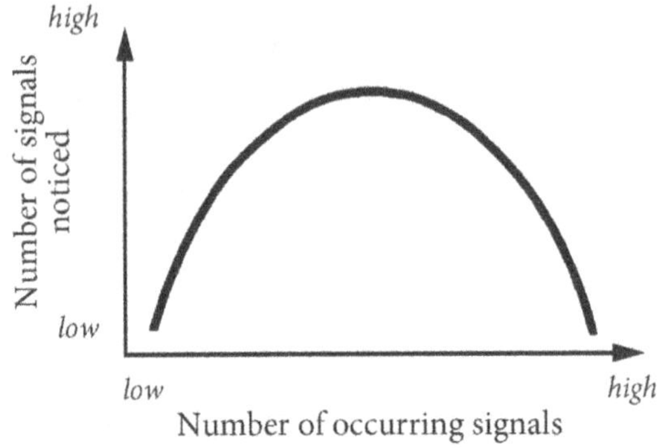

FIGURE 11.4 Relative relationship between numbers of signals occurring and being noticed.

to the early 1900s ideas of Frederick Taylor*, created monotonous and repetitive tasks with the same, repeated demands on the operator's dexterity. Consequently, with practice, the worker became highly skilled and could perform the tasks at high speed. However, there are serious problems with this Tayloristic approach: while some people enjoy repetitive work, others are bored by drudgery and monotony. Furthermore, excessive repetition can lead to injury to the human musculoskeletal system, as discussed below.

Satisfaction with one's work

Individuals' preferences and aspirations differ, yet almost everyone desires the fulfilling experience of mastering their work tasks and having control over how to perform their jobs, especially if they can implement better methods and improve outcomes. *Detail* (or *instrumental*) *control* pertains to the task level, involving decisions about workplace arrangements, tools, procedures, and the immediate environment. *Conceptual* (or *managerial*) *control* operates at a higher level and can include decisions about design, work organization, priorities, and administrative policies. Having detail and/or conceptual control over one's job fosters feelings of power and significance, while a lack of control can hinder performance and cause social stress. Feedforward and feedback among workers and management are essential components of participatory teamwork, as detailed in Section IV of this book. Active communication can significantly enhance job satisfaction and create a social climate that promotes high productivity. Recognition, rewards, and incentives following successful work foster satisfaction, personal interest, and ego involvement*, all of which improve performance.

11.3 Designing for suitable postures at work

No "static work"

In light and moderate work, fatigue often results from the requirement to maintain static positions of the limbs or the whole body for extended periods*. Sustaining a posture typically requires the involved muscles to maintain constant contraction, which compresses their own tissues. This pressure hinders blood flow through the muscle (as discussed in Chapter 10), and the obstructed blood flow cannot effectively remove metabolic byproducts. As these byproducts accumulate, the muscle "fatigues" and needs to relax to recover. During this time, blood flow is restored, and metabolic waste is removed. To avoid such fatiguing and debilitating postures, various often simple ergonomic solutions can be applied.

Avoiding fatiguing, awkward postures

Many tasks still require tiring body postures*. Standing, stooping, and kneeling increase metabolic costs compared to sitting or resting, as Figure 11.5 shows. Other examples of fatiguing conditions include twisting the trunk, bending close to the ground, reaching up, working with extended arms, standing for hours (such as at sales counters), and prolonged sitting with eyes on displays and hands on keyboards (such as in computerized offices). Tasks should avoid severe twisting of the trunk, as shown in Figure 11.6, to move a heavy packet onto a shelf; instead, use mechanical lifting assistance or, at a minimum, move your feet. Figure 11.7 illustrates that thoughtlessly placing a large box on the floor requires a person to "dive" into it to sort and remove the contents. Simply placing the box on its side atop a support allows the worker to maintain a more upright posture (see also Chapter 21 for a discussion of material handling).

FIGURE 11.5 Increases in metabolic cost associated with body posture, relative to the reference of lying down and resting.

FIGURE 11.6 Avoid twisting the trunk.

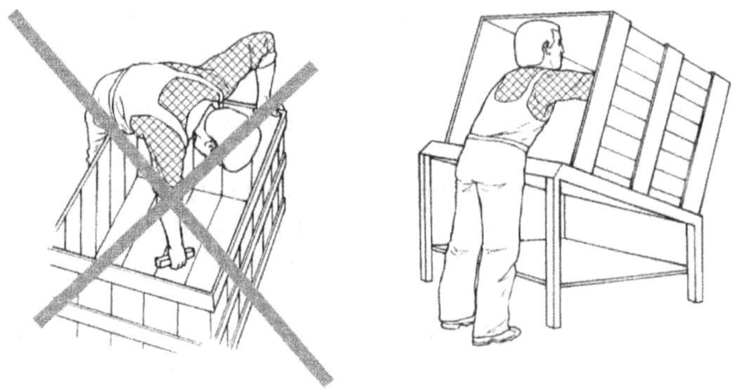

FIGURE 11.7 Avoid a deep "diving" reach, and instead elevate and tilt the large box on its side.

Sitting at work

Sitting is less tiring than standing, as shown in Figure 11.5 and confirmed by our own experience. However, when working at a workstation (whether at home or in the office), there should be enough space for the legs and provisions for comfortable sitting. Not every operator prefers to fold their legs beneath their body, as the woodcarver does in Figure 11.8. One could argue that this traditional approach to carving, where the toes help hold the workpiece, is what many local craftsmen have long been accustomed to. (Note that the carver still has to hunch over to see and use the tools.) However, such reasoning does not apply to the grinder workplaces in Figure 11.9. This drawing indicates that the grinders must tuck their feet under, bend

FIGURE 11.8 Wood carver sitting on his work piece.

FIGURE 11.9 Workers at grinders in miserable and fatiguing work postures.

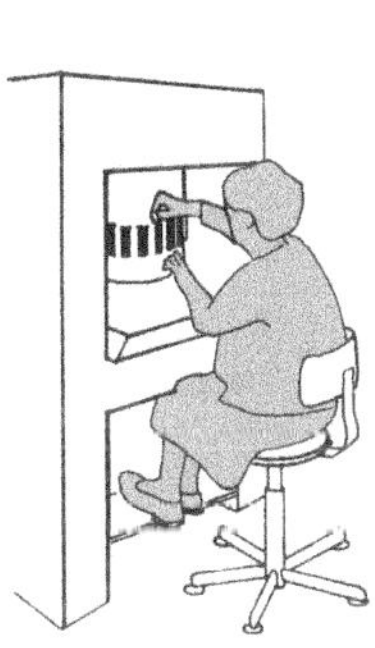

FIGURE 11.10 Operator loading and unloading parts with extended arms.

forward, and work with extended arms—one can only hope that proper safety measures are in place at these miserable workstations. The operator in Figure 11.10 has some legroom and even a footrest, but the edge of the machine seems to press against her knees; furthermore, she has to perform much of her work with her arms stretched upward and forward without support.

Too much sitting Depending on the occupation, sitting at work can be prolonged. Examples of nearly intolerably long and immobile sitting can be found in long-distance transportation. A truck driver covering long distances may need to stay seated with hands on the wheel, feet on the pedals, and eyes on the road for many hours before standing up and stretching. A military aircraft pilot on a long mission faces an even worse situation, as there may be no opportunity to get up and move. One solution used in 1940s military airplanes involved separate, air-filled sections in the pilot's seat that would automatically inflate and deflate: this pulsating body support alleviated some of the constant pressure on the body. Individuals who cannot move, such as some healthcare patients, often develop pressure ulcers ("bedsores") on the areas of the body that continuously bear the most weight. Moving instead of remaining still is essential for well-being and comfort.

11.4 Designing for accurate, fast, and skillful activities

Exact manipulations In light work, especially if it requires frequent and precise movements, the work area for the hands should be at about waist height and in front of the trunk. This preferred work area is shown in Figure 11.11 (and in Figure 2.15). It facilitates fast and accurate hand motions, achieved by inward rotation of the whole arm at the shoulder joint and by raising and lowering the forearm. Primary, secondary, and tertiary workspaces are indicated,

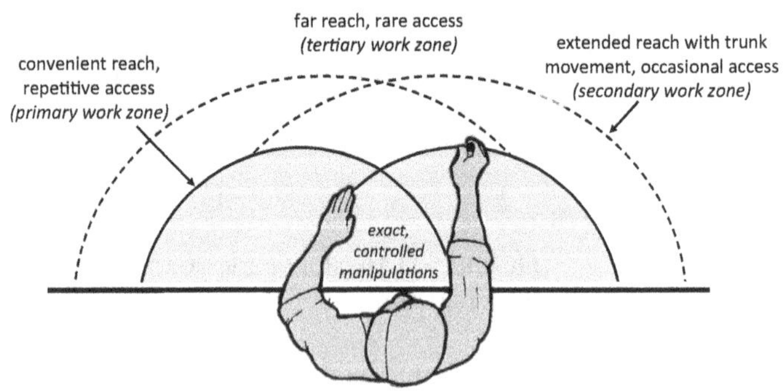

FIGURE 11.11 Space for fast and accurate light work with hands.

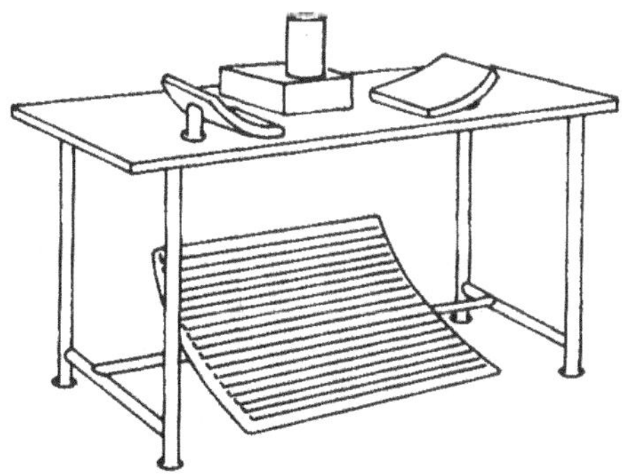

FIGURE 11.12 Workplace with supports for the elbows and the feet.

although trunk movement may still be necessary for the furthest and less frequent reaches. If the work requires particularly precise manipulation, armrests, as shown in Figure 11.12, might help stabilize the upper body. Note that this workplace ideally should allow for adjustments in surface height to accommodate different (seated or standing) operator heights.

Seeing what we are doing

Being forced to keep the neck bent or twisted can lead to discomfort; such an awkward posture is often caused by having to look at a work object that is not easily visible. Work with fixed loupes or microscopes has always been challenging because, to maintain the eyes in the proper positions at the ocular lenses, neck and trunk muscles must hold strict and static contractions. Electronic displays can alleviate this issue because they project the image onto a monitor, which the operator can adjust and view from various positions of the eyes and head.

Describing head posture

Describing head and neck posture using the so-called Frankfurt Plane (still described in older textbooks) used to be tedious, but it is simple when employing the ear–eye (EE) line, shown in Figure 11.13 (which duplicates Figure 5.4). The EE line runs through two anatomical markers on the head: the ear hole and the meeting point of the eyelids. Its angle relative to horizontal defines the head position as viewed from the side. The angle of the line of sight (LOS) against the EE line, called LOSEE in Figure 11.13, defines the direction of the LOS.

LOS

The LOS connects the eye to the visual target. When the target is far ahead, most people tend to look slightly downward below the horizon. The preferred angle of the sight line relative to the EE line (LOSEE in Figure 11.13) is about 10°–15°; however,

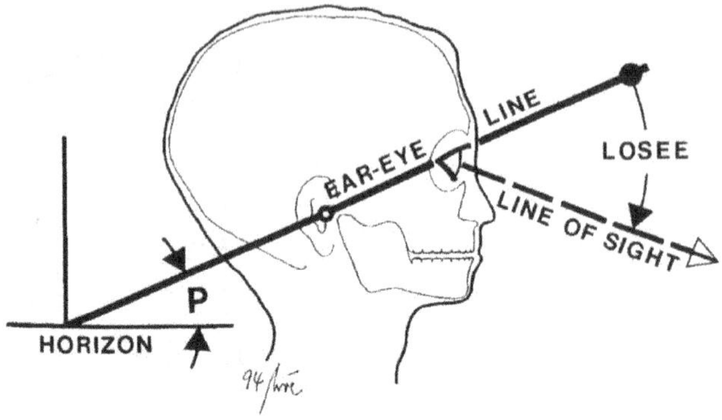

FIGURE 11.13 LOS is usually angled down from to the EE line. LOSEE is the angle between LOS and EE; P is the position angle between EE and the horizon. LOS, line of sight; EE, ear–eye.

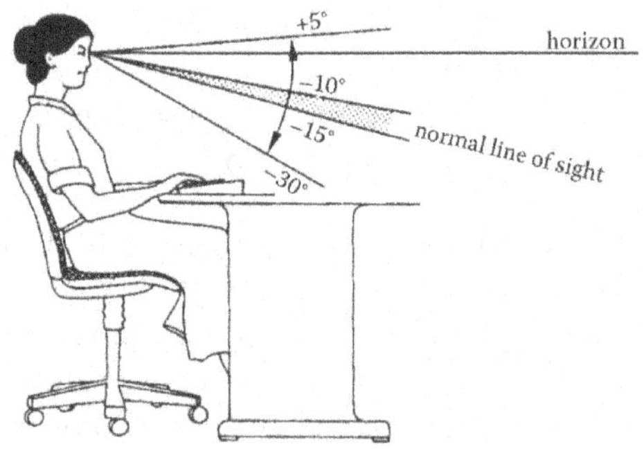

FIGURE 11.14 Visual cone of easy sight.

this angle increases as the object to be focused on comes closer. At "reading distance," about forearm's length (around half a meter) from the eyes, the preferred angle is about 45°. However, there are significant variations in preferred angles and distances among individuals, and wearing corrective eyewear (especially glasses) can also play a role. Eye movements within about 15° up, down, left, or right of the average LOS angle remain comfortable. Thus, common viewing tasks should be within a 30° cone around the normal LOS, as shown in Figure 11.14.

Better tools

Contorting, bending, and twisting the body into awkward positions often results from improper task or tool design, such as mispositioning controls on machinery. Figure 11.15 illustrates such design flaws in machine tools that can be avoided by considering body sizes and reaches. Even traditional tasks like gardening and fieldwork can be improved through simple

FIGURE 11.15 Machine tool operators forced into bent body postures by poor tool and workstation design.

FIGURE 11.16 Stooped posture during seeding or fertilizing by hand can be addressed with simple tool design.

means; Figure 11.16 shows an example for hand seeding and fertilizing. Variations in basic field-working tools, such as hand-held hoes, can make work easier or harder: in loose soil, the open hoe shown in Figure 11.17 requires less energy to use than a conventional blade and also performs well in compacted earth.

Repetitive work

Much skillful work consists of frequently practiced tasks. Repetition-related injuries to muscles, connective tissues, and joints of the human body have become a significant concern in occupational safety and health over the past few decades. This should not be surprising, as even in 1713, physician Bernardino Ramazzini* described occupation-related diseases and injuries from physically demanding jobs. He noted that "secretaries" (scribes) suffered from muscle pains and cramps due to their repetitive tasks, while "chairworkers" (such as cobblers

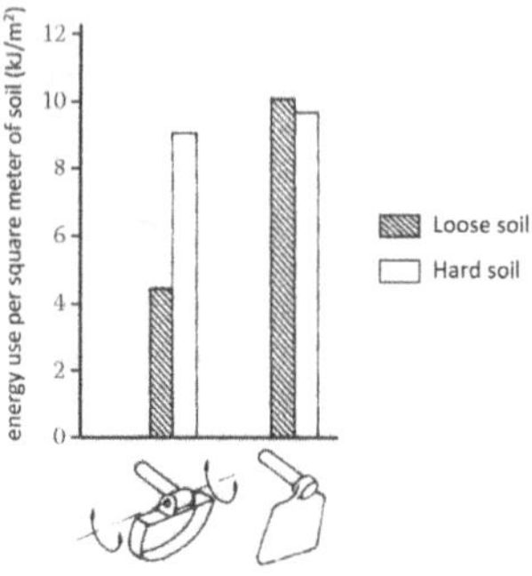

FIGURE 11.17 Energy use to move loose or hard soil with two kinds of hoes (open or fixed).

and tailors) developed curvature and back pain. Forceful exertions, hard physical contact, vibrations, and sustained or awkward postures can lead to repetitive strain injuries*, but light work can also cause them. Although each activity alone may not result in injury, frequent repetition over long periods can be harmful. The accumulation of mini-traumas can lead to a final injury, described by the term *cumulative trauma disorders*, which refers to excessive and repeated wear and tear on tendons, muscles, joints, and sensitive nerve tissue.

The human body is not designed for repetitive actions, whether at work or leisure, as Table 11.2 illustrates. The reasons for the listed health problems are clear from the descriptive names of the injuries. Hands, wrists, elbows, shoulders, and the lower back are the primary locations for *overuse disorders* caused by repetition.

Rest breaks

Breaks in physical work offer time for rest and recovery, benefiting both physiological and psychological health. Multiple short breaks are more effective than a few longer interruptions since recovery is most significant at the start of a break (as discussed in Chapter 10, Figures 10.3 and 10.4). Interrupting the flow of work, particularly if it is repetitive, monotonous, or demanding mechanically or conceptually, helps prevent physical and mental overload. Additionally, breaks provide an opportunity for social interaction while the body's energy supplies are replenished with food and drink. Therefore, offering frequent rest breaks, chosen by the worker if possible, is essential for improving well-being, job satisfaction, and overall performance across all types of work, whether mental or physical, heavy or light.

Table 11.2 Descriptive names of repetition-related disorders

Traditional descriptive names	Contemporary descriptive names
• Writer's cramp or scribe's palsy • Goalkeeper's, seamstress's, or tailor's finger • Bowler's, gamekeeper's or jeweler's thumb • Bricklayer's hand • Telegraphist's or pianist's wrist • Stitcher's, tobacco primer's, meat cutter's, or washerwoman's wrist • Carpenter's, jailer's or student's elbow • Jackhammerer's or carpenter's arm • Porter's neck • Weaver's bottom • Shoveller's hip • Housemaid's or nun's knee	• Typist's myalgia • Typist's tenosynovitis • Texter's thumb • Baseball catcher's hand • Knitter's or crocheter's wrist • Typist's or cashier's wrist • Letter sorter's wrist • Yoga wrist • Keyboarder's carpal tunnel syndrome • Golfer's or tennis player's elbow • Keyboarder's mouse elbow • Letter carrier's shoulder • Nurse's back • Carpet layer's knee • Ballet dancer's or nurse's foot

Source: Adapted from Kroemer (2001).

Summary

Light and moderate work do not impose heavy burdens on our physical capabilities; instead, the primary demands are to work with diligence and attention to detail. As a result, many workers with sedentary jobs choose to increase their daily activities through demanding leisure activities and sports. Whether at work or during leisure, the easily measurable HR can be a useful indicator of physical strain; it responds to both dynamic work and static muscle loading, such as holding the body or its segments in one position. This sensitivity of HR also extends to mental reactions such as excitement or frustration, reflecting more than just the physical demands of work. Another way to assess work conditions is to gather people's opinions and judgments, either informally or more commonly using prepared questionnaires and scaled responses.

Many individuals enjoy jobs that include diverse tasks that challenge both mental and physical capabilities. However, others are attracted to jobs that consist of the same or similar tasks. For most people, performance—and job satisfaction—is highest with a job that is neither overly complex nor too simple and that allows some control over task execution, workplace arrangements, and the immediate environment.

Physical fatigue can result from the requirement to maintain limb positions or body posture for extended periods. While sitting is generally less tiring than standing, excessively long periods of immobile sitting (or other fixed postures) occur in some jobs. Being forced to keep the body bent or twisted can be painful. Moving around instead of remaining still is essential for well-being.

Work that requires frequent and precise hand movements should be done at about waist height, in front of the trunk. Typically, a viewing task should be within a narrow cone around the normal LOS, which is inclined forward and downward. However, there are significant differences in preferred viewing angles and distances from person to person.

Some skilled work consists of frequently practiced tasks. However, repetitive activity using the same muscles, tendons, and joints can lead to injury, whether at work or during leisure. Breaks in physical work allow time for rest and recovery, benefiting both physiological and psychological well-being.

Steps to fit the human

Step 1: Assess the type and magnitude of body strength necessary to perform a specific task. Make work tasks easy to perform; avoid fatiguing postures and highly repetitive tasks.

Step 2: Ask people for their opinions and judgments about their work conditions. Consider whether the task involves diversity or monotony, needs vigilance, is repetitive and/or fatiguing, and requires tiring body postures.

Step 3: If the task causes boredom but requires alertness and/or causes pain, redesign the task and signals (if used).

Step 4: Design, then test. Modify and iterate as necessary.

Notes

The text contains markers, *, to indicate specific references and comments, which follow.

11.1 Basics of work physiology and psychology

Measuring workload effects: Karwowski (2006), Kumar (2007), Kumar and Mital (1996), Stanton, Hedge, Brookhuis et al. (2005).

Study affects the outcome: The so-called Hawthorne effect recognizes that people change their behavior when they know they are being observed, or that productivity increases due to heightened motivation when interest is shown in workers (to the delight of any ergonomic consultant); see also Chapter 13.

11.2 Addressing tiredness, boredom and alertness at work

Mound (inverted U) curve: This up–down curve (and its mirror image, the down–up "valley" or "U") is often used to describe human physical and physiological behavior; for another example, see Figure 8.7 on climate perception.

Alertness: Proctor and Van Zandt (2008), Stanton, Hedge, Brookhuis et al. (2005), Wickens, Lee, Liu et al. (2004).

Frederick Taylor: Mechanical engineer FW Taylor (1856–1915) was one of the first management consultants specializing in production efficiency. "Taylorism" is associated with a focus on optimized efficiency, time-and-motion studies, and, often, ruthless cost-cutting. See also Bjoerkman (1996), Lepore (2009).

Ego involvement, work performance: Bailey (1996); Kroemer Elbert, Kroemer, and Kroemer Hoffman (2018).

11.3 Designing for suitable body postures

"Static work": Because there is no displacement in an isometric (static) muscle contraction, this muscular effort does not produce "work" as defined in physics: work is a force acting over a distance (the integral of force and displacement). Hence, the term "static work" is a misnomer but is often used (see also Chapters 3 and 10).

Tiring body postures: Chaffin, Andersson, and Martin (2006), Delleman, Haslegrave, and Chaffin (2006), Kroemer (2001).

Ramazzini: Italian physician Bernardino Ramazzini (1633–1714) is often called the father of occupational medicine.

Repetitive work: Kroemer (2001), Kroemer Elbert, Kroemer, and Kroemer Hoffman (2018), Kumar (2001, 2004, 2007).

References and further reading

Bailey RW (1996). *Human Performance Engineering: Designing High Quality, Professional User Interfaces for Computer Products, Applications, and Systems*, 3rd ed. Upper Saddle River, NJ, Prentice Hall.

Bjoerkman T (1996). The rationalisation movement in perspective and some ergonomic implications. *Applied Ergonomics*, 27(2):111–117. https://doi.org/10.1016/0003-6870(95)00065-8

Chaffin DB, Andersson GBJ, Martin BJ (2006). *Occupational Biomechanics*, 4th ed. New York, Wiley.

Delleman NJ, Haslegrave CM, Chaffin DB (2004). *Working Postures and Movements*. Boca Raton, FL, CRC Press. https://doi.org/10.1201/9781482265095

Karwowski W (2006b). *International Encyclopaedia of Ergonomics and Human Factors*, 2nd ed. Boca Raton, FL, CRC Press. https://doi.org/10.1201/9780849375477

Kroemer Elbert K, Kroemer HB, Kroemer Hoffman AD (2018). *Ergonomics: How to Design for Ease and Efficiency*, 3rd ed. London, UK, Elsevier. https://doi.org/10.1016/C2016-0-02516-0

Kroemer KHE (2001). Keyboards and keying: an annotated bibliography of the literature from 1878 to 1999. *Universal Access in the Information Society UAIS*, 1(2):99–160. https://doi.org/10.1007/s102090100012

Kumar S (2001). Theories of musculoskeletal injury causation. *Ergonomics*, 44(1):17–47. https://doi.org/10.1080/00140130120716

Kumar S (2004). *Muscle Strength*. Boca Raton, FL, CRC Press. https://doi.org/10.1201/9780203503591

Kumar S (2008). *Biomechanics in Ergonomics*, 2nd ed. Boca Raton, FL, CRC Press. https://doi.org/10.1201/9780849379093

Kumar S, Mital A (1996). *Electromyography in Ergonomics*. London, UK, Francis & Taylor. https://doi.org/10.1201/9780203758670

Lepore J (2009). Not so fast. *The New Yorker*, 85(32):114–122.

Proctor RW, Van Zandt T (2008). *Human Factors in Simple and Complex Systems*, 2nd ed. Boca Raton, FL, CRC Press. https://doi.org/10.1201/9781420063608

Ramazzini B (1713). *De morbis artificum diatriba* [diseases of workers]. Translated in *American Journal of Public Health*, 2001, 91(9):1380–1382. https://doi.org/10.2105/AJPH.91.9.1380

Stanton N, Hedge A, Brookhuis K, Salas E, Hendrick HW (2004). *Handbook of Human Factors and Ergonomics Methods*. Boca Raton, FL, CRC Press. https://doi.org/10.1201/9780203489925

Wickens CD, Lee J, Liu Y & Gordon-Becker S (2004). *An Introduction to Human Factors Engineering*, 2nd ed. Upper Saddle River, NJ, Prentice-Hall/Pearson education.

CHAPTER TWELVE

Task load and stress

What is stress? The physician Hans Selye* introduced the term "stress" as a medical concept in the 1930s. He used it to describe the reactions of humans to good situations (*eustress*) and to bad situations (*distress*). Today, the emphasis is almost entirely on the negative condition: for example, we say we feel "stressed out" by a stressor.

Stressor causes stress Selye's choice of terms was unfortunate; he did not realize that engineers define *stress* as an external load, pressure, thrust, or impact (such as from a heavy truck on a bridge or from a sudden blow on the back) that causes *strain* (a relative displacement) in the receiving structure (in the framework of the bridge or in the musculoskeletal components of the trunk). These two conflicting meanings of stress—either considered the "cause" (in engineering and physics) or the "result" (in psychology)—have led to much confusion. This book uses the convention that a *stressor causes stress.* We need to understand how we perform and respond to task load and stress to fit the task to the human.

12.1 Basics of task load

Human reactions to task loads Heavy physical work is typically accompanied by high energy consumption and significant demands on the heart and lungs. Energy consumption and cardiac effort often limit the ability to perform physically strenuous tasks (see Chapter 10); therefore, these bodily functions can indicate the severity of a physical task load. Similarly, the severity of a mental task load is usually assessed by the demands that associated stressors place on the mind in terms of stress. Stressors can be viewed as discrete

DOI: 10.1201/9781032674575-16

events that may disrupt normal psychological functioning. The resulting stress manifests as mental (cognitive and emotional) and physiological (biological) reactions.

Blue- and white-collar work?

Well into the mid-20th century, the working population was often categorized as either blue-collar workers, who performed physical labor, or white-collar workers, who primarily used their intellect. Today, this simplistic division no longer holds, as work tasks and job requirements have evolved (along with clothing colors and styles). An example of this shift is the widespread use of electronic devices (such as computers) for information retrieval and storage, even in occupations traditionally associated with physical labor, such as automobile repair. However, some professions predominantly rely on mental capabilities; for example, airplane pilots, medical personnel, process control operators, teachers, and journalists, as well as most office workers. It is reasonable to assume that such jobs impose unique demands on mental capabilities. Therefore, we should be able to measure both the mental task load and the related mental capabilities.

"Computer adaptation syndrome"

The introduction of computers into offices in the late 1900s exemplifies the workload stress experienced by a large group of people. Many experienced office workers were concerned about their ability to handle the new demands of using computers, especially if they feared that any lack of performance might lead to job loss. There was anxiety that computers would require extraordinary technical knowledge; in addition, there were concerns about potential radiation health risks associated with the cathode ray tubes in the monitors then in use. Fortunately, the stress related to computer-use* diminished within a relatively short time. One primary reason was that smarter software, coupled with well-designed training and instructions, made it easier than anticipated to learn how to use computers. As many older office workers became proficient, they found that computers actually made certain tasks easier and others more interesting and challenging. Younger office personnel, having grown up with computerized devices as everyday gadgets, had no qualms about using computers at work. Thus, the so-called computer adaptation syndrome completely vanished within a decade.

What is mental task load?

Surprisingly, a generally accepted definition of mental workload does not exist even today. In 2005, Megaw repeated the 1989 statement by Paul Linton: "The simple fact of the matter is that nobody seems to know what workload is. Numerous definitions have been proposed, and many of them seem complete and intuitively "right." Nevertheless, current definitions of workload all fail to stand the test of widespread acceptance or quantitative validation." Even ISO 10075, which addresses ergonomic principles related to mental workload, does not provide a convincing definition.*

Demand and resource

To understand the concept of workload, we can use the resource construct: it assumes a certain quantity of human capability from which the task demands a portion. Figure 12.1 depicts the "partially filled cup" analogy. Workload is the portion of resource expended in performing a given task. If less is required than is available, a reserve remains, as indicated by the excess capacity in Figure 12.1. If more is demanded than is provided, overload occurs (continuing the cup analogy, the workload overfills the capacity). In this analogy, the size of the cup varies both between individuals and over time: humans are neurodiverse, leading to different responses to workload and other situations, which can result in varying levels of reserve capability.

Overload and underload

In the case of an overload, task performance remains incomplete or impossible, and the operator is likely to suffer (physically, psychologically, or both) from overload stress. Measuring performance outcomes provides clues about the type and quantity of overload. In the case of an underload, the operator is capable of performing more or better, based on remaining reserve capacity; satisfactory performance of the primary task does not require all of the operator's capabilities, allowing the

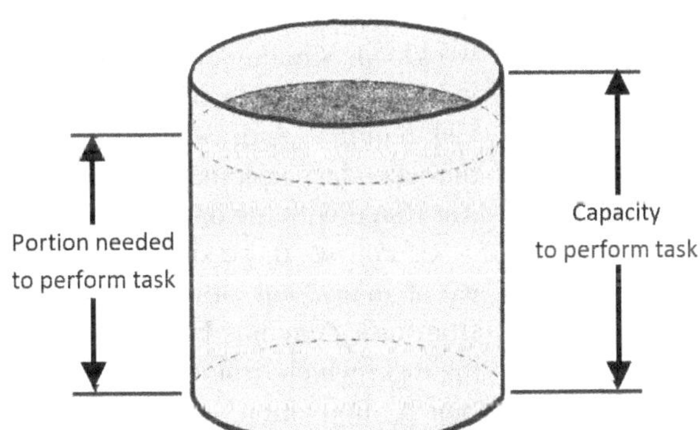

FIGURE 12.1 Resource model with "partially filled cup" analogy.

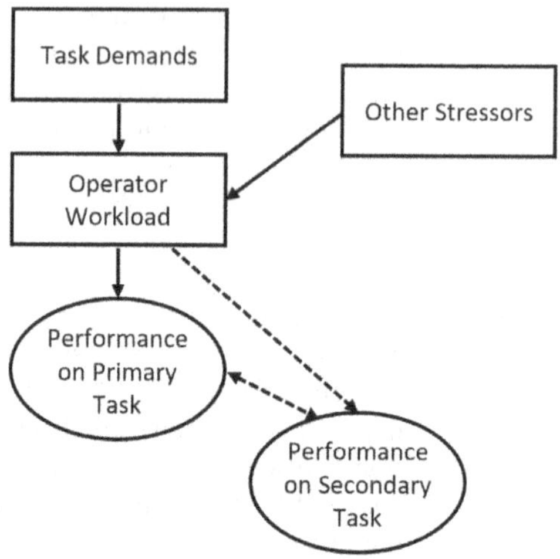

FIGURE 12.2 Model of workload and task performance.

person to perform a secondary task simultaneously. Measuring performance on the secondary task is one way to assess the workload of the primary task, as discussed further below.

Task performance Figure 12.2 schematically shows the relationship between task demands, operator workload, and task performance. Task demands may be defined in terms of intensity, complexity, time, or other measures of difficulty. The operator's workload depends on the person's task-related capabilities at the moment of demand and is influenced by other stressors. Performance on the primary task impacts, and is likely impacted by, any secondary task.

Multitasking (or maybe more accurately, rapid task switching) often allows us to assess task demands and operator workload. Simultaneous performance on a secondary task can indicate how much of the operator's capabilities the primary task requires. A drawback of this approach is that performing the secondary task may affect execution of the primary task (as illustrated with the dashed line in Figure 12.2).

Of course, these simple concepts only hint at the complexity of many real situations. Workload stress results not only from task demands but also from how the operator perceives the task, which, in turn, depends on the operator's skill, experience, motivation, and other factors at and outside of work. These circumstances may determine the strategies the operator adopts: the most effective strategy for the ergonomist is to reduce task load, which diminishes the stressor and, hence, the stress. However, the individual perception of task load (and the person's task performance) depends on task-related capabilities, aptitude, attitude, motivation, arousal, and fatigue. While working, the operator may gain skill or ways to simplify the task; both outcomes modify the resulting task load.

12.2 Designing for mental workload

Stress causes physiological reactions

Selye discovered that stress triggers physiological responses in both positive situations (eustress) and negative situations (distress). A positive stress response, eustress, results from challenging yet achievable situations that are enjoyable or fulfilling (such as embarking on a difficult but important project at work). A negative stress response, distress, occurs in intense or prolonged situations perceived as painful or threatening (such as a demanding project at work facing insurmountable technical or financial challenges). Both stress responses lead to similar reactions in the endocrine system and the autonomic nervous system. Common responses include increased secretion of hormones from the adrenal glands, especially adrenaline and noradrenaline, which put the entire organism in a state of heightened alertness.

Stress at work and leisure

The mind–body interaction functions in both directions: psychological factors can contribute to the onset or aggravation of various physical disorders, while physical problems can affect a person's thinking or mood. At work, the intensity of stress and its effects depend on how the individual responds to specific conditions in the work environment; stressors outside of work can also complicate the ability to manage work-related stress. Individuals have differing emotional responses to the same stressors, and their bodily reactions differ as well.

Is stress always harmful?

Clearly, stress is an unavoidable part of daily life. It is a necessary condition for all living beings that must respond appropriately to new situations. A life without stressors would not only be unnatural but also boring. Paracelsus, a 16th-century physician, coined the famous dictum, "Solely the dose determines that a thing is not a poison," meaning that the quantity establishes whether something is toxic (or dangerous) or not. This principle applies to stress as well; the amount of stress determines whether it adversely affects well-being or enhances a person's ability to cope with life events. The boundary between healthy and pathological stress varies among individuals and between situations. Often, challenging tasks appear to us as engaging and interesting, and achieving them brings satisfaction: rewards, promotions, accomplishments, compliments, pleasant surprises, joyful events, comical situations, unexpected acts of kindness, and friendly words make us feel alive and energized. Thus, positive stressors and healthy stress are essential components of a fulfilling life.

12.3 Addressing stress and distress

Stress as an emotion

Today, psychologists assert that the basic experience of stress is emotional*. One's state of mind can cause joy or anxiety, either of which can trigger the body's endocrine and autonomic nervous systems. Emotions provoked by distress can lead to muscle tension, resulting in pain in the head, neck, back, or elsewhere. Healthwise, distress may cause disturbed sleep, gastrointestinal complaints, cardiovascular diseases, musculoskeletal impairments, and weakened immune functions.

Behavior under stress

Intensive or sustained stress often changes how people behave and feel. Irritability, general dissatisfaction, impaired attention, damaged interpersonal relationships, anxiety, and depression are common symptoms. Stress may also lead to maladaptive behaviors, such as excessive smoking and alcohol or drug abuse. Work-related stress can negatively impact the organization and reduce its functioning through increased staff absence, poor work attitudes, and decreased productivity and quality of work. Thus, stress can cause individual suffering and may also be costly for the employing organization*.

Coping with stress

Distress, anger, anxiety, and even depression may arise when a person cannot adequately cope with demanding conditions. Because stress is a subjective experience, techniques to manage it must be tailored to each individual. Dealing with stress involves activating cognitive and behavioral strategies. These strategies may directly target the causes by eliminating stressors or by modifying the stressors and their perceived demands through techniques such as time management, workstyle adaptation, assertive communication, and setting limits. Coping strategies may also focus on individual emotions, including cognitive re-evaluation of the situation (reframing the problem) and the use of humor, relaxation exercises, off-work activities, and hobbies. The person's physical health, nutrition, and habits (including smoking or alcohol consumption) may require attention and possibly a change in lifestyle. A radical solution is to quit the stressful job and seek work that is more suitable; this is more feasible if another, better job is available.

Eliminating stressors at work

Job stress primarily results from a mismatch between job demands and a person's ability to meet them. Instead of (or in addition to) trying to adapt the individual, it often makes more sense to reduce or eliminate stressors at work—that is, fit the tasks to the human. The following list contains examples of stressors in the work environment:

1. Job content;
2. Intense job demands;
3. Complex task loading;
4. Repetitive, monotonous tasks;
5. Lack of control over one's job;
6. Excessive responsibilities of the job;
7. Lack of recognition and insufficient rewards for achievement;
8. Environmental factors such as noise, unpleasant climate (indoors or outdoors), crowded office space, and technical issues including frequent equipment malfunctions;
9. Lack of job security and fear of losing one's job;
10. Lack of social support from management and peers;
11. Unreasonable, unfriendly, and unfair supervisor;
12. Threats or bullying from the boss or coworkers.

Stress metrics

Workload, as a general term for "stressor(s)," lacks a precise psychometric definition. Nonetheless, various techniques are available to categorize a person's perception of stressful workloads and to scale the resulting stress. Physiological stress metrics have been derived from the electroencephalogram, electrocardiogram, transcranial Doppler sonography, eye tracking, and heart rate variability. In the medical field, additional health assessments exist, as do various approaches in industrial engineering and management.

Psychologic metrics

Common psychologic metrics for perceived mental task load* include the NASA Task Load Index (NASA-TLX), the Subjective Workload Assessment Technique (SWAT), the Multiple Resource Questionnaire (MRQ), and other related task load indices that combine specific measures. Further assessments include well-being surveys, stress arousal checklists, mood assessments, emotional intelligence questionnaires, and evaluations of coping capabilities. New and more refined approaches continue to emerge in the psycho-social literature. The sensitivity of workload indices to the cognitive demands of tasks varies significantly, as do the relationships between indices. Usually, stress indices are specific to both the task and the individual. Clearly, such measurements require well-trained experts to select stress metrics, evaluate test results, and determine stress relief strategies—this is not for the layperson.

12.4 Addressing underload and overload

Too little, too much

Most people feel stressed due to task overload, which demands too much from them in terms of intensity, complexity, diversity, time pressure, and similar factors. However, some jobs require too little effort, leaving workers idle, bored, and underutilizing their capabilities. These observations have led to the notion that we can describe the relationship between task demands and experienced workload as a "valley" function* resembling the letter U: in its center, an appropriate balance exists between demand and load that is just right. However, requiring either too high a task load or too little leads to worker overload or underload, respectively. Task performance suffers under both conditions. Figure 12.3 shows a postulated U-function; other theories propose an S-shaped (sigmoid) relationship. What is appropriate depends on the situation and the individual operator.

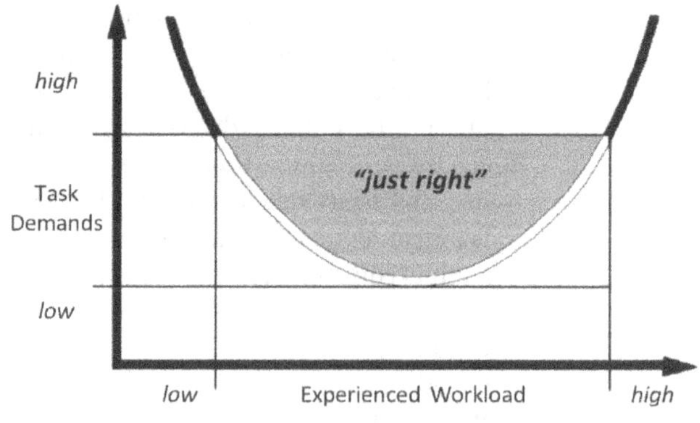

FIGURE 12.3 Assumed "U" function (valley function) between task demands and the workload experienced by the worker.

Monotony and boredom

Monotony is a characteristic of an environment that remains unchanged over time or where changes occur in a repetitive and predictable manner over which the operator has little or no control. A varied environment usually provokes interest and even excitement, while monotonous conditions produce boredom. *Boredom* at work is an emotional state resulting from a dull, uninteresting job environment. A bored person often complains of feeling tired or fatigued, frequently accompanied by yawning, the same response expected from someone exhausted by heavy physical work.

12.5 Assessing task load psychophysically

The understanding and definition of task load depend on two factors: the intensity of the task and the related capability of the person to perform the job. This relationship between demand and ability is further influenced by the person's perceived stress and willingness to perform. There are straightforward cases in which we can measure both workload and the operator's resulting physical strain, such as a baggage handler loading an airplane cargo hold or a mason carrying bricks up a ladder at a construction site. Here, an investigator may determine, in physical terms, the energy the job requires, and in physiological terms, how this impacts the person's metabolism and circulation, as discussed in Chapter 10.

Complex jobs, complex assessments

However, such simple cases with only a few easily assessable components are rare. Usually, the task demands are complex, and environmental conditions often contribute intricately to these demands. Furthermore, it is often impossible to measure a person's body and mind responses to the demands using just a few established physiological techniques, which assess only one of several strains on the body (such as oxygen uptake) and whose apparatus often hinders task performance. The literature reflects a long-running discussion* about the usefulness of so-called objective (analytical, quantitative) measures compared to assessments that rely on subjective (qualitative, empirical) judgments made by either an observer or the worker. Table 12.1 lists a sampling of methods useful for task load assessments.

Listen to the worker

The last entry in Table 12.1 lists psychophysical measures: these integrate elements of biomechanical, physiological, and psychological assessments in the form of subjective judgments made either by an observer or the operator. The underlying premise

Table 12.1 Some typical methods of task load assessments

Methods	Techniques
Archives and records	Description, statistics, trend
Indirect observations	Interview, discussion, survey
Direct observations	Chart, checklist, critical incident review
Expert analyses	Walk-through, judgment, scoring
Task analysis	Flow/time chart, time-and-motion study
Simulation	Mathematical/physical mock-up
Modeling	Task network, control processing
Performance measures	Work rate, output in amount and quality
Biomechanical measures	Posture, force/moment/work analysis
Physiological measures	Oxygen consumption, heart rate
Psychological measures	Perceived exertion, attitude, rating
Psychophysical measures	Subjective assessments *(see text)*

is that an experienced person can perceive and merge all existing loadings and then summarize their combined effects in the form of a judgment. The reliability of judgments increases when they are formulated within a standardized and controlled framework, such as pre-selected words or phrases or numbers that reflect rankings (less than, more than) or mathematical relations (such as double or triple as much) as a reference.

Borg scales

Borg's psychometric scales* incorporate such organized judgments in an interval scale. Table 12.2 contains the rating of perceived exertion (RPE) and the Category-Ratio-10 (CR-10) scales. The steps are "verbally anchored" with the words listed in Table 12.2 and are arranged to follow mathematical formulas. The RPE ratings are in equidistant steps, which are numbered so that, when multiplied by 10, they correspond to heart rates measured at these work levels. The CR-10 Scale has ratings in steps set to constant ratios (and hence, it is called a ratio scale).

Like any other measurements, these Borg scales require rigidly controlled procedures to yield credible results. For example, his CR-10 Scale, from Table 12.2, is presented to the person who will rate the perceived intensity of work with these instructions:

> "You do not have to specify your feelings, but do select the number that most correctly reflects your perception of the job demands. If you feel no loading, you should answer zero, nothing at all. If you start to feel something just noticeable, your answer is 0.5, extremely weak. If you have an extremely strong impression, your answer will be 10. So, the more you feel, the higher the number you are choosing. Keep in mind that there are no wrong numbers; be honest, do not overestimate or underestimate your ratings."

Table 12.2 Borg scales for perceived exertion

1960 Borg RPE scale (modified 1985)	Borg general (CR-10) scale
6 – no exertion at all	0 – nothing at all
7 – extremely light	0.5 – extremely weak (just noticeable)
8	1 – very weak
9 – very light	2 – weak (light)
10	3 – moderate
11 – light	4 – somewhat strong
12	5 – strong (heavy)
13 – somewhat hard	6
14	7 – very strong
15 – hard	8
16	9
17 – very hard	10 – extremely strong (almost maximal)
18	11 or higher, the individual's maximum
19 – extremely hard	
20 – maximal exertion	

RPE, rating of perceived exertion; CR-10, Category-Ratio-10.
Source: Adapted from Borg (2001, 2004).

Borg claimed that his scales have high reliability (meaning that a repeated test yields the same results as the initial test) and high validity because the test results correlate well with measures of heart rate. These scales have been used for over 40 years in occupational health and ergonomics.

Such psychophysical assessments (of which there are many, including more complicated ones*) have several practical advantages: they are relatively easy and inexpensive to conduct, do not usually interfere with task performance, and provide a comprehensive summary of the effects of all stressors at work. However, they need to be administered carefully; otherwise, their results may be misleading.

Summary

Work-related stress results from stressors at work. This stress is an individual's biological and emotional reaction to aspects of work demands, work environments, and work organizations. While people react differently to job conditions, in general, stress results from a mismatch between demands and a person's ability to meet them. The reaction to stressors occurs whether the situation is perceived as positive or negative, and whether the worker feels overloaded or unloaded. Instead of trying to adapt the person, it often makes more sense to change the demands.

Steps to fit the human

Step 1: Assess the mental and physical demands required to perform a specific task. Reduce excessive task demands and adverse conditions.

Step 2: Ask people for their opinions and judgments about their workload conditions. When people complain of overload underload, or boring tasks, redesign those tasks.

Step 3: Listen to and observe the people doing the work. The workers themselves can tell you what's working and what needs improvement.

Step 4: Design, then test. Modify and iterate as necessary.

Notes

The text contains markers, *, to indicate specific references and comments, which follow.

Hans Selye (1907–1982) has been called the founder of stress theory and wrote *The Stress of Life* (1956, 1978) and *Stress without Distress* (1974). He was not the first to notice the impact of stressors on the body; physician and physiologist Walter Bradford Cannon (1871–1945) also researched the physiological impact of adverse external environments and coined the term "fight or flight response" (now sometimes stated as "fight, flight, freeze or fawn").

12.1 Basics of task load

Computer-use stress problem: Grandjean (1987).

Mental task load: Per ISO 10075-1:2017, "mental workload is regarded as an umbrella or generic term, … and does not have a specified or standardized meaning of its own within the document." See also Longo, Wickens, Hancock et al. (2022).

12.2 Designing for mental workload

Stress is an emotion: Cox and Griffiths (2005), Carayon and Lim (2006).

12.3 Addressing stress and distress

Stress is costly to the organization: Cox and Griffiths (2005).

Stress metrics: Rubio, Diaz, Martin et al. (2004), Megaw (2005), Finomore, Shaw, Warm et al. (2013), Matthews, Reinerman-Jones, Barber et al. (2015).

Valley function: The inverse of the "U" or "valley" curve is the "mound" (inverted U) function, which is also discussed in Chapters 8 and 11.

12.5 Assessing task load psychophysically

Discussion of measures: see also *J. Ergonomics*, Volume 45, issue 14, 2002; Finomore, Shaw, Warm et al. (2013), Matthews, Reinerman-Jones, Barber et al. (2015), Wilson and Corlett (2005).

Borg Scales: developed by Gunnar Borg, see Borg (2001, 2004).

Complicated scales: Wilson and Corlett (2005).

References and further reading

Borg G (2001). Rating scales for perceived physical effort and exertion. In Karwowski W (ed), *International Encyclopaedia of Ergonomics and Human Factors.* London, UK, Taylor & Francis, pp. 358–541.

Borg G (2004). Scaling experiences during work: perceived exertion and difficulty. Chapter 11 in Stanton N, Hedge A, Brookhuis K, Salas E, Hendrick HW (eds), *Handbook of Human Factors and Ergonomics Methods.* Boca Raton, FL, CRC Press, pp. 768. https://doi.org/10.1201/9780203489925

Carayon P, Lim SY (2006). Psychosocial work factors. Chapter 5 in Marras WS, Karwowski W (eds), *The Occupational Ergonomics Handbook: Interventions, Controls, and Applications in Occupational Ergonomics*, 2nd ed. Boca Raton, FL, CRC Press, pp. 848. https://doi.org/10.1201/9781420003642

Cox T, Griffiths A (2005). The nature and measurement of work-related stress: theory and practice. Chapter 19 in Wilson JR, Sharples S (eds), *Evaluation of Human Work*, 4th ed. Boca Raton, FL, CRC Press, pp. 1031. https://doi.org/10.1201/b18362

Finomore VS, Shaw TH, Warm JS, Matthews G, Boles DB (2013). Viewing the workload of vigilance through the lenses of the NASA-TLX and the MRQ. *Human Factors*, 55(6):1044–1063. https://doi.org/10.1177/0018720813484498

Grandjean E (1987). *Ergonomics in Computerized Offices.* London, UK, Taylor & Francis. https://doi.org/10.1201/b12564

ISO 10075-1 (2017). *Ergonomic Principles Related to Mental Workload: General Issues and Concepts, Terms and Definitions.* Geneva, CH, International Organization for Standardization.

Longo L, Wickens CD, Hancock G, Hancock PA (2022). Human mental workload: a survey and a novel inclusive definition. *Frontiers in Psychology*, 13:883321 with erratum in *Frontiers in Psychology*, 13:969140. https://doi.org/10.3389/fpsyg.2022.883321

Matthews G, Reinerman-Jones LE, Barber DJ, Abich J (2015). The psychometrics of mental workload: multiple measures are sensitive but divergent. *Human Factors*, 57(1):125–143. https://doi.org/10.1177/0018720814539505

Megaw T (2005). The definition and measurement of mental workload. Chapter 18 in Wilson JR, Sharples S (eds), *Evaluation of Human Work*, 3rd ed. Boca Raton, FL, CRC Press, pp. 1031. https://doi.org/10.1201/b18362 (4th edition in 2015)

Rubio S, Díaz E, Martín J, Puente JM (2004). Evaluation of subjective mental workload: a comparison of SWAT, NASA-TLX, and workload profile methods. *Applied Psychology*, 53:61–86. https://doi.org/10.1111/j.1464-0597.2004.00161.x

Selye H (1956, revised 1978). *The Stress of Life.* New York, McGraw-Hill.

Selye H (1974). *Stress without Distress*. Philadelphia, PA, Lippincott.

Wilson JR, Corlett N (2005). *Evaluation of Human Work*, 3rd ed. London, UK, Taylor & Francis.

SECTION **IV**

Organizing and managing work

How individuals feel about their company, and hence about their work, largely depends on the organizational set-up. The design of the work organization and the way it is managed can be considered *macroergonomic* issues, in contrast to the more detailed concerns about specific human factors, such as the workspace for hands.

The fourth section of this book addresses the key factors that determine how people interact with each other at work and, as a result, how much effort they invest in their work. We then discuss another set of managerial responsibilities related to the organization of work itself: the number of hours worked per day, the best times for work during the day, and how to effectively organize shift work. We recognize and address differences in how people work and work together so that we can design tasks, gear, equipment, and procedures to fit their specific needs. Therefore, the following chapters provide relevant information for ergonomic design:

- Chapter 13, Working with Others
- Chapter 14, Working in an Organization
- Chapter 15, Working Hours and Sleep
- Chapter 16, Working in Shifts

DOI: 10.1201/9781032674575-17

CHAPTER THIRTEEN

Working with others

A working story

What follows is an example* of an office move that shows how the whole organization, including the layout of the workspace for individuals and work teams, can strongly affect people's work attitudes and output. We need to understand how we work with others and in teams to better fit the task to the human.

> *A few years ago, a well-respected professional services company in Chicago moved from an older office tower downtown into a newly renovated building. The company, with a rich, multi-decade history, employed thousands of people. Employees often described themselves as part of a "family," with extensive interaction among all levels of the corporate hierarchy and a strong prevailing team spirit. The old offices had sentimental value for many employees; this is where many had begun their careers. However, the new office building was easier to access via public transportation, allowed for increased company growth, and was more reflective of the company's success, with luxurious interior appointments.*
>
> *Once the relocation was complete, employees realized that the move with its office redesign had some unintended and unfortunate consequences. Formerly, executives' offices and cubicles were interspersed among staff cubicles; now all executives' offices were located on two separate floors. This new lay-out was intended to encourage easier communications among the now co-located directors, and also to reward executives' performance by providing especially posh office suites. However, now there were few casual interactions between executives and employees: no chance encounters at the coffee machine or in the hallways, no "management by walking around." Now, these interactions were limited to business-only discussions at formal meetings, which led to anxiety and reduced trust between employees and executives. This physical separation between employees and executives led to an emotional separation: employees felt disconnected and disengaged*

DOI: 10.1201/9781032674575-18

from leadership and from the company as a whole. Executives also felt out-of-the-loop from the employees. The previous team atmosphere transitioned into more of an "us-versus-them" philosophy. The company culture had changed.

A few years later, the company decided it needed to move again. This time, the company wanted to adapt the office to a post-2020 "hybrid" work environment: on any given day, some employees worked in-person at the office while others worked from remote locations (from their homes or other business locations). In the newest office space, employees would no longer have an assigned desk and could sit at any open desk when they were on-site. The company could save money by supporting less (physical) office space, and also democratizing the space it had. The intent was to help the company culture become more team-focused and help employees re-engage with each other and with the company. This move and re-design also had unintended consequences: although the flexibility in work location helped employees manage their work and personal lives and reduce commuting expenses, it also had disadvantages. Having a sizable number of employees working off-site on any given day made it harder to onboard new employees, trickier for managers to lead and supervise their teams, less likely for employees to cultivate relationships to form an internal professional network, more difficult for employees to learn (and shape) the company's unique culture and really feel a part of the broader company and not just their immediate work teams. Once again, the company culture had changed.

13.1 Basics of getting along with others

The example above refers to many types of personal interactions, such as chance encounters in shared spaces like a coffee (or tea) room or the hallway, where employees can chat and connect as individuals, outside of their work teams. Indeed, the most basic and perhaps most important everyday interactions occur between individuals, often informally.

Personal space

Expectations for personal interactions vary in different regions of the earth and in different cultural traditions. Social interactions typically occur at culturally determined distances that people unconsciously have learned and seek to maintain. The way individuals manage their personal space, particularly the physical distances they keep from others, conveys psychological and social messages. Broadly speaking, people from high-contact cultures (including South/Central Americans, Southeastern Europeans, and Middle Easterners) have smaller personal spaces than those from low-contact cultures (such as North Americans, Northern Europeans, Chinese, and Japanese)—although contact distances and interpretations of who is a close friend can vary widely among individuals and in different situations (such as on a crowded public bus).*

Social distance Most of us perceive a certain volume around us that we consider as our own sphere. When another person enters that space, it can evoke strong emotions. The so-called social distance typically extends to three or four meters. Where individuals position themselves within this personal space reflects the nature of their relationship.

Personal distance Most in-person interactions occur within an extended arm's length (the *personal distance*), whereas a narrower *intimate distance* often involves physical contact. Friends and good acquaintances usually move in closely (within about a meter), or even closer (within 0.5 m) for very dear friends and family. More casual friends, work colleagues, and groups generally maintain a personal distance of up to 1.2 m. Interactions between unacquainted individuals and more formal business transactions typically take place within a social distance of about two or three meters. Beyond that distance, interactions are generally formal. The *public distance*, extending beyond about four meters, may require raised voices for communication. Of course, local customs and even a person's immediate attitude influence these preferred interpersonal distances.

Using personal space Using personal space can be complex. When someone who, for personal or social reasons, is not supposed to be close nevertheless invades our personal space, we may experience discomfort, anxiety, and even anger. Conversely, if coworkers, good acquaintances, and friends remain too distant, disappointment and alienation are likely, and formerly close relationships may deteriorate. At work, when group members are expected to cooperate as teams, performance is optimal when they are in close proximity. However, competition tends to be fiercer when people stay outside the distances of personal relationships.

Owning an area Some people display behavior patterns called *territoriality*, which is the inclination to occupy and control a defined physical space, such as property, a workplace, or even a section of a workbench. Preventive or reactive defenses may keep perceived intruders at bay; an "invader" may be met with offensive behavior, or the space may be abandoned. Often, the "owner" of an area personalizes it and marks it in various ways, such as with plants or photographs; these measures convey a sense of control, possession, and security.

Crowding

Crowding is the negative experience of too many people within a given space. Like other feelings associated with personal space, the perception of crowding depends on individual characteristics, the assessment of one's coping capabilities, and the physical and social settings. Individual reactions to crowding may include aggression or withdrawal.

Teamwork

Working in groups is a common requirement at work, whether in offices, in planning and design, in production and assembly, or in maintenance and repair tasks. Teamwork is essential in high-pressure situations such as emergency rescue, firefighting, and military operations. *Teams* consist of two or more individuals who pursue a common goal by performing their tasks in interdependent, complementary, or distinct manners*. Team members must communicate with each other, adapting and adjusting their work to perform their tasks in a timely and integrated fashion. At the core of teamwork are sets of interrelated knowledge, skills, attitudes, and behaviors, including formal and informal organizational aspects. In general, team members must be willing to communicate and be flexible in adapting to changing allocations of functions within the team. Not all individuals are willing or able to subordinate their work habits and specific skills to the rules of teamwork; furthermore, some regional cultures emphasize collaboration, while others stress individual work.

Remote and in-person teams

Coming out of the early 2020s, many workplaces had to adjust to new ways of working. For office-based workers, the global COVID-19 pandemic led to the widespread (and often forced) adoption of remote work instead of in-office duties. Data from the United States in 2024 show that about one-quarter of remote-capable jobs are fully remote, over half are hybrid, and the rest are on-site (in-person, as we used to think of most office work); 2019 data reflected an opposite trend, with fewer than 10% of remote-capable jobs being remote and 60% being on-site. Estimates indicate that in 2023, full-time workers in the United States, Canada, the United Kingdom, New Zealand, and Australia worked remotely an average of 1.4 days per week, compared to 0.8–0.9 days per week for European and Latin American countries*. Overall, many global companies have implemented policies that allow for a hybrid work schedule, balancing off-site and on-site work for those whose job duties permit it. The long-term implications of teams who do most or all of their work without having met each other in person are unclear, but may include feeling less connected to their colleagues and their organization's culture, as well as having impaired working relationships. Team members generally need opportunities to form interpersonal relationships to perform effectively, and this is more challenging (and psychologically taxing) with only virtual communication.

13.2 Designing for motivation and behavior

The field of study that seeks to understand and address interpersonal and managerial challenges in the workplace is called *organizational behavior.* The relatively recent "lean and mean" business years, marked by cost-cutting and layoffs, have led many employees around the world to fear for their jobs and feel uncertain about their professional futures. In response, some companies actively strive to retain their employees and keep them happy, satisfied, and motivated. This is not solely out of altruism; there is also a recognition that improving employee satisfaction enhances work efficiency and profit.

Happy employees Happy employees are more productive; they provide better customer service, work harder, and take fewer sick days. Moreover, they are more likely to stay with the organization, which reduces one of the most significant costs—employee turnover related to dissatisfaction. Employee attrition can generate substantial expenses for recruiting and training new staff: replacing a valuable employee can cost anywhere from half to several times a year's pay. It takes time, training, and supervision to bring a newly hired employee's performance up to that of an experienced worker. Additionally, there is the less tangible cost of unhappy employees who remain on the job; they tend to exhibit low productivity and poor customer service, suffer personally from stress (see Chapter 12), and may develop serious stress-related illnesses as they trudge through a dreary or demoralizing work routine. Their negative attitudes can also adversely affect their coworkers.

Psychosocial work factors How we feel about ourselves as members of a work organization depends on a variety of psychosocial work factors*, listed in Table 13.1. These factors relate to job demands, the tasks we perform and how we perform them, our cooperation with peers and superiors in the organization, and our expectations regarding our professional and employment futures

Table 13.1 Psychosocial job factors

Category	Factors
Job demands	Physical workload
	Variations in workload
	Work pressure
	Cognitive demands
Job content	Challenges, interest
	Repetitiveness, monotony
	Development and use of personal skills
Job control	Control over work pace
	Control over physical environment
	Task/instrumental control
	Organizational control
Social interaction	Interactions with colleagues
	Interactions with supervisors and senior management
	Dealing with clients and customers
Job future and career issues	Ambiguity about job future
	Fear of losing job
Organizational and management issues	Employee participation
	Management style

Source: Adapted from Carayon and Lim (2006).

Division of labor

Since the early 1900s, Frederick Taylor's* concept of the division of labor has led to work mechanization and specialization. Although many of Taylor's ideas have been frequently (and often justifiably) criticized, his concepts also emphasized motivating individuals and sharing responsibilities between management and labor. He asserted that a worker's interests should be placed on par with those of the employer. Taylor's approach generated efficient work systems through time studies, careful work design, provision of proper tools, and the selection of capable individuals. A worker's task was often divided into small components, which reduced skill requirements and facilitated performance evaluations. Unfortunately, this also had the consequence of many jobs becoming simplified, standardized, repetitive, and monotonous.

Quality of work life

Half a century later, as the workforce became more educated and liberated, it also started to place more importance on working conditions and the quality of work life. This raised issues of recognition of personal value and improvement of working conditions. Consequently, theories were developed regarding individual workers' motivation, productivity, and well-being. Individual needs and wants were emphasized, and workers' behavior and attitudes towards work were considered important.

Motivation, ability, performance

Motivation* incites, directs, and maintains behavior towards goals. Motivation and job performance are, of course, related: a motivated person who desires to do well at work is willing to expend effort to achieve that. One way to express this is to consider that performance is the product of motivation and ability, where motivation includes personal commitment and drive, and ability includes personal aptitude along with organizational training and resources. Performance is moderated by situational constraints at work, which are factors that can hinder or enhance performance; examples include the work climate (both psychological and physical) and up-to-date equipment.

Needs hierarchy

In the 1950s, Abraham Maslow and later his disciples developed the *Hierarchy of Needs**, depicted schematically in Figure 13.1. According to this model, motivation is a function of meeting personal needs, which start with basic physiological necessities and then progress to higher order aspirations. Thus, we must first work to meet basic needs such as food and shelter (at the base of the triangle in Figure 13.1). Once these are met, we strive to satisfy our safety needs, which focus on economic and physical security. The next step concerns our social needs for a sense of belonging and love; above these are our needs for esteem, where we gain self-confidence through

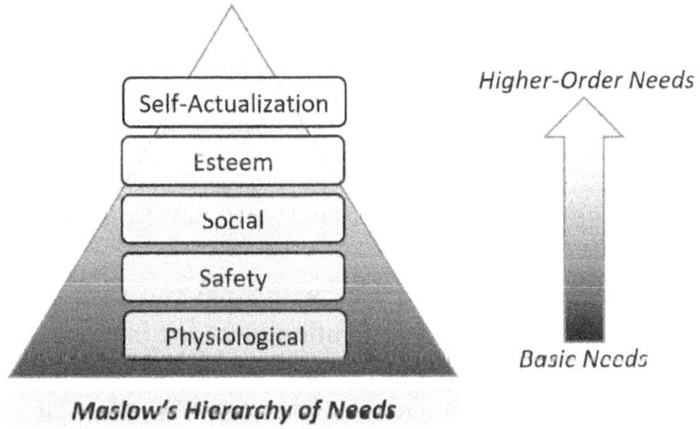

FIGURE 13.1 Maslow's hierarchy of needs theory.

recognition, appreciation, and respect. The fifth step fulfills our highest-order needs: self-actualization, where we achieve our full potential and capabilities.

Changing needs

Maslow's hierarchy of needs theory posits that behavior is motivated by the urge to satisfy the needs and wants of increasingly higher orders. This tenet has been modified in various ways to explain observations of people moving back and forth among needs: what are steps according to Maslow's theory may actually be a continuum where needs and desires change, overlap, and merge. By 1972, Clayton Alderfer proposed another approach known as the *ERG theory**; this concept explains shifts among our concerns that relate to

- *Existence*: food and essential supplies, compensation, working conditions.
- *Relatedness*: relationships and interactions with family, friends, colleagues.
- *Personal Growth*: the desire for personal development, advancement, recognition.

Expectancy theory

Another individual-focused approach to motivation is *expectancy theory*, which gained prominence in the field of motivation research in the 1960s and is still widely applied today. Expectancy theory states that a person's motivation (and resulting satisfaction) depends on the difference between what is expected and what is offered. Motivation is considered the product of the individual's expectation that work effort will lead to performance, the expectation that performance will lead to reward, and the value of the reward. This theory assumes that the individual is rational about consequences and has a desired reward. The reward itself is under the control of the organization: it may be positive (recognition, pay, promotion) or negative (loss of employment). To maintain motivation, the organization should demonstrate the link between worker performance and outcomes with rewards that are important to the worker. This sustains the expectancy (and thus the performance) on the part of the worker.

Job satisfaction

Satisfaction with one's job, or dissatisfaction, can greatly influence motivation and task performance. As usual, there are significant differences in individual perceptions of what makes a job satisfying or not. For example, someone might value personal growth and recognition at work over monetary rewards; this person would be satisfied with a relatively low-paying job as long as there is appreciation of their performance and success in professional challenges. Another person, however, might prioritize job security and good pay, feeling satisfied even if the position is considered of lower rank.

Two-Factor theory

In 1966, Frederick Herzberg proposed another concept of job satisfaction with his *two-factor theory**. This theory focuses on work conditions (which should be within the control of the employer and amenable to ergonomic design) rather than the individual. It assumes that positive *content* factors of a job mostly explain satisfaction while negative *context* factors lead to dissatisfaction.

As Table 13.2 illustrates, Herzberg and his disciples isolated five *content* factors that act as satisfiers or motivators, including promotion, opportunities for growth, and responsibility. If these factors are present and positive, they can create job satisfaction and motivation to excel at work; their absence would leave an employee feeling neutral or indifferent.

On the opposite side are *context* factors (also called *hygiene* factors), which include management, compensation, and work conditions and relationships. If these context factors are present and positive, they can help prevent negative consequences, such as employees becoming dissatisfied and losing motivation. Furthermore, a job lacking good context factors would make an employee dissatisfied, while the absence of context factors would result in a neutral or indifferent employee. (The following Chapter 14 addresses some of these issues.)

APCFB model

Additional models have been developed to explain the cognitive processes linking external events to employee behavior. One of these is the *APCFB* model. This model posits that a person's closely held *assumptions* shape their *perceptions* of a given

Table 13.2 Herzberg's two-factor theory: content satisfiers and context dissatisfiers

Content factors that (when positive) promote satisfaction and motivation (*intrinsic needs, motivators*)	**Context factors** that (when positive) prevent dissatisfaction (*extrinsic needs, hygiene factors*)
Promotion opportunities	Supervision quality
Personal growth opportunities	Pay and benefits
Recognition	Company policies
Responsibility	Interpersonal relationships
Achievement	Working conditions
Interesting and engaging work	Job security

event, leading to highly individual *conclusions*. These conclusions, in turn, evoke *feelings* that result in *behaviors*. Since everyone has a different set of assumptions, we perceive events differently and draw our own conclusions, which shape our individual feelings and actions. Simply put, one common event can lead to widely divergent individual behaviors and reactions.

Basic understanding

We can use the concepts of Maslow, Herzberg, and others regarding behavior, motivation, and satisfaction to gain fundamental insights into why people act the way they do on the job. Newer research continues to refine and redirect our knowledge and understanding.

13.3 Designing task demands and job rewards

Motivating work conditions

Many researchers and theorists have considered the work conditions that motivate people to perform their tasks efficiently. History has shown that, under austere conditions and for short periods of time, even negative factors such as fear or greed can drive people to outstanding task performance. However, when we consider ourselves as a society of valuable individuals, we strive for a high quality of work life; everyone should have the right to a fulfilling, rewarding, safe, and secure job. This goal aligns with the business intent to prosper, as job satisfaction is linked to critical revenue-impacting variables like employee turnover, absenteeism, and job performance.

Enlarging and enriching jobs

In contrast to earlier work specialization based on Tayloristic principles, the concepts of job enlargement and job enrichment form the foundation of many contemporary job design theories. *Job enlargement* refers to having a greater variety of tasks and activities, typically at the same professional level. *Job enrichment* involves expanding work skills and increasing responsibilities, often taking on tasks at higher professional level.

Job characteristics

How jobs are designed, both generally and specifically, can significantly affect our attitudes and hence our performance. One approach to explaining how job design influences behavior is the *job characteristics model**, which proposes five core features:

1. *Skill variety* is the number of different talents and activities that the job requires.
2. *Task identity* is the specifics of the work that a person completes from beginning to end.

3. *Task significance* is the impact of one's work on others, procedures, and products.
4. *Autonomy* is the degree of independence in planning, controlling, and determining work procedures, helping to foster a sense of responsibility.
5. *Feedback* is information about how one's performance is evaluated and perceived.

These five job dimensions provide the jobholder with a sense of meaningfulness, responsibility for work outcomes, and understanding of results. These experiences generate high motivation and satisfaction for the employee and, importantly for management, lead to higher quality work, lower absenteeism, and reduced turnover.

Hawthorne effect

Sociologists and industrial psychologists have conducted numerous experiments intended to change working conditions and observe their effects, if any, on work attitude and output. Among the best-known experiments are those from the 1920s at the Hawthorne Works* near Chicago. Here, the intervention involved adjusting the lighting conditions in a manufacturing assembly task. Each increase in lighting level resulted in improved output; surprisingly, when the illumination was eventually lowered, performance still improved. The explanation, in simplified terms, is that paying attention to the workers, taking their comments and activities seriously and listening to them—in short, treating them as important—led to improved output regardless of the magnitude or even the direction of the overt work intervention.

The term "Hawthorne effect" has become proverbial, describing a work situation where an introduced change increases productivity, not because of the change itself but because workers feel they are being observed and given attention, which encourages motivation and extra effort. These effects are usually short-lived, but they highlight the effectiveness of positive engineering and managerial measures.

Setting goals and rewards

The rewards inherent in a job must be meaningful and valuable to the employee; given individual differences, no universal reward system exists. Goal-setting theories suggest that people set targets and then purposefully pursue them. The premise is that conscious ideas underlie our actions, and the goals we set motivate us, direct our behavior, and help us determine how much effort to invest in our work. Within limits, more challenging goals foster higher levels of commitment. The more specific the goal, the more focused an individual's efforts will be to achieve it.

Motivation and work behavior

In spite of how compelling these existing theories are, there is no clear-cut answer to the question of what motivates people. Instead, various theories apply to different individuals under different circumstances. Motivation is both intrinsic and extrinsic: factors within us and external to us drive our behavior. Most likely, we consider our own needs and wants and either consciously or subconsciously determine our goals; then we act to increase our chances of obtaining what we desire. We know from the theories of Maslow, Herzberg, and others that we strive first to fulfill basic survival needs like securing food and shelter; beyond this, needs are still real but vary widely among individuals in terms of priority and strength. People around us influence and shape our motivation and behavior because we are all, to some degree, social creatures. How hard we work depends on what our efforts will yield: if we feel rewarded in ways that are meaningful and valuable to us, and if we perceive a definite link between the strength of our efforts and performance, we will work strenuously to perform well and gain those desired rewards.

Motivators other than money

What motivates people can shift. In Europe and North America, for example, pay and stability used to be among the most important job motivators; now, with an increasingly diverse, lifestyle-conscious workforce, there are many non-financial ways to reward employees and keep them motivated and happy. On-site childcare, flexible work hours, the ability to work remotely from alternate locations (including a home office), time off for volunteer activities, and paid vacation time are a few examples of the benefits that employees might seek. Some workers prefer close teamwork, while others may be more introverted and perform better on individual tasks. Most people place high value on having friendly coworkers and a supportive corporate atmosphere: feeling that the company is “a good place to work” and that “my coworkers are like family” is important to many.

Summary

Today’s educated and liberated workers seek pleasant conditions at work where they can use and expand their skills; they view work as an important part of their quality of life. Careful design of work, provision of the right tools, and a suitable work environment are basic ergonomic measures that must be augmented by an organizational culture (see Chapter 14) that encourages personal development.

Every person holds beliefs about how an employer should operate. We each have our own value system and believe that a suitable job should meet our corresponding expectations. While we are willing to make some adjustments, we acknowledge that to satisfy us, the job should motivate us to achieve our personal professional goals. Behavior theories suggest that individuals strive to satisfy certain needs, which in turn drives their behavior. Expectancy theories indicate that an individual's motivation (and resulting satisfaction) depends on the difference between what the work environment offers and what the worker expects. Motivation incites, directs, and maintains behavior toward goals.

An individual's motivation, capabilities, well-being, conditions, and productivity are interconnected: performance results from motivation and ability. Performance is influenced by situational constraints at work, which are factors that can hinder or enhance performance. A motivated person who wants to do well at work is willing to expend effort to succeed. How hard a person works depends on what that effort yields: if there is a clear link between exertion and performance, and if one feels rewarded in ways that are personally meaningful and valuable, individuals will generally work diligently to perform well and attain those desired rewards.

Steps to fit the human

Step 1: Determine what work needs to be done and what work environment and conditions are suitable.

Step 2: Provide opportunities for developing individual skills and fulfilling aspirations.

Step 3: Assess the workplace culture, and ensure good management, relationships, and work conditions.

Step 4: Be innovative, flexible, adaptive.

Step 5: Monitor conditions, and modify the work and iterate as necessary.

Notes

The text contains markers, *, to indicate specific references and comments, which follow.

13.1 Basics of getting along with others

Example: Adapted and expanded from Kroemer and Kroemer (2017).

Personal space: See more on proxemics in Proctor and Van Zandt (2008), Gokmen, Turen, Erdem et al. (2021).

Teams and team work: A group is a collection of individuals who coordinate their efforts, while a team is a group of people who share a common goal. For more on teamwork, see Muchinsky and Culbertson (2016), Stanton, Hedge, Brookhuis et al. (2004).

Global working trends: These numbers will continue to change; for 2023 data, see Aksoy, Barrero et al. (2023).

13.2 Designing for motivation and behavior

Psychosocial work factors: Smith (2007).

Frederick Taylor: Mechanical engineer FW Taylor (1856–1915) was one of the first management consultants specializing in production efficiency, see, for example, his book *The Principles of Scientific Management* (1911).

Motivation: see Landy and Conte (2006), Muchinsky and Culbertson (2016) and many other related articles and books.

Maslow's hierarchy of needs theory: Maslow (1943, 1954).

Alderfer's ERG theory: in effect compressing Maslow's five needs into three, see Alderfer (1972).

Herzberg's two-factor theory: Herzberg (1966, 1968).

13.3 Designing task demands and job rewards

Job characteristics model: Hackman and Oldham (1980).

Hawthorne effect: Roethlisberger and Dickson (1939), Parsons (1974), Levitt and List (2009), McCambridge, Witton, and Elbourne (2014).

References and further reading

Aksoy CG, Barrero JM, Bloom N, Davis SJ, Dolls M, Zarate P (2023). *Working from Home Around the Globe: 2023 Report*. https://wfhresearch.com/wp-content/uploads/2023/06/GSWA-2023.pdf (retrieved June 11 2024).

Alderfer CP (1972). *Existence, Relatedness and Growth: Human Needs in Organizational Settings*. New York, Free Press.

Carayon P, Lim SY (2006). Psychosocial work factors. Chapter 5 in Marras WS, Karwowski W (eds), *The Occupational Ergonomics*

Handbook: Interventions, Controls, and Applications in Occupational Ergonomics, 2nd ed. Boca Raton, FL, CRC Press, pp. 848. https://doi.org/10.1201/9781420003642

Gokmen Y, Turen U, Erdem H, Tokmak I (2021). National preferred interpersonal distance curbs the spread of COVID-19: a cross-country analysis. *Disaster Medicine and Public Health Preparedness*, 15(6):e20–e26. https://doi.org/10.1017/dmp.2020.295

Hackman JR, Oldham GR (1980). *Work Redesign.* Reading, MA, Addison-Wesle.

Herzberg F (1966). *Work and the Nature of Man.* New York, Thomas.

Herzberg F (1968). One more time: how do you motivate employees? *Harvard Business Review*, 81:87–96; reprint 87507 HBR 1987 and 2003.

Kroemer AD, Kroemer KHE (2017). *Office Ergonomics: Ease and Efficiency at Work*, 2nd ed. Boca Raton, FL, CRC Press. https://doi.org/10.1201/9781315368603

Landy FJ, Conte JM (2006). *Work in the 21st Century: An Introduction to Industrial and Organizational Psychology.* Malden, MA, Blackwell (7th edition in 2024 from Wiley)

Levitt SD, List JA (2009). Was there really a Hawthorne Effect at the Hawthorne Plant? An analysis of the original illumination experiments. National Bureau of Economic Research, Cambridge, MA, Working Paper 15016. https://doi.org/10.3386/w15016

Maslow AH (1943). A theory of human motivation. *Psychological Review*, 50(4):370–396. https://doi.org/10.1037/h0054346

Maslow AH (1954). *Motivation and Personality.* New York, Harper & Row.

McCambridge J, Witton J, Elbourne DR (2014). Systematic review of the Hawthorne effect: new concepts are needed to study research participation effects. *Journal of Clinical Epidemiology*, 67:e267–e277. https://doi.org/10.1016/j.jclinepi.2013.08.015

Muchinsky PM, Culberson SS (2016). *Psychology Applied to Work*, 11th ed. Summerfield, NC, Hypergraphic Press.

Parsons HM (1974). What happened at Hawthorne? *Science*, 18:922–932. https://doi.org./10.1126/science.183.4128.92

Proctor RW, Van Zandt T (2008). *Human Factors in Simple and Complex Systems*, 2nd ed. Boca Raton, FL, CRC Press. https://doi.org/10.1201/9781420063608

Roethlisberger FJ, Dickson WJ (1939). *Management and the Worker.* Cambridge, MA, Harvard University Press. https://doi.org/10.4324/9780203503010

Smith TM (2007). *Job Satisfaction in America: Trends and Socio-Demographic Correlates.* Available at https://www-news.uchicago.edu/releases/07/pdf/070827.jobs.pdf

Stanton N, Hedge A, Brookhuis K, Salas E, Hendrick HW (2004). *Handbook of Human Factors and Ergonomics Methods.* Boca Raton, FL, CRC Press. https://doi.org/10.1201/9780203489925

Taylor F (1911). *The Principles of Scientific Management.* New York, Norton.

CHAPTER FOURTEEN

Working in an organization

For the smooth and successful operation of any business, a "good fit" is necessary not only between workers and their work tasks but also among workers and their employing organization. Other chapters of this book primarily discuss the ergonomics of the individual and the task—a type of microergonomics that focuses on optimizing the details (physical and cognitive) of the workspace. This chapter addresses the bigger picture: the macroergonomics of individuals within an organization. An organization can be considered a network of related elements that must work together to enable the whole organization to function well. Of course, a family running a small farm requires a different scale of organization compared to managing a firefighting or space flight team, running a hospital, or setting up a global manufacturing business. This chapter discusses various kinds of organizations and how to create a good place to work. We need to understand how we work within organizations in order to fit the task to the human.

Human–system integration

*Macroergonomics** focuses on how we design work systems and interactions within the organization to optimize workplaces and enhance the effectiveness of companies. Macroergonomics is also a sociotechnical approach that considers the shifting value system of workforces: in North America and Western Europe, for example, people now expect to have control over the planning and pacing of their work, and they want to exercise decision-making responsibilities. Many employees enjoy broadly defined jobs that provide feelings of responsibility and

DOI: 10.1201/9781032674575-19

accomplishment. The so-called human–system integration (HSI) approach acknowledges that, compared to conventional organizations, new work structures are less formalized and less controlled by defined and written procedures, rules, and detailed job descriptions; they are also less centralized than traditional bureaucracies because tactical decisions are delegated to lower-level supervisors and workers. These represent profound changes from traditional work systems.

14.1 Basics of the human-centered organization

Fitting together employees, work, and organization leads to smooth and efficient performance. Humans are at the center of the organization's operations because companies cannot exist solely by their strategy, structure, and machinery; instead, they are collections of living, working, and interacting people. Without humans, the organization is lifeless. People influence the organization and its functioning, and in turn, the organization affects these individuals. Understanding why people act the way they do—and how to keep them satisfied and interacting productively at work—is a fascinating puzzle that has engaged psychologists and behaviorists* for many decades. Some explanations were discussed in Chapter 13; others follow in this chapter.

The individual in the organization

The individual is the innermost and most critical asset of an organization. When a company's market value is assessed, "hard" assets like property, plant, and equipment are part of the equation, alongside sales and financial liabilities. However, companies with similar financial balance sheets regarding tangible assets may have vastly different market values, often due to "soft" human attributes such as customer base, reputation, and, importantly for this chapter, employee engagement. Human resources are a company's most valuable assets. However, this workforce consists of unique individuals. Familial upbringing, social environment, life experiences, and personalities all contribute to our differences. It is a complex proposition to assemble individuals into groups in a work environment and expect them to collaborate according to company directives, given the inherent tensions of a job: demanding or monotonous duties, long hours, and imposed relationships with colleagues. The organization has limited opportunities to select and train individuals so that, as employees, they will fit in; hence selection and training are common tasks in industrial psychology.

Rather than relying solely on these individual approaches, greater success can be achieved by structuring the organization and designing work tasks and conditions to accommodate and leverage various individual capabilities and aspirations. The organization can be tailored to fit the human.

Social contracts

Throughout history, the relationship between individual workers and their employers has fluctuated between extremes: from employees demonstrating dedicated devotion ("I'll work hard for them until I die") to those doing the minimum ("good enough for government work"), and from organizations providing near-parental nurturing to heartlessly exploiting. In some cases, these social contracts (not always written but well understood by all involved) have included appropriate pay, a secure workplace with no layoffs, and access to health and retirement care in exchange for reliable, thorough, diligent work, once aptly summarized by expressions like "A Krupp man forever" or "Once IBM, always IBM." For employees, these conditions fostered a sense of security and belonging, recognition, and appreciation, which in turn generated motivation to excel in their work, toiling carefully and thoughtfully in the interest of the company. For organizations, it could be costly to underwrite care for the employees, particularly in cases of illness or disability and after their working years, but loyalty was generally rewarded.

Organizational components

The schematic for a human-centered organization* shown in Figure 14.1 depicts the basic components of the organization that determine the roles of its human players. In this case, five main elements interact to define a company: its *strategy* (the

FIGURE 14.1 A basic human-centered organization. Adapted from Kroemer and Kroemer (2017).

company's plan for success), its *structure* (corporate hierarchies), its *conduits* (channels for allocating, controlling, and tracking corporate resources), its *guidelines and rules* including policies and procedures, and its *social climate* (employees' feelings about the company) *and culture* (behaviors and feelings within the company). All these elements affect one another, and changes in one will impact the others. They also all influence individual employees: their thoughts, feelings, behaviors, and, thus, their performance within and for the organization.

Strategy

14.1.1 Organizational strategy The diagram in Figure 14.1 shows strategy as one of the main components of an organization. *Strategy* is the plan—stated or implicit—that the company has adopted to achieve its goals: an organizational master plan. It guides the company's operations and determines specific tactics that the company will use. Ideally, all employees are aware of and aligned with this strategy. Examples of organizational strategies include a repair service that pledges to dispatch a plumber within 2 hours of a request, a grocery store that aims to offer a wider selection of fresh produce and gourmet items than competitors, and a manufacturer intending to produce the most fuel-efficient airplanes.

Structure

14.1.2 Organizational structure *Structures* outline the hierarchies within an organization. In larger traditional companies, they are usually depicted in detailed diagrams linking elements in the organizational chart: this can be on a functional level (showing which group has what responsibilities) or an individual level (which manager supervises whom). An organization's structure determines accountability and authority within its ranks; essentially, it defines the official relationships that exist between employees. Each level in a structured organization has its own degree of authority and responsibility; moving up in the hierarchy increases one's authority and responsibility.

Down/up/sideways relations

In its most basic interpretation, the structure of an organization determines each employee's position: the ways in which instructions come down from the top, how reports move upward, and how workers at the same level communicate with each other. In a family-run shop or in small groups, formal organizational structures are usually unnecessary because the divisions of labor and responsibilities develop naturally according to personal preferences and skills. However, formal down/up/sideways relations are crucial in more complex organizations, such as manufacturing plants, airlines, community hospitals, or city police departments.

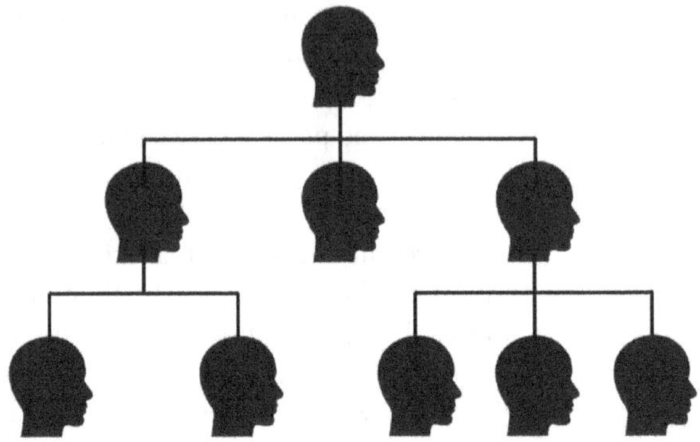

FIGURE 14.2 A basic layered organization with clear hierarchy.

Fixed structure

Volumes have been written on organizational designs*, ranging from rigid structures (traditional in the military) to loose conglomerates. The simplest and clearest structure follows the "unity of command" principle, where each employee is accountable to one boss, who (organizationally) is at the top of all decisions and functions. Figure 14.2 shows a basic example of such a conventional layered organizational chart. It depicts an organizational structure with several divisions or departments containing individuals at the same hierarchical level who perform specific tasks. Each department is headed by one person; department heads report to division heads; and finally, all division chiefs report to the top boss. Among the main advantages of this rigid structure is the clarity of communication up and down. However, employees and work projects may not thrive in this firm, hierarchical structure but rather may perform better in a more flexible network.

A common modification of the layered, hierarchical organization is to overlay a matrix structure, as illustrated in Figure 14.3. In a matrix set-up, workers may effectively have two (or more) bosses: one for their work function (such as financial management or production) and one for a project (such as developing a product, supporting a geographic region, or improving a manufacturing line). This allows the company additional flexibility to form work teams across functions and provides employees opportunities to learn new skills and gain insights. Disadvantages include potential confusion regarding roles and responsibilities and conflicting priorities between an employee's two (or more) bosses.

Overlapping assignments

Teamwork that includes job enlargement and enrichment (see Chapter 13) often does not thrive under a conventional, rigid,

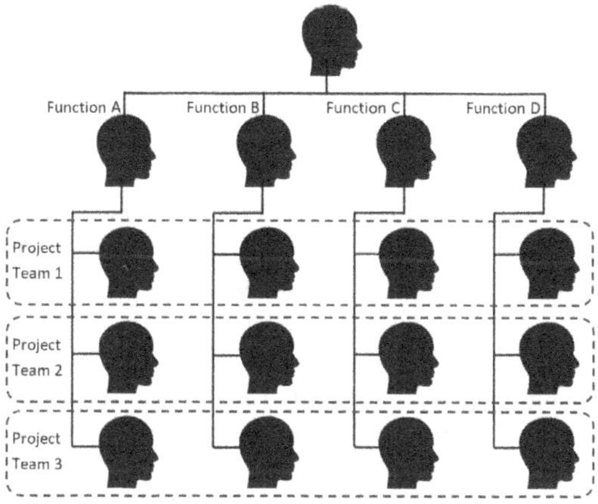

FIGURE 14.3 A matrix structure work organization.

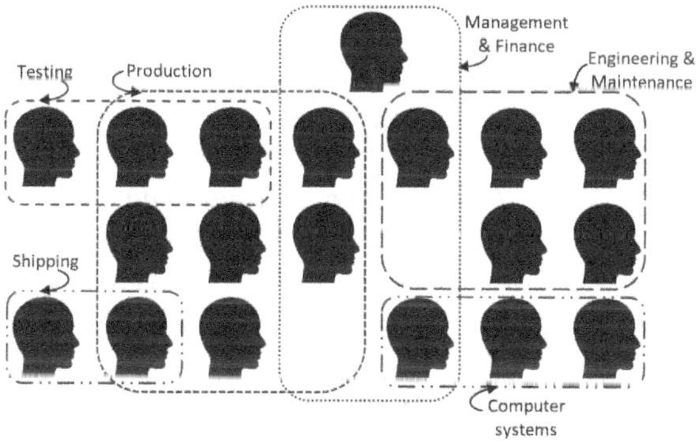

FIGURE 14.4 A flexible work organization with overlapping worker teams. Adapted from Nagamachi (2001).

layered setup. For example, originally an electrician was responsible exclusively for maintaining all electrical equipment; but as the company evolved into a more flexible, team-oriented organization, the electrician acquired additional tasks such as testing and designing new equipment, developing software, planning work, and even handling some financial budgeting and accounting. This illustrates how an employee can take on multiple roles that overlap the previously separate areas of maintenance, equipment, production, delivery, strategic planning, and financial management. Overlapping and multiple tasks can lead to a complicated amoeba-type organizational structure, sketched in Figure 14.4, which provides flexibility but complicates supervisory assignments.

Flexible organization

Loose and flexible organizational set-ups are necessary for autonomous teams, which self-organize to solve varying problems and meet changing tasks. This so-called employee empowerment can largely dissolve the traditional organization, eliminate managers, and abolish job titles. Self-governance* can represent a radical new system by transferring substantial authority to each employee, who must be well-informed, knowledgeable, skillful, and engaged to follow the organizational strategy set by the "big boss."

Employee self-governance

A few companies have decentralized their management and use self-organizing teams to distribute authority and decision-making: in this type of *holacracy*, the aim is to improve transparency and agility. This brings about its own set of challenges: for example, some managers may now feel disenfranchised, and some employees may be uncertain about how to exercise their new authority. Issues to be addressed include how to establish, communicate, administer, and implement company goals; how to prevent task overload; how to evaluate performance; and how to compensate each employee.

Conduits

14.1.3 Organizational conduits *Conduits* are the dispatch systems through which information flows within the organization. These channels ideally align with official structures that facilitate the allocation, control, and tracking of people, money, and resources (machines, equipment, supplies). Conventional organizations have clearly defined conduits that enable predictable asset distribution. However, these channels may become informal (sometimes termed "grapevines") if the organizational structure loosens. For example, in a self-governed business, who assesses employee performance and who allocates travel funds?

Guidelines and rules

14.1.4 Organizational guidelines and rules *Guidelines and rules* govern an organization's conduct. These policies and procedures prescribe how employees and supervisors interact. Examples include the types of medical and disability benefits offered, available retirement plans, the number of paid vacation days an employee receives, and the expected code of conduct for employees. In larger companies, policies are typically documented in an employee handbook. Some aspects of guidelines and practices may not be formally documented yet remain widely understood; for instance, how often emails are checked each day or whether employees may occasionally arrive late or leave early.

Conduits and rules interact

Breaking the organizational structure into components, as done in this chapter, helps specify topics and allows for separate consideration; however, in reality, these components interact. This is particularly true for organizational conduits and rules. The tighter the organization, the more regulated the communications and procedures, especially down and up within the hierarchical structure. Furthermore, the available means of interaction also influence the frequency and manner of communication. Consider the shift from many years ago to today: previously, a carefully edited, formally typed and signed report was delivered in person to the Herr Direktor by an appropriately dressed underling who had navigated through the antechamber of the esteemed Executive Secretary; today, an electronic memorandum, complete with inputs from involved colleagues and updated with links to the latest data, is sent with the push of a button to all who need to know, any of whom can review and comment instantly. Significant changes have arisen from portable computers and, especially, from ubiquitous mobile phones facilitating (and even provoking) constant use of communication media. Intraorganizational conduct, formally and informally, is largely determined by the available means of communication.

Social climate and culture

14.1.5 Organizational culture The norms, beliefs, unwritten rules, rites, rituals, behaviors, and practices among the members of an organization establish its *climate and culture*. These elements stem from underlying values that often remain submerged but surface in the behaviors of the system's members. A friendly social climate and culture have become increasingly important, especially in bureaucracies, as major determinants of employees' overall happiness and of organizational performance. (This was well-known but often disregarded by the "lean and mean" management style prevalent in the United States around the 2000s.) Today, with casual work environments, telecommuting, job-sharing, and an emphasis on individuality and participatory work, company cultures vary widely and appear more important to people than ever before.

The structural design of a company strongly influences its social climate and culture. Climate and culture are group phenomena that relate to the shared values, behaviors, beliefs, customs, and ideas held by employees collectively. This, in turn, affects individuals' emotional states and how they feel about the company, their coworkers, and their jobs.

Different cultures

Consider two advertising agencies: one is a large, highly traditional company headquartered in a high-rise building in* *a big city, while the other is a small "boutique" agency in a smaller town. The first agency, founded many decades ago, has many long-standing blue-chip clients and is highly structured, with formal dress codes, office procedure manuals, and written policies for all conceivable situations. The second agency, ten years old, has only a handful of young partners who employ creative individuals in an entirely unstructured office. Casual attire is standard, no handbooks exist, and decisions are made quickly and with virtually no bureaucracy.*

Clearly, although both companies operate in the same industry, they have very different cultures with widely diverse values and behaviors, leading to different climates. They also attract (or retain) employees with different personalities. Employees who begin working at either organization quickly learn about its culture and climate; over time, they either adopt the culture and become socialized into the broader group, or they become outsiders—perhaps even outcasts—and typically leave the company.

The culture defines the social rules within the organization that govern behavior, collaboration with others, and the performance of individual and team tasks. One might be tempted to believe that under harsh or high-stress conditions, such as working in a foundry or fighting a fire, social climate and culture would be unimportant; however, organizational culture may be even more crucial in critical functions to ensure smooth, safe, and efficient workflows. For example, if the co-pilot of a commercial airliner is inhibited by social climate and culture from speaking up to warn the pilot of a danger, or is disregarded by the pilot due to the co-pilot's lower position in the cockpit hierarchy, a terrible accident can result. Social interactions vary across different industries and organizations but remain vital to an individual's overall feelings toward the organization, their assessment of working conditions, and consequently, their motivation to work.

14.2 Addressing individual thoughts, feelings, and behavior

Chapter 13 discusses basic concepts of individual job behavior, motivation, and satisfaction. Maslow, Herzberg, Alderfer, and many subsequent researchers offer insights into why people act the way they do at work. This understanding provides guidance for creating organizational conditions that foster positive attitudes and behaviors, benefiting both the individual and the company.

Motivation and performance

Every worker holds beliefs about how their employer or workplace ought to function. We each have our own value system and prefer a job that meets our corresponding expectations. While we are willing to make some adjustments, we also recognize that to satisfy us, the job should motivate us to achieve personal professional goals. As discussed in Chapter 13, many behavioral theories hold that individuals strive to satisfy certain needs, and that this quest drives their behavior. Expectancy theories posit that an individual's motivation (and resulting satisfaction) is related to the gap between the worker's expectations and what the work environment offers. Motivation incites, directs, and maintains behavior towards goals. Motivation and job performance are interconnected; a motivated person who aims to succeed at work is willing to expend effort to do so. Performance is affected (and sometimes limited) by situational constraints at work, which are factors that can enhance or hinder performance. These situational constraints include the organization's climate (both psychological, as discussed above, and physical) and up-to-date equipment.

Satisfaction

Job satisfaction closely correlates with motivation, and several theories explain the degree of pleasure an employee derives from their job. Some approaches suggest that job satisfaction is determined within the individual, achieved when a person's physical and physiological needs are met, while other concepts focus on external factors of social comparison. Work conditions strongly influence job satisfaction; Herzberg's two-factor theory (see Chapter 13) suggests that certain job conditions (*context* or *hygiene factors*) like pay and the work environment can lead to dissatisfaction if perceived negatively. However, according to Herzberg, simply making these factors positive (such as providing good pay) does not result in satisfaction; only positive work *content factors* like achievement (and the subsequent recognition and pay increase) produce job satisfaction.

Stress

In the 1930s, Hans Selye introduced the term *stress** (discussed in Chapter 12). In his theory, stress can have a positive side: on the job, it is a condition in which a person feels compelled to improve a situation, enhance performance, and achieve greater satisfaction. Today, we typically refer to *distress*, as Selye called it: a strained personal condition that is nerve-racking and exhausting, caused by demands that exceed our coping abilities. Such stressors and the resulting stress they cause can occur at all organizational levels and in all jobs. Situations outside of work can cause stress, which then interacts with job-related stress.

Stress is an individual's physiological and psychological response to excessive demands. It can have severe consequences for both physical and mental health. Addressing negative stress is critical for employee well-being and, consequently, for the ongoing operational health of an organization.

Stressors

Since individuals react differently to demands, stressors vary in their impact on people. Generally, the biggest organizational stressors include task attributes, work environment, lack of recognition and support from supervisors, and unfair or bullying behavior from peers and bosses (as discussed in Chapter 12). Proper organizational policies and prudent actions by those in authority can eliminate many sources of stress.

14.3 Designing a good place to work

Satisfaction theories

As discussed in Chapter 13, needs-based and value-based theories propose that job satisfaction is an attitude that is specific to each individual. Every person has physical and physiological needs that they strive to fulfill to achieve satisfaction. A satisfying job meets physical needs (such as food and shelter) through appropriate income and safety, and meets psychological needs (such as self-esteem and intellectual stimulation) through job opportunities and professional recognition.

Other theories focus on social comparison. They postulate that people assess their own feelings of job satisfaction in relation to others; by inferring others' feelings about their jobs, they compare themselves to colleagues in similar roles. Work conditions are the focus of the third group of theories that attempt to explain job satisfaction, with Herzberg's two-factor theory being the most well-known.

Ultimately, the various theories contribute to an overall understanding of what kind of job is necessary to keep employees satisfied.

The best place to work

As a summary of the various factors underlying motivation and job satisfaction, it may be best to consider comments from workers themselves. One US-based company that continually makes the list of the 100 Best Companies to Work For has offered these reasons for its success: "This is a positive place. People are friendly … they feel challenged. They feel respected and valued. And they respond with loyalty." The company's philosophy includes: "Treat others the way they want to be treated. Strive for mutual respect and foster an atmosphere that makes people proud to work here. Provide career opportunities. Say thank you for a job well done."

Quality of life at and off work

Striking a balance between life at work and life outside the job is, of course, of primary importance to the individual; but this has become an issue for employers as well. A leanly staffed company in a tight labor market is likely to create a situation in which employees are overloaded with work. Technological developments that connect employees to work around the clock, such as mobile phones and handheld computers, also contribute to the burden many employees feel. Employers recognize that overworking employees may lead to reduced productivity and increased employee turnover; consequently, organizations may take measures to prevent employees from feeling overwhelmed at work. These measures include limiting work hours, providing time-off free of work tasks and phone calls, encouraging lengthy and uninterrupted vacations, offering employee sabbaticals, limiting meetings, and allowing flexible schedules: a work–life balance for employees is essential for a company's continued success.

Macroergonomic principles

Applying macroergonomic principles can improve productivity and work–life quality. This requires consideration of interactions between psychosocial, cultural, and technical factors within the overall organizational system. Here are some macroergonomic concepts from a company in Japan*:

- Body movement and thinking on the job stimulate alertness.
- A complex job enriches skills and motivation and generates satisfaction when completed.
- Flexibility and freedom of choice in task execution result in efficiency and productivity.
- Opportunities for decision-making create a sense of responsibility that leads to performance-driven motivation and satisfaction.
- An organization that promotes participation enhances its members' self-development and self-realization, thereby advancing its own success.

Summary

An organization cannot function solely on its strategy, structure, and machinery; it is powered by its working, thinking, aspiring, and interacting employees. People influence the organization; they determine how it operates—and, in turn, the organization impacts these individuals. A "good fit" between individuals, work, and the organization allows the system to run smoothly. Motivating employees is a complex and highly desirable goal. Many theories seek to explain motivation, which can be roughly divided into two categories: one focuses on individual traits, while the other emphasizes the immediate environment and the overall work organization.

Today's educated and informed employees expect to be acknowledged and valued as skilled and resourceful individuals. They seek recognition and respect for their efforts. They anticipate better-designed jobs, greater participation in decision-making, and the freedom to work in a less formal organization. They desire a pleasing environment and wish to work collaboratively with colleagues as peers and managers at various levels within the organization. These expectations foster a HSI that is less hierarchical, less formalized, and less centralized than in the past.

A human-centered design for the whole organization and an ergonomic design of the work environment can shift attitudes away from the aloof and disengaged "I just work here" to the excited and all-in approach "I want to do my best, this job needs me."

Steps to fit the human

Step 1: Determine the organization's plan for work to be done, in accordance with its overall organizational strategy.

Step 2: Structure the organization to utilize people's skills and aspirations. Create policies and procedures that provide opportunities for employees to develop skills and pursue their ambitions.

Step 3: Encourage a work climate where the members of the organization can work the way they "like to do," as this supports the strategic goals. Ensure good management, relationships, and work conditions.

Step 4: Be innovative, flexible, and adaptive.

Step 5: Monitor conditions, modify the work (or the organization), and iterate as necessary.

Notes

The text contains markers, *, to indicate specific references and comments, which follow.

Macroergonomics: Hendrick and Kleiner (2005).

14.1 Basics of the human-centered organization

Psychologists and behaviorists: Greenberg (2010), Landy and Conte (2006), Stanton, Hedge, Brookhuis et al. (2005), Vercruyssen and Hendrick (2011), Wilson and Corlett (2005), Wilson and Sharples (2015).

Schematic for human-centered organization: Based on material in Kroemer and Kroemer (2017).

14.1.1 Organizational strategy

Organizational designs: Booher (2003), Pew and Mavor (2007).

14.1.2 Organizational structure

Comprehensive self-governance: Use (2015).

14.1.5 Organizational culture

Advertising agencies: Based on material in Kroemer and Kroemer (2017).

14.2 Individual thoughts, feelings, and behavior

Selye's original stress theory: Selye (1956, 1978). Note that in this theory, a stressor causes stress.

14.3 Designing a good place to work

Concepts from Japan: Paraphrased from Nagamachi (2002).

References and further reading

Booher HR (2003). *Handbook of Human Systems Integration.* New York, Wiley.

Greenberg J (2010). *Behavior in Organizations: Understanding and Managing the Human Side of Work*, 10th ed. Upper Saddle River, NJ, Prentice-Hall.

Hendrick HW, Kleiner BM (2005). *Macroergonomics: Theory, Methods, and Applications*, 2nd ed. Boca Raton, FL, CRC Press.

Kroemer AD, Kroemer KHE (2017). *Office Ergonomics. Ease and Efficiency at Work*, 2nd ed. Boca Raton, FL, CRC Press. https://doi.org/10.1201/9781315368603

Landy FJ, Conte JM (2006). *Work in the 21st Century: An Introduction to Industrial and Organizational Psychology*. Malden, MA, Blackwell (7th edition in 2024 from Wiley).

Nagamachi M (2002). Relationships among job design, macroergonomics, and productivity. Chapter 6 in Hendrick HW, Kleiner B (eds), *Macroergonomics. Theory, Methods, and Applications.* Bora Raton, FL, CRC Press, pp. 428. https://doi.org/10.1201/b12477

Selye H (1956, revised 1978). *The Stress of Life*. New York, McGraw-Hill.

Stanton N, Hedge A, Brookhuis K, Salas E, Hendrick HW (2004). *Handbook of Human Factors and Ergonomics Methods*. Boca Raton, FL, CRC Press. https://doi.org/10.1201/9780203489925

Use J (2015). Are bosses necessary? *The Atlantic*, October issue:28–32.

Vercruyssen M, Hendrick HL (2011). *Behavioral Research and Analysis*, 4th ed. Boca Raton, FL, CRC Press.

Wilson JR, Corlett N (2005). *Evaluation of Human Work*, 3rd ed. London, UK, Taylor & Francis.

Wilson JR, Sharples S (2015). *Evaluation of Human Work*, 4th ed. Boca Raton, FL, CRC Press. https://doi.org/10.1201/b18362

CHAPTER FIFTEEN

Working hours and sleep

We have all experienced how work affects sleep and how sleep affects work. Parents with young children are busy all day and often during the night. Students may struggle to complete school assignments late into the evening. Farm work has always been associated with long working hours, typically starting at sunrise and ending at sunset or later. During the industrial revolution around 1800 in Europe, both home-based handwork and factory work often lasted 10 to 12 hours a day, six or even seven days a week. Today, in communities around the world, emergency responders like firefighters and medical personnel are on-call for 12 hours a day, although much of that time is—they hope—spent waiting; they must be ready for any emergency.

The human body changes its physiological functions throughout the 24-hour day. During waking hours, the body is naturally prepared for physical and mental work; during the night, it relaxes and sleeps. Attitudes and behaviors also change regularly during the day. New time signals and unusual periods of activity and rest (such as when starting shift work, see Chapter 16) can disrupt the daily rhythm and affect work performance and health. We need to understand how we perform and respond to working hours and sleep (or lack thereof) to better fit the tasks to the human.

15.1 Basics of circadian body rhythms

Body rhythms

Daily rhythms are natural temporal programs that regulate the lives of animals and plants. In humans, functions such as body temperature, heart rate, and blood pressure follow a set of daily recurring fluctuations called *circadian** (or *diurnal*) rhythms,

DOI: 10.1201/9781032674575-20

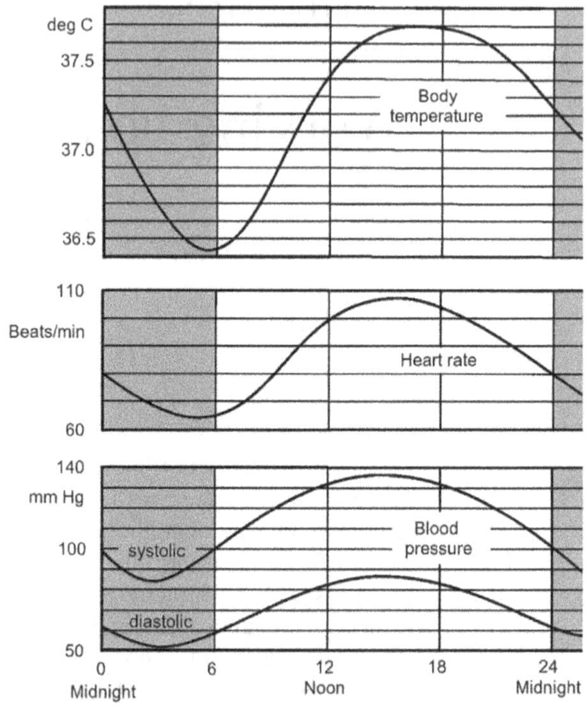

FIGURE 15.1 Typical variations in nominal body functions over the day. Adapted from Colligan and Tepas (1986).

as illustrated in Figure 15.1. These rhythms show higher values during the day and lower values at night. Several rhythms intertwine with each other, such as core body temperature, blood pressure, and sleepiness.

Daily rhythms

For most of us, during the day, we are awake and active; we eat and drink. At night, we sleep and fast. Physiological events do not exactly follow this general pattern. For example, core body temperature falls even after we have been sleeping for several hours, usually reaching its lowest point before six o'clock in the morning. Core temperature then rises quickly when we get up and continues to increase, with some variations, until late in the afternoon, before starting to fall again as we prepare to sleep. This illustrates that body temperature is not just a passive response to our daily activities, such as waking up, eating meals, working, and enjoying social interactions. This circadian physiological rhythm is established and self-regulated. Another daily rhythm is skin temperature, which increases with the onset of sleep, regardless of when that occurs. Hormones also fluctuate on a daily cycle: melatonin (associated with sleepiness) rises as bedtime approaches, whereas cortisol (which regulates metabolism) peaks as we are waking up.

24-hour cycles

When a person is completely isolated from external time markers and events, such as the darkness of night and the light of day, and no regular activities are set, the internal body rhythms are said to be "running free" and under internal control* only. Many experiments have consistently shown that circadian rhythms continue while running free; however, their time periods are slightly longer than the regular 24-hour duration: earlier research estimated 25 hours, while more recent studies suggest 24.2 hours. These experiments indicates that body rhythms can persist independently of external stimuli and follow their own internal clocks. However, when subjected to daily 24-hour activities and associated time markers, the internal rhythms synchronize to a common 24-hour cycle.

Individual differences

Some people have consistently shorter (or longer) free-running circadian rhythm periods than others. Statistically, those with shorter periods are likely to be "morning types" who rise early from sleep, while those with longer internal rhythms are probably "evening types" who prefer to stay up late. Women may have a slightly shorter average intrinsic circadian period than men. Rhythm amplitudes, such as body temperature, usually decrease with age, which could explain the shift towards early morning activity among the elderly. Additionally, oscillatory controls seem to lose some of their power as we age, indicating a greater susceptibility to rhythm disturbances with increasing age.

External and internal events

Under regular circumstances, there is a well-established phase coincidence between external activities and internal circadian events. For example, during the night, low values of physiological functions, such as core temperature and heart rate, are primarily due to the body's diurnal rhythm; however, they are also influenced by the body's nighttime inactivity and fasting. During the day, peak activity usually coincides with high values of the internal functions. Thus, the observed diurnal rhythm is typically the result of external (*exogenous*) and internal (*endogenous*) events that largely coincide in time. If that balance of concurrent events is severely disturbed, well-being, health, and performance decline.

Synchronizing biological clocks

The regulation of alertness, wakefulness, sleepiness, sleep, and many physiological functions appears to be under the control of two internal clocks*: one regulating sleep and wakefulness, and the other regulating physiological functions such as body temperature. Under normal conditions, these internal clocks are linked so that body temperature and other physiological activities increase during wakefulness and decline during sleep. However, this congruence of the two rhythms may be disrupted by night-shift work, where one must be active during nighttime and sleep during the day (more on shift work in Chapter 16). As such patterns continue, the physiological clocks adjust to the external requirements of the new sleep–wake regimen, leading to a flattening of the formerly well-established physiological rhythm. Within a period of up to 2 weeks, the rhythm re-establishes itself according to the new sleep–wake schedule.

Resetting biological clocks

Strong and persistent external events can reset the regular circadian rhythms. This commonly occurs when we travel quickly and far to the east or west. Upon arrival at a new location, we initially experience "jet lag" (also called *desynchronosis*): our body continues to function according to its typical timing for our previous time zone, which now differs from our new location. The local time settings for daylight and darkness, wakefulness and sleep, working hours and meal times all act as time markers. Their regular repetition resets our internal biological clocks, regulating and resetting our diurnal rhythms.

Airline crews and other short-term visitors to a different time zone may decide to simply stay on their home time setting until they return to their regular time zone. However, this is impractical for longer-term moves to a new time zone, where realignment is preferrable. This readjustment usually takes at least three days: the larger the time difference, the longer the adjustment period needed to overcome the rhythm lag.

The body needs sleep?

15.2 Basics of sleep

While it is obvious that humans need sleep (after all, we spend about one-third of our lives asleep), exactly why our bodies need sleep is not clear*. We know that sleep affects our health, and that a chronic lack of sleep increases the risk of health problems including heart disease, diabetes, and depression. There is a general opinion that sleep has recuperative benefits, allowing some sort of restitution or repair of tissue or the brain following the "wear and tear" of wakeful activities. However, what is meant by restitution or repair is usually not clearly expressed nor fully understood. Certainly, sleep is accompanied by rest and, consequently, by energy conservation. However, when not forced to be active, a human can attain similar relaxation during wakefulness without actually sleeping. Regarding the restitution of the body, it is an everyday experience that muscle aches resulting from a day's effort are gone the next morning; yet, they might disappear just as well during a long rest while remaining awake. Limited sleep deprivation does not impair the physiological ability to perform physical work. Apparent reductions in physical capability, blamed on sleep deprivation, may be mostly due to reduced psychological motivation. The lack of experimental findings regarding the physical benefits of sleep is somewhat surprising because this challenges common experience.

The brain needs sleep

In contrast, the restorative benefits of sleep for the brain are well established. The brain needs sleep in order to consolidate memories. Partial sleep deprivation affects attention and vigilance, whereas total sleep deprivation affects memory, attention, and decision-making. Two or more nights of sleep deprivation result in psychological performance detriments, particularly reduced motivation to perform (but apparently not a reduction in inherent cognitive capacity), behavioral irritability, suspiciousness, slurred speech, lack of attention, drowsiness, memory problems, and other performance reductions. Lack of sleep diminishes mental performance less on stimulating and motivating tasks than on boring, repetitive, and unrewarding tasks (see also Chapter 16). More than two nights of sleep deprivation reduce all task performances. These performance degradations indicate impairments of the function of the central nervous system due to sleep deprivation, hence a need for the brain to sleep. The effects of sleep deprivation are also influenced by gender and aging, as well as large differences between individuals.

Sleep phases

The brain and muscles are the human organs that show the largest changes from sleep to wakefulness: their electrical activities have allowed for well-established observation techniques. Electrodes attached to the surface of the scalp can pick up the electrical activity of the brain (encephalon). Thus, the name of this measuring technique is electroencephalography (EEG). The EEG signals provide general information about brain activity. Another common technique records the electrical activity associated with the muscles that move the eyes and those in the chin and neck regions. The recording of eye activity is called electrooculography (EOG). Both techniques (already mentioned in Chapter 9) are often applied together to record and describe events during sleep.

EEG during sleep

The amplitude of EEG signals is measured in microvolts; the amplitude rises as consciousness falls from alert wakefulness through drowsiness to deep sleep. EEG frequency is measured in cycles per second, called hertz (Hz). The frequencies observed in human EEGs range from 0.5 to 25 Hz. Frequencies above 15 Hz are called fast waves; frequencies below 3.5 Hz are called slow waves. Frequency decreases as sleep deepens; it is the slow-wave sleep (SWS) that is of particular interest to researchers.

Certain EEG frequency bands are labeled with Greek letters. The main bands of interest are typically described as follows:

- *Beta*, fast waves above 14 Hz: These fast waves of low amplitude (under 10 μV) occur when the cerebrum is alert or even anxious.
- *Alpha*, between 8 and 14 Hz: These frequencies occur during relaxed wakefulness when there is little information input to the eyes, particularly when they are closed.
- *Theta*, between 3.5 and 8 Hz: These frequencies are associated with drowsiness and light sleep.
- *Delta*, slow waves under 3.5 Hz: These are waves of large amplitude (often over 100 μV) and occur more often as sleep becomes deeper.

Additionally, certain occurrences in the EEG waves have descriptive names, such as vertices, spindles, and complexes, which are regular occurances associated with specific sleep characteristics.

REM and non-REM sleep phases

EOG outputs from the eye muscles often serve as the main descriptors of sleep episodes. Accordingly, sleep is divided into two types: sleep associated with rapid eye movements (REM) and non-REM sleep. Non-REM sleep is further subdivided into three stages according to their associated EEG characteristics*. These sleep stages are listed in Table 15.1. We cycle through these stages of sleep several times each night. Each cycle generally takes about 90 minutes, but this can vary widely between individuals and sleep cycles. We also tend to experience longer

Table 15.1 Stages of sleep

Sleep stage	Condition	Brain EEG	Sleep percentage
Wake	Awake	Alpha and beta waves	–
REM	Sleeping, often dreaming	Alpha and beta waves	25%
Stage 1 Non-REM	Transitional light sleep	Theta waves; loss of alpha; sharp vertex waves	5%
Stage 2 Non-REM	Deeper sleep	Theta waves; few delta waves; sleep spindles; k-complexes	45%
Stage 3 Non-REM	Deepest sleep	Delta waves; slow-wave sleep	25%

and deeper REM sleep closer to the morning. As we age, we spend less time in stage 3 and REM sleep and more time in stage 2 sleep.

- *REM sleep*: this is when most dreaming occurs. Eyes move rapidly under closed eyelids. Heart rate and EEG increase to near waking levels, and breathing may be fast and irregular.
- *Stage 1 non-REM sleep*: the drowsy transition from being awake to light sleep, usually lasting just a few minutes. Heart rate and breathing slow, and muscles relax. EEG begins to slow.
- *Stage 2 non-REM sleep*: the period of deeper sleep, where we spend most of our time while asleep. Heart rate and breathing continue to slow, muscles relax further, and body temperature drops. Brain electrical signals (EEG) slow but show brief bursts of activity.
- *Stage 3 non-REM sleep*: the deep sleep we need to feel refreshed upon waking. Heart rate and breathing are at their slowest. EEG slows even more (Figure 15.2).

Normal sleep requirements

Certain age groups exhibit fairly regular sleeping hours: newborns typically sleep 16–18 hours a day, young children 12 hours, and young adults sleep on average about 7.5 hours (though they may need more). Some adults feel well-rested after about 6 hours of sleep, while others habitually require 9 hours or more. Although some people can function on just a few

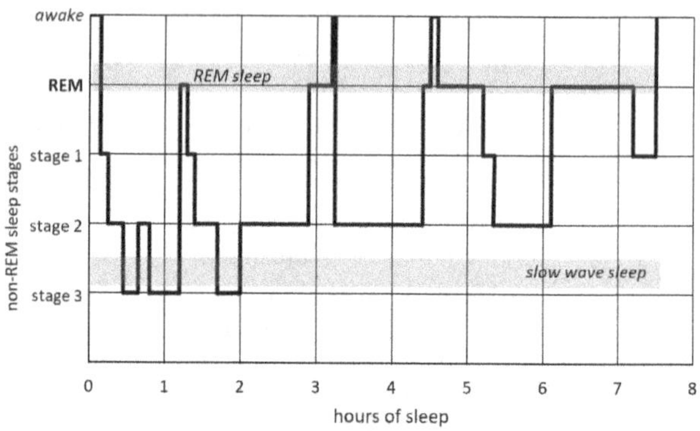

FIGURE 15.2 Typical sleep stages during an undisturbed night's sleep.

hours of sleep per day, many are able to maintain their performance levels even if the total sleep time attained is shorter than desired. This lower limit on sleep seems to be around 5 hours per day, but even shorter periods can still be somewhat useful. The amount of SWS (stage 3) in both short and long sleepers is about the same, but the time spent in REM and non-REM sleep periods differs considerably. Individuals who naturally sleep less than 3.5 hours are very rare among middle-aged people, with no true non-sleepers having ever been found among healthy individuals.

Sleep loss and tiredness

The brain assumes a physiological state during sleep that is unique to this state and cannot be attained during wakefulness. Muscles, for example, can rest during relaxed wakefulness, but the cerebrum remains in a condition of "quiet readiness," prepared to act on sensory input without any decrease in responsiveness. Only during sleep do cerebral functions show marked increases in thresholds of responsiveness to sensory input. In the deep sleep stages associated with slow-wave non-REM sleep, the cerebrum appears to be functionally disconnected from subcortical mechanisms. The brain needs sleep to recuperate, a process that cannot occur sufficiently during waking relaxation. However, it seems not all the regular full sleep time is essential for brain recuperation. Following sleep deprivation, usually not all lost sleep is reclaimed; a sleep period that is up to 2 hours shorter than usual can be endured for many months without negative consequences. Evidently, it is the first 5–6 hours of regular sleep (which contain most of the slow-wave non-REM sleep and at least half the REM sleep) that are necessary to maintain psychological performance at normal levels.

Sleep deprivation

Traveling through several time zones, then being ready to get some sleep but forced to stay awake, can put us in a difficult situation: we need sleep. Although it's not clear why sleep is necessary for the body, we all know that performing everyday tasks can be very difficult while deprived of sleep.

There are occasions that demand continuing work for long periods, such as a full day or longer. This not only involves being busy for extended period of times, with the associated mental and physical fatigue, but also encompasses loss of sleep. Hence, the negative effects on operator well-being and performance associated with extended working shifts result partly from the prolonged work itself, but they are also due to the lack of sleep.

Performing tasks while sleep-deprived

Sleepiness and wakefulness appear in unpredictable waves; thus, work performance may be fine for a while and then become difficult and faulty. Sleepiness has little effect on the performance of demanding work tasks that are interesting and appealing, especially if they last less than half an hour, even if they involve making complex decisions. Accuracy in performing a job may still be quite good even after losing sleep, but it takes longer to complete the job. Self-paced work deteriorates less with sleepiness than externally paced tasks.

However, sleep loss negatively affects performance on work that must be done over longer time periods, particularly if the task is repetitive, monotonous, or complex. Decision-making takes longer if the task is disliked and unappealing. Long-term and short-term memories degrade when people are required to stay awake for extended periods of time.

15.3 Addressing rest pauses and time off work

Given the natural circadian rhythms in physiological functions, one expects corresponding changes in attitudes and activities throughout the day. The daily organization of getting up, working, eating, relaxing, and sleeping also strongly affects attitudes, work habits, and work performance. In controlled experiments, it is possible to separate the effects of internal and external daily rhythms, but this is generally not feasible in our everyday lives.

For practical purposes, it is reasonable to consider the overall results of the combined internal and external factors as they affect workload and work output.

Performance changes during daylong work

For many people, the morning hours are optimal for strenuous activities, both mental and physical, because "fatigue" resulting from work already performed may reduce performance as the day progresses. Often, performance shows a pronounced decrease early in the afternoon; this phenomenon is frequently referred to as a "post-lunch dip." However, there appear to be no similar changes in physiological functions; for example, body temperature remains unchanged at that time of day. Therefore, it seems reasonable to speculate that the interruption of activities by a noon meal, along with subsequent digestive processes and reduced psychological arousal, expectation, and habits, would account for this reduction in performance. In contrast, in medium to heavy physical work, no such post-lunch dip occurs except when food and beverage intake is very heavy and when true physiological fatigue has built up during the pre-meal activities.

Long working hours

Some individuals are capable of and even enjoy sustained heavy physical labor, while others prefer lighter physical tasks or mental activities throughout the workday; the types and intensities of activities impose varying demands on the worker. Section III of this book (Chapters 10–12) discusses how to assess actual workloads and consequently how long people can perform such work. If metabolic, circulatory, or respiratory work demands exceed sustainable limits, even with interspaced rest periods, the work must be terminated. At lower levels of physical demand, work can proceed smoothly, even over extended periods of time, as long as rest pauses can be freely chosen.

Breaks needed

Mental activities such as visual searches or information processing, and tasks requiring concentrated attention, may be interesting and generate alertness—or they may be overtaxing and fatiguing—depending on their intensity and duration, as well as how they align with individual traits and motivation. Individuals tasked with performing monotonous mental tasks that are low in novelty, interest, and incentive, or that are high in complexity, need breaks every half hour or so; otherwise, performance declines sharply. Accordingly, tasks that are monotonous or complex should not be assigned to workers over long periods of time. Varying or new tasks are more resistant to deterioration if they require decision-making and problem-solving or deal with information that is particularly interesting or rewarding.

Rest pauses

In both physical work and mental activities, suitable breaks allow us to literally "catch our breath" and refocus our minds. Pauses and off-work time enable relaxation, recovery, and reloading. The working human body needs to maintain a balance between energy consumption and energy replenishment: put simply, a balance between work and rest. This "supply–use–supply" process is an integral part of the operation of muscles, heart, brain, and of the organism as a whole. Rest pauses are indispensable physiological and psychological requirements if performance is to be maintained.

There are four types of work pauses:

1. Spontaneous pauses,
2. Disguised pauses,
3. Work-caused pauses, and
4. Management-prescribed pauses.

Spontaneous pauses are taken by workers on their own initiative to interrupt the flow of work in order to "take a break and recharge"; this may include napping*. These pauses are in response to effort and are needed for recuperation. As discussed in Chapters 10 and 11, as a rule, more frequent and shorter pauses provide greater recuperative value than fewer, longer ones.

Disguised pauses are those during which the worker engages in an activity other than the main job; for example, replacing work tools or tidying the workplace in between performing the primary work tasks. Many of these disguised pauses serve to cover a necessary spontaneous pause. It would be better to take these pauses openly so that the necessary recovery from work can be fully realized.

Work-caused pauses are interruptions that arise from the work itself, such as waiting for a machine to complete an operation or for a tool to cool down, or they may stem from a breakdown in the flow of work. These breaks are beyond the worker's control, even if they occur with regularity, and they still provide time to recuperate.

Management-prescribed pauses are wisely implemented so that workers can step away from the job and rest openly, for example, to allow time for eating and other personal needs. As a rule, during shifts lasting up to 8 hours, one long break of 30 minutes or longer should be provided in the middle of the work period, and two shorter breaks, about 15 minutes each, should break up the two halves of the working time.

Pauses are physically or psychologically helpful (or even necessary) to maintain performance throughout the entire work shift. Management should encourage taking spontaneous pauses; disguising a break with another activity diminishes the recovery value of the time off. Since not all work activities will include work-caused pauses, implementing management-prescribed breaks is not only a good idea but is often required by local employment laws.

15.4 Designing daily and weekly working time

One prerequisite for our well-being is the maintenance of our circadian rhythms in spite of external disturbances. This state of balanced control, called *homeostasis*, regulates our body functions, makes us sleepy or awake, enables us to work, establishes our inclinations to engage in certain activities, and influences our social behavior.

Days on/off work

Most people are accustomed to working during daylight. Physiologically, we should be able to do so every single day, provided that we have suitable working conditions, including rest breaks. However, tradition and societal customs have led us to expect regular days free from work: in many regions, this occurs every seventh day. In most of Europe, for example, Sundays have been designated as off-work days since at least the mid-1800s; work-free Saturdays followed in the mid-1900s. Since then, a 4-day workweek has been widely discussed. The conventions of specific sequences of working days followed by days off have spurred various theories and experiments* to determine suitable numbers of working hours per day in relation to working days during the week (discussed below) and regarding shift work (discussed in Chapter 16).

History of working time

Weekly and daily working times are closely related. In Switzerland, the Factory Act of 1877 stipulated a 65-hour workweek, with 11 hours of work each weekday and 10 on Saturday. In 1914, a legal amendment shortened the weekly work to 48 hours. In the United States, non-agricultural jobs averaged around 70 hours per week in the 1850s, which shortened to around 50 hours per week by 1940. Although working time in industry has been around 40 hours per week since the early 1960s, in the United States nearly 20% of working adults work 48 hours per week, and 7% work at least 60 hours per week. Global trends are shown in Figure 15.3, with the caveat that data gathered before and after 1950 use different assumptions*. However, the

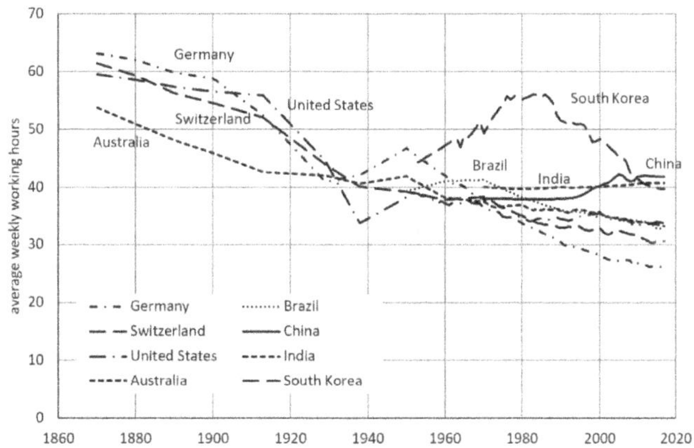

FIGURE 15.3 Non-agricultural weekly working hours in countries around the world. Adapted from https://ourworldindata.org/grapher/annual-working-hours-per-worker and Huberman and Minns (2007) and Penn World Table PWT 9.1 (2019).

trend is clear: for most of us (except in post-World-War-II South Korea), the average number of hours we work each week is less than that of our (great-) great-grandparents. About 40 hours of labor in a working week are common in most industrialized regions on Earth, although a 6-day workweek is still in place in some countries.

40-hour workweek

The reduction of the weekly working time to 40 hours, and even less in some countries, was often associated with a changeover from a 6-day workweek to 5 or even 4 days of work each week. Employees and employers generally both welcomed that reduction in workdays; it often came with a slight increase in productivity and a reduction in absenteeism. It is of socio-economic interest that this reduction in working hours and days allowed, in some countries, companies to hire more workers and thus reduce widespread unemployment.

8-hour workday

It seems that dividing 24 hours into three, approximately equal sections of time fits the current lifestyle in many societies. This set-up is commonly accepted because it offers about 8 hours for work, reserves 8 hours for sleep, and still provides 8 hours for family and social interactions (as well as all the chores and errands needed for daily life). As mentioned above, it is appropriate to interrupt the 8-hour work shift about halfway with a long break of at least 30 minutes, typically used for taking in a meal. A truly long break could be used for an extended "siesta" around noon, as is traditional in some countries with hot climates. Each of the two blocks of working time should be split by another scheduled break lasting perhaps 15 minutes, and workers should be encouraged to take additional pauses as desired and needed (as discussed as part of management-prescribed pauses above).

The 8-hour work shift of has become widespread: it suits organizational purposes in trade and industry, especially because the 8-hour work period allows for the establishment of a 3-shift schedule that covers the 24-hour day; however, a 2-shift arrangement with 12-hour work periods could also achieve this—see more discussion of shift arrangements in Chapter 16.

Brief work periods

Work that is very demanding, mentally or physically, may be difficult to maintain for the customary 8 hours of working time a day, even with suitable breaks for recovery; instead, 6 hours of work may be more acceptable. Even in such shortened work arrangements, a substantial break around mid-time is generally required, and each of the two halves of the working period should be interrupted by one scheduled and management-prescribed break, with additional pauses taken as needed.

Long work periods

Long work shifts may be scheduled for various reasons: for example, big-city firefighters spend much of their 10- or 12-hour work schedules waiting for the call of duty. In automated plants, supervisory presence ensures that the process runs smoothly with occasional human intervention as needed. Twelve-hour shifts may be desired to fill the 24-hour day (more about shift arrangements in Chapter 16). However, working periods of 10 or 12 hours are not suitable if they require continuous effort from the worker. Humans cannot perform physical or mental work at a consistently high rate over long periods of time without experiencing physical or mental fatigue, along with its accompanying negative effects on well-being and work performance: lapses in attention are unavoidable, mistakes are likely, and accidents are possible.

Flextime

Flexible working hours, often called *flextime*, are a recent form of work organization. Its specific feature is that a person can distribute the required work time, say 40 hours per week, over the official workdays as long as certain requirements are met, including completing assigned duties or being available during prescribed core hours (e.g., from 9 to 12 in the morning). *Sliding time* (or *gliding time*) is a subgroup of flexible work hours in which the start of the daily working hours can be slid (adjusted) by the worker. As with weekly flextime, certain requirements must be met, such as being present at pre-defined times for meetings. Flextime is often offered as part of a hybrid work arrangement comprising both remote and in-person (at the worksite) work. Flextime arrangements have generally been popular with employees and employers, though not all types of work are suitable for flexible arrangements. Furthermore, in nearly every organization, there are groups of employees who must be present at the worksite during all conventional hours, such as receptionists, cafeteria workers, and manufacturing assembly workers organized in groups.

Emphasis on performance

Working at home, especially in a home office usually associated with computer use, offers the most flexible arrangements. The individual controls the number and scheduling of working hours; the primary requirement is to complete the assigned work on time. These flextime arrangements abolish much of the traditional control over working hours by the employer and put more responsibility on the individual worker. The emphasis is on performance, not on hours spent on the job. Yet, overly ambitious individuals might spend more time working than they would if they were on regular schedules. For example, companies that offer "unlimited" time off have often found that workers take fewer days off than if they had a set number of days, possibly due to uncertainty about how much vacation is really meant to be taken.

Compressed workweek

Flextime is sometimes associated with *compressed workweeks* that have more working hours per day but fewer workdays per week. For example, the customary number of weekly work hours is 40, divided into 5 days with 8 hours of daily work. This can be consolidated into 4 days per week, with each day consisting of 10 hours of work. This allows the worker to have three free days instead of two each week. This arrangement is attractive to many employees and employers for several reasons: there are more work-free days (preferably taken together in one block), the number of commuting trips to and from work is reduced, and there are fewer set-ups and shutdowns at work. However, there are also concerns about increased fatigue due to long workdays and reduced performance and safety (see Tables 15.2 and 15.3).

The type of work to be performed primarily determines whether compressed workweeks can and should be used. Long shifts have mostly been employed in cases where one is

Table 15.2 Benefits and consequences of flex time

Positive (benefits)	Negative (consequences)
• Appeals generally to employees and employers • Makes work-free available time at the employee's choosing • Does not result in reduced employee pay • Reduces commuting and traffic problems • Less fatiguing for workers • Increases job satisfaction • Recognizes and utilizes employees' individual differences • Reduces tardiness • Reduces absenteeism • Reduces employee turnover • Increases performance	• Makes it difficult to cover jobs at all required times • Makes it difficult to schedule meetings or training sessions • Reduces communication within the organization • Increases energy and maintenance costs • Requires more sophisticated planning, organization, and control • Requires adjustment in management objectives • Requires special recording of work time • Requires additional supervisory personnel

Source: Adapted from Knauth (2007), Kroemer Elbert, Kroemer, and Kroemer Hoffman (2018).

Table 15.3 Benefits and consequences of extended workdays/compressed workweeks

Positive (benefits)	Negative (consequences)
• Appeals generally to employees and employers • Makes available more work-free days for family and social life • Provides more time per day for scheduling meetings or training sessions • Generates fewer start-up and warm-up expenses • Does not result in reduced employee pay • Reduces commuting and traffic problems	• Decreases job performance • May require overtime pay • Tends to fatigue workers • Increases tardiness and early departure from work • Increases absenteeism • Increases on-the-job and off-the-job accidents • Increases energy and maintenance costs • Requires adjustment in management objectives

Source: Adapted from Knauth (2007), Kroemer Elbert, Kroemer, and Kroemer Hoffman (2018).

"on standby" much of the time and thus not engaged in mentally or physically fatiguing work for the entire shift. Additionally, activities that require only minimal physical effort, especially if they are diverse and interesting yet fall into routines, have been handled in long shifts. Examples of compressed workweeks include nursing, clerical and administrative work, technical maintenance, computer supply operations, and supervision of automated processes. There is not much data on the use of long shifts in manufacturing, assembly, and machine operations. Furthermore, the existing information mostly stems from subjective statements of employees, results of psychological tests, and scrutiny of performance and safety records in the industry.

12-hour workdays?

Working very long shifts, such as 12 hours, is likely to lead to fatigue, cause drowsiness, reduce cognitive abilities and motor skills, and generally decrease performance during the course of the work shift. This effect becomes more pronounced as the workweek progresses. There is potential for a fatigued worker to take careless shortcuts to complete a job. Thus, work practices may become less safe in highly repetitive tasks or those that are tedious due to high cognitive or information-processing demands.

Summary

Human physical and psychological functions, as well as behaviors, follow a set of regular and predictable daily fluctuations: the circadian (diurnal) rhythm. Under regular living conditions, this temporal program is well established and persistent. Disrupting the rhythm with strong external time markers and events, such as working in a different time zone, can lead to reduced performance and well-being.

Regular work shifts should ideally be scheduled during daylight. The suitable length of the working time per day primarily depends on

- The type and intensity of work and how these relate to worker capacities, and
- The ways work is organized within societal and social customs.

A working time of 8 hours a day is generally acceptable for psychological, physiological, social, and organizational reasons. This workday set-up is commonly accepted because it offers about 8 hours for work, reserves 8 hours for sleep, and still provides 8 hours for personal, family, and social interactions. During the 8-hour work shift, scheduled and freely taken rest periods are necessary. Under certain circumstances, shorter or longer daily work times are reasonable, as long as the fatiguing aspects of the work tasks are considered.

Steps to fit the human

Step 1: Determine what work is to be done and what the suitable length of the workday is.

Step 2: Consider whether fixed or flexible arrangements are desirable.

Step 3: Provide for breaks.

Step 4: Monitor conditions, modify the work schedule, and iterate as necessary.

Notes

The text contains markers, *, to indicate specific references and comments, which follow.

15.1 Basics of circadian body rhythms

Daily rhythms: Called *circadian* from the Latin *circa*, about, and *dies*, the day; *diurnal* from *diurnus*, of the day.

Rhythms "running free" under internal control: Folkard and Monk (1985).

Model of two internal clocks: Horne (1988, 2006), Foster and Kreitzman (2004).

15.2 Basics of sleep

Why we need sleep is not clear: Horne (2006), Siegel (2003)

Importance of REM and non-REM sleep: Horne (1988, 2006). Note that sleep stages are now typically classified as REM and three stages on non-REM sleep; some older references list a fourth stage of non-REM sleep that has now been included within stage 3, collectively called SWS and which is the deepest stage of sleep.

15.3 Designing rest pauses and time off work

Napping: Driskell and Mullen (2005).

15.4 Designing daily and weekly working time

Theories and experiments to determine suitable working hours: Discussed by Hornberger, Knauth, Costa et al. (2000), Folkard, Lombardi, and Tucker (2005), Foster and Kreitzman (2004), Popkin, Howarth, and Tepas (2006), Refinetti (2016), Silva and Costa (2023).

Weekly working hours: Global data from online (internet) resources such as OurWorldInData.org (https://ourworldindata.org/grapher/work-hours-per-week) which references Huberman and Minns (2007), Penn World Table PWT 9.1 (2019); US numbers are also available from Centers for Disease Control and Prevention NHIS (2015). Different sources use different definitions and criteria (e.g., including part-time work and time-off work in the calculations of the average of hours worked).

References and further reading

Centers for Disease Control and Prevention (2015). NIOSH, National Health Interview Survey (NHIS) Report dated 5 Aug 2015. https://wwwn.cdc.gov/NIOSH-WHC/chart/ohs-workorg?OU=*&T=OU&V=R

Colligan MJ, Tepas DI (1986). The stress of hours of work. *American Industrial Hygiene Association Journal*, 47(11):686–695. https://doi.org/10.1080/15298668691390485

Czeisler CA, Gooley JJ (2007). Sleep and circadian rhythms in humans. *Cold Spring Harbour Symposia on Quantitative Biology*, 72:579–597.

Driskell JE, Mullen B (2005). The efficacy of naps as a fatigue countermeasure: a meta-analytic integration. *Human Factors*, 47:360–377.

Folkard S, Lombardi DA, Tucker PT (2005). Shiftwork: safety, sleepiness and sleep. *Industrial Health*, 43(1):20–23. https://doi.org/10.2486/indhealth.43.20

Folkard S, Monk TH (1985). *Hours of Work*. Chichester, UK, Wiley.

Foster R, Kreitzman L (2004). *Rhythms of Life: The Biological Clocks that Control the Daily Lives of Every Living Thing*. London, UK, Profile.

Hornberger S, Knauth P, Costa G, Folkard S (2000). *Shiftwork in the 21st century*. Frankfurt, DE, Lang.

Horne J (1988). *Why We Sleep – The Functions of Sleep in Humans and Other Mammals*. Oxford, UK, Oxford University.

Horne JA (2006). *Sleepfaring: A Journey through the Science of Sleep*. Oxford, UK, Oxford University Press. https://doi.org/10.1016/j.eeh.2007.03.002

Huberman M, Minns, C (2007). The times they are not changin': days and hours of work in old and new worlds, 1870–2000. *Explorations in Economic History*, 44(4):538–567.

Knauth P (2007). Extended work periods. *Industrial Health*, 45(1):125–136. https://doi.org/10.2486/indhealth.45.125. Erratum in: *Industrial Health*. 2007 Apr;45(2):375.

Kroemer Elbert K, Kroemer HB, Kroemer Hoffman AD (2018). *Ergonomics: How to Design for Ease and Efficiency*, 3rd ed. London, UK, Elsevier. https://doi.org/10.1016/C2016-0-02516-0

Popkin SM, Howarth HD, Tepas DI (2006). Ergonomics of work systems. In Salvendy G (ed), *Handbook of Human Factors and Ergonomics*, 3rd ed. Hoboken, NJ, Wiley, pp. 761–800.

Refinetti R (2016). *Circadian Physiology*, 3rd ed. Boca Raton, FL, CRC Press.

Siegel JM (2003). Why we sleep. *Scientific American*, November issue, 92–97.

Silva I, Costa D (2023). Consequences of shift work and night work: a literature review. *Healthcare*, 11(10):1410. https://doi.org/10.3390/healthcare11101410

CHAPTER SIXTEEN

Working in shifts

Day and night

Physiological functions of the human body wax and wane rhythmically throughout the 24-hour day, as described in Chapter 15. During daylight, the body is prepared for physical and mental work; at night, sleep is expected. Our attitudes and behavior also fluctuate throughout the day, as does our social life. New time signals and unusual periods of activity and rest (such as when we have shifting work schedules) can disrupt our normal daily rhythms. To avoid negative health and social effects and to prevent reduced work performance, work schedules should be arranged to least disturb physiological, psychological, and behavioral rhythms.

Human physiological and psychological functions, as well as behaviors, follow regular and predictable daily fluctuations: the *circadian* rhythms. They are easy to observe in body temperature, heart rate, and blood pressure, as shown in Figure 15.1. These natural temporal programs keep our body and mind healthy*; they enable us to work and function in our social environment. The circadian programs intertwine: for example, core temperature and blood pressure are coupled with sleepiness and wakefulness. Self-sustained "internal clocks" control the rhythms, which run on a 24-hour cycle synchronized by our daily time markers: getting up, eating, working, and going to sleep.

The circadian rhythms show a high persistence under varying external conditions. However, strong and repeated external events and time markers can disrupt the regular circadian rhythms and disturb the daily balance of wakefulness and sleep, activities and rest, alertness and drowsiness. Such balance is a prerequisite for human performance and individual health, both physical and mental. Getting enough sleep is part of that requirement, as discussed in Chapter 15.

DOI: 10.1201/9781032674575-21

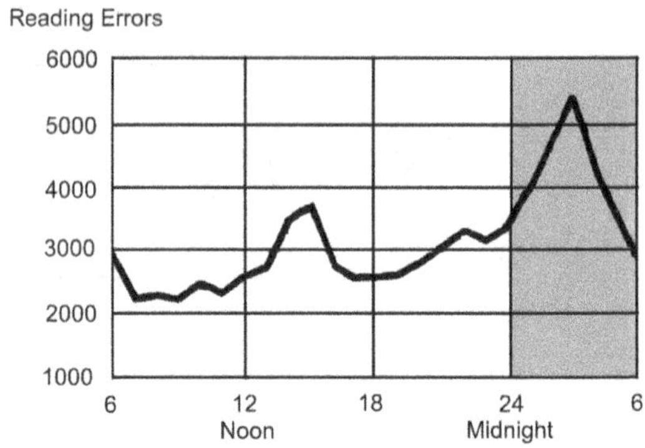

FIGURE 16.1 Number of errors in gas meter readings. Adapted from Lehmann (1953).

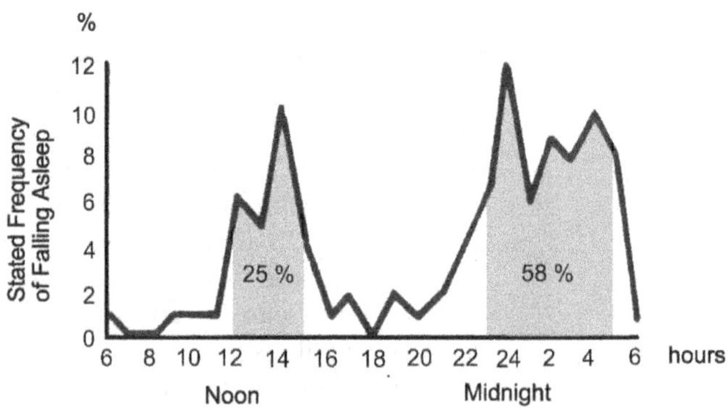

FIGURE 16.2 Frequency at which truck drivers fell asleep while driving. Adapted from Lehmann (1962).

Job performance at night

Performance levels are generally lower during nighttime activities, when the body and brain usually rest, than during the daytime. This has long been known: Figure 16.1 shows a summary graph of errors in the reading of gas meters recorded between 1912 and 1931. During that time, about 175,000 readings of gas meters were taken around the clock. The number of errors was by far highest around two o'clock in the morning. Another peak in errors, though noticeably lower, occurred after lunch. Similar relationships between mishaps and the time of day have been reported ever since. Figure 16.2, with data from the mid-1950s, illustrates the frequency at which truck drivers fell asleep at the wheel. This occurred most often after midnight and, again, around lunchtime. Sleep deprivation and its resulting drowsiness likely aggravate any naturally occurring fluctuations in performance. We need to understand how we perform and respond to shift work in order to fit the task to the human.

16.1 Basics of shift work

Defining shift work

Shift work involves two or more individuals, or teams of individuals, working in sequence at the same workplace. Often, each worker's shift repeats in the same pattern over a number of days. For the individual, shift work means either attending the same workplace regularly at the same time (continual shift work) or at regularly recurring times (discontinuous, rotating shift work). Operating with more than one staggered shift allows a facility to function outside of traditional business hours, including around the clock: for example, a housekeeping service that cleans an office at night, operators that keep a factory running 24 hours a day, or nurses that provide much-needed care for patients at all hours.

Important factors that define shifts and shift patterns* include

- The starting and ending time of a shift;
- The number of workdays in each week;
- The hours worked in each week;
- The number of free days per week or per rotation cycle;
- The number of consecutive days on the same shift (*cycle length*), which may be a fixed or variable number;
- The schedule by which an individual worker either works or has free days;
- The number of free days per week and per year; and
- The number of shift teams.

History of shift work

Shift work is not new. In ancient Rome, a decree required that deliveries be made at night to relieve street congestion. Bakers have traditionally worked through the late night and early morning hours to have bread ready for morning customers. Guards and firefighters have always been assigned night and day shifts. With industrialization, long working days became common, with teams of workers relieving each other to maintain blast furnaces, rolling mills, glass works, and other workplaces where continuous operation is desired. Covering the 24-hour period with either two 12-hour work shifts or three 8-hour work shifts became common practice. In the first part of the 20th century, the then-traditional 6-day workweeks with 10-hour shifts were shortened, as discussed in Chapter 15. Since the middle of the 20th century, many work systems use a shift arrangement of 8 hours per day during five workdays per week. Data from the US Bureau of Labor Statistics show that in 2017–2018, over 16% of the working population worked full time on evening shifts, night shifts, rotating shifts, or other employer-arranged irregular schedules. This is consistent with data from other industrialized societies, where approximately 15%–20% of the labor force works shifts*.

Health concerns

With shift work, the matter of circadian rhythm adjustments and the fear of consequent negative health outcomes is an important topic in industrial medicine. The first large surveys of the consequences of shift work came from Norway around 1960. They showed that shift workers had significantly more digestive ailments and nervous disorders than their colleagues on regular day shifts. Many workers disliked shift arrangements and tried to return to regular day work.

Half a century later, these early concerns about the negative health effects of shift work are still valid. Working at unusual times, especially during the night, working long shifts, such as 10 or 12 hours, and changing the timing of work: these can all cause health problems, especially if these shift work conditions combine to cause extraordinarily long working hours and significant deficits in rest and sleep. Prominent on the list of concerns for human well-being are as follows:

- Interruption of natural physiological and psychological rhythms,
- Fatigue, drowsiness, or sleepiness,
- Increased risk of accidents, both on and off the job,
- Sleep disorders,
- Gastrointestinal disorders,
- Cardiovascular diseases,
- Nervous disorders,
- Cognitive decrements*, and
- Disturbances of family and social life.

Clearly, the main problems are physiological and psychosocial: *physiological*, in that such work needs to be performed at times when the body is set to sleep and rest; and *psychosocial* because work must be done at times that are usually reserved for family and social interactions and for leisure activities.

Ideally, it is best to schedule job activities to follow entrained body rhythms and to schedule work during daylight. If this is not possible, evening shifts are preferable to night shifts.

Societal expectations of free time

People working permanent evening or night shifts establish their own personal schedules for sleep, leisure time, and social networking. These schedules often conflict with societal habits that place a higher value on free time in the evening and on weekends than on free time at other times of the day or week. This creates a problem: if during the week you work nights and sleep during the day but on the weekend you try to spend time with family and friends during the day and sleep at night, you will be out of sync with your work shift rhythm. Hence, many overnight shift workers need to decide what schedule they, along with their families and friends, will follow.

16.2 Basic solutions for shift work

There are three basic solutions that help to avoid serious complications resulting from working shifts during the evening or night, or at other unusual times. The common aim of all three is to not upset the circadian rhythm and thus mitigate the negative consequences of shift work.

Just one odd shift The first solution is to insert just one odd work shift and then return to the normal schedule. This means muddling through that one unusual work shift (perhaps two shifts) and thereafter taking a good rest before returning to the regular work schedule. However, as discussed in the foregoing Chapter 15, getting through the unusual work period is fatiguing and strenuous and may seriously reduce work performance.

Permanent shift assignments The second solution is to adopt a permanent work arrangement at the "unusual" time. Often, there are people who volunteer to work continual evening or night shifts; they are willing to readjust their circadian rhythms and their habits of private and social interactions permanently. Such volunteers include, for example, bakers and night guards of olden times; today, they are firefighters, nurses and physicians, drivers of trains, buses, trucks and taxis, workers on offshore oil rigs or road construction, and many others who agree to be on evening or night shifts for weeks, months, even years on end. These individuals readjust their physiological and social rhythms fully and permanently; therefore, they are less likely to suffer from disorders related to rhythm disturbances. Figure 16.3 sketches the schedules for permanent shifts starting in the morning, evening, or at night.

Shift–pause–shift The third solution entails a limited number of work shifts (say, 5 workdays a week) at the odd time, followed by an interruption

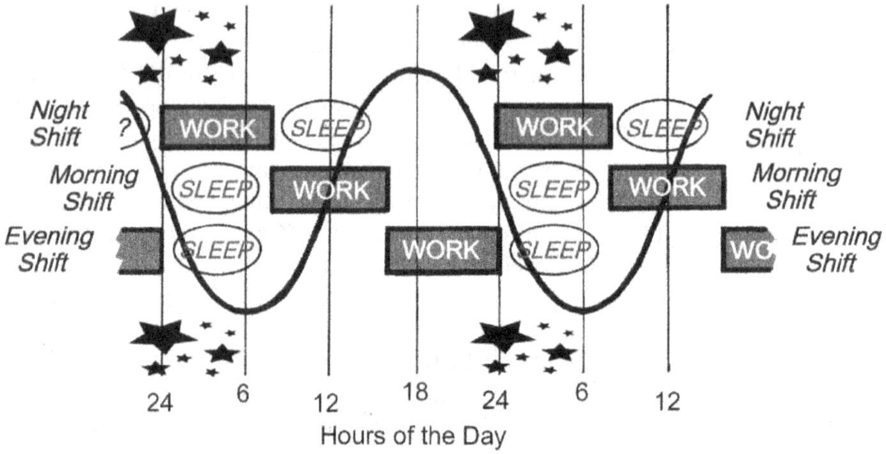

FIGURE 16.3 Typical work and sleep patterns with permanent shifts. Adapted from Kroemer, Kroemer, and Kroemer-Elbert (2020).

(often the weekend) and then continuing again with a limited number of work shifts at the same or another timing. Although this schedule is often used, it has two major repeating disadvantages: the interruption is likely to disrupt the adjustments in body rhythms and individual habits attained during the preceding work period, and the new set-up requires making additional adjustments. So, while seemingly "fair" when all involved workers are being shifted in the same manner, each of them suffers the same negative consequences.

16.3 Designing shift patterns

For practical reasons, shift work is difficult to avoid in many organizations. Shift work appears in several basic patterns, but any selected shift system may comprise aspects of several patterns. Five particularly important considerations are listed below:

- Do shifts extend into hours that would normally be spent asleep?
- Do shifts cover the entire 7-day week?
- What days of rest, such as a free weekend, do shifts include?
- How long is each shift?
- Do shift crews rotate, or do they work the same shift permanently?

All of these aspects are of concern regarding the welfare of the shift workers and their work performance, and they affect organizational scheduling.

Rotating shifts

In terms of organizing the work staffing schedule, it is easiest to establish a permanent set-up or a schedule that rotates weekly. The literature discusses many such solutions, most of which rely on traditional societal expectations, especially regarding weekends, that originated in Europe and North America. In most systems used there today, the same 8-hour shift is worked for 5 days, usually followed by two free days on the weekend. This regimen, however, does not cover all 21 (7 days × 3 8-hour workdays) shift periods of the week, thus additional crews are needed to work on weekends.

One approach to cover all of the 21 shift periods of the week is called the metropolitan rotation. This system uses three shifts a day (day D, evening E, night N) with four teams. The shift pattern (for each team) is D-D-E-E-N-N-free-free, yielding a ratio of six work days to two free days per cycle, and a cycle length of 8 days. Another approach is the *continental rotation*, which also assumes three shifts per day and four crews. This shift pattern has the sequence D-D-E-E-N-N-N, free-free-D-D-E-E-E, N-N-free-free-D-D-D, E-E-N-N-free-free-free, yielding a work/free day ratio of 21/7 and a cycle length of 4 weeks. Note that for this rotation, each set of night shifts is followed by free days. This is consistent with the finding that one long night's sleep usually restores performance to a normal level, even after extensive sleep deprivation.

Table 16.1 shows a schedule for rotating shifts that seems to be particularly suitable for Europeans. Each person works the same shift twice, then rotates forward to the next later shift for

Table 16.1 Sample 6:4 rotation shift schedule, where D represents a day (morning) shift, E represents evening shift, and N represents night shift

	Day of the week							Day of the week						
Week	Mo	Tu	We	Th	Fr	Sa	Su	Mo	Tu	We	Th	Fr	Sa	Su
1, 2	D	D	E	E	N	N	*Free*				D	D	E	E
3, 4	N	N	*Free*				D	D	E	E	N	N	*Free*	
5, 6	*Free*		D	D	E	E	N	N	*Free*				D	D
7, 8	E	E	N	N	*Free*				D	D	E	E	N	N
9, 10	*Free*				D	D	E	E	N	N	*Free*			

Source: Adapted from Knauth (2007b).

two periods, then rotates forward again to the next set of two shifts. This is followed by four free days, which, however, do not always include weekend days. The system has a work/free day ratio of 6/4, and its cycle length is exactly 10 weeks; on average, each week has 33.6 work hours.

16.4 Designing suitable shift systems

There are two sets of criteria by which to judge the suitability of shift plans: one set focuses on the worker, while the other focuses on the organization. Humans are by nature accustomed to daytime activity, with the night reserved for rest. Working evenings or at night (and possibly on weekends and holidays) generates physical and psychosocial stress, which may be mild or severe, depending on the circumstances and the individual. Organizational criteria include the number of shifts per day, the length of each shift, overall coverage of the day and week by shifts, work on holidays, and work performance. Clearly, any shift work schedule is inherently a compromise among various, and occasionally conflicting, expectations.

Selecting shifts

The following points are considerations in the selection of one of the many possible shift patterns:

- Daily work duration (i.e., each shift) normally should not exceed 8 hours.
- Either minimize or maximize the number of consecutive evening or night shifts. In the first case, use the fewest possible number of consecutive evening or night shifts; it is best to intersperse only one late shift between day shifts. Alternatively, maximize the number of consecutive evening or night shifts by remaining permanently on the same late shift.
- A full day of free time should follow every single night shift or series of night shifts.

Each shift plan should include consecutive work–free days, preferably encompassing the weekend.

Making late shifts easier

Performance is influenced by four major interrelated factors: the body's internal diurnal rhythm, the daily organization of work activities, the worker's subjective motivation and interest in the work, and the type and conditions of work. If late work shifts are required, the employer can assist workers by recognizing these factors. Work can be organized so that physically or mentally demanding or critical tasks are performed during day or evening shifts, as physical and mental performance are particularly weak during the "low" periods of the circadian cycle, such as in the early morning hours. Work can also be structured to accommodate the fact that performance deteriorates during prolonged periods of work, especially if the worker is suffering from a lack of sleep.

During evening or night shifts, management should provide high illumination levels at the workplace to suppress the production of the hormone melatonin, which causes drowsiness. Furthermore, environmental stimuli can help to keep the worker alert and awake, such as occasional lively music and the provision of snacks and (possibly caffeinated) beverages. Interaction with co-workers, and even taking short naps, should be encouraged as the type of work allows. The work tasks should be interesting and rewarding, as boring and routine tasks are difficult to perform efficiently and safely during the night hours.

Coping strategies The shift worker can use coping strategies for setting the biological clock, obtaining restful sleep, and maintaining satisfying social and domestic interactions. For example, sleep should be taken immediately after a night shift, rather than in the afternoon just before the night shift. Sleep time should be regular and uninterrupted. Shift workers should seek their family's and friends' understanding of their rest needs. Certain times of the day should be set aside specifically and regularly to spend with family and friends.

Summary

Human body functions and human social behavior follow circadian rhythms. Shift work required at unusual times can disrupt these synchronized rhythms and the associated behaviors of sleep (usually during the night) and activities (usually during the day). Resulting sleep loss and tiredness can negatively affect the shift worker's health, diminish family and social life, and reduce work performance.

Shift work is often desired for organizational and economic reasons. When possible, work should be scheduled during daylight hours; if this is not possible, evening shifts are preferred over night shifts. Recommendations for acceptable regimes of shift work include limiting a shift to 8 hours; for evening or night shifts, either work only one evening or night shift before returning to day work or stay permanently on the same shift. Night shifts should be followed by work–free days.

Task performance is influenced by four major interrelated factors: the internal diurnal rhythm of the body, the external daily organization of work activities, subjective motivation and interest in the work, and the type and conditions of work. Work schedules and tasks can be arranged to consider and accommodate the expected deterioration in physical and mental performance due to body rhythms, prolonged work, and sleep loss. By providing interesting and rewarding work, interaction with co-workers, breaks (naps), snacks and beverages, and environmental stimuli such as bright illumination, the negative effects of shift work can be mitigated.

Steps to fit the human

Step 1: Determine what work is to be done, and seek to schedule this work only in daytime shifts.

Step 2: If two or three shifts a day are necessary, attempt to find volunteers for permanent shift assignments.

Step 3: If shift rotations are necessary, carefully lay out schedules that suit workers' preferences.

Step 4: Consider limiting work to 8-hour shifts and provide interesting and rewarding work.

Step 5: Monitor conditions, modify the work schedule, and iterate as necessary.

Notes

The text contains markers, *, to indicate specific references and comments, which follow.

These natural temporal programs keep our body and mind healthy: Costa (2010), Folkhard, Lombardi, and Tucker (2005), Folkard and Tucker (2003), Foster and Kreitzman (2004), Monk (2006), Refinetti (2016).

16.1 Basics of shift work

Other important identifiers of shifts and shift patterns: Costa (2010), Folkard and Tucker (2003), Knauth (2006, 2007a, 2007b), Popkin, Howarth, and Tepas (2006), Silva and Costa (2023).

Shift workers: US Bureau of Labor Statistics for 2017–2018 https://www.bls.gov/news.release/flex2.nr0.htm, Sweileh (2022).

Cognitive decrements: Long-term shift work impairs cognition, as shown by Marquie, Tucker, Folkard et al. (2013).

References and further reading

Costa G (2010). Shift work and health: current problems and preventive actions. *Safety and Health at Work*, 1(2):112–123. https://doi.org/10.5491/SHAW.2010.1.2.112

Folkard S, Lombardi DA, Tucker PT (2005). Shift work: safety, sleepiness and sleep. *Industrial Health*, 43(1):20–23. https://doi.org/10.2486/indhealth.43.20

Folkard S, Tucker P (2003). Shift work, safety and productivity. *Occupational Medicine (London)*, 53(2):95–101. https://doi.org/10.1093/occmed/kqg047

Foster R, Kreitzman L (2004). *Rhythms of Life: The Biological Clocks that Control the Daily Lives of Every Living Thing*. London, UK, Profile.

Knauth P (2006). Workday length and shift work issues. Chapter 29 in Marras WS, Karwowski W (eds), *The Occupational Ergonomics Handbook: Interventions, Controls, and Applications in Occupational Ergonomics*, 2nd ed. Boca Raton, FL, CRC Press. https://doi.org/10.1201/9781420003642

Knauth P (2007a). Extended work periods. *Industrial Health*, 45(1):125–136. https://doi.org/10.2486/indhealth.45.125. Erratum in: *Industrial Health*. 2007 Apr;45(2):375.

Knauth P (2007b). Schicht- und Nachtarbeit (shift work and night work, in German). In Landau K (eds), *Lexikon Arbeitsgestaltung*, Stuttgart, DE, Gentner.

Kroemer KHE, Kroemer HJ, Kroemer-Elbert KE (2020). *Engineering Physiology: Bases of Human Factors Engineering/Ergonomics*, 5th ed. Heidelberg, Springer. https://doi.org/10.1007/978-3-030-40627-1

Lehmann G (1953). *Praktische Arbeitsphyiologie* (2nd ed. 1962), Stuttgart, DE, Thieme.

Marquié JC, Tucker P, Folkard S, Gentil C, Ansiau D (2015). Chronic effects of shift work on cognition: findings from the VISAT longitudinal study. *Occupational and Environmental Medicine*, 72(4):258–264. https://doi.org/10.1136/oemed-2013-101993

Monk TH (2006). Shiftwork. Chapter 32 in Marras WS, Karwowski W (eds), *The Occupational Ergonomics Handbook: Fundamentals and Assessment Tools for Occupational Ergonomics*, 2nd ed. Boca Raton, FL, CRC Press. https://doi.org/10.1201/9781420003635

Popkin SM, Howarth HD, Tepas DI (2006). Ergonomics of work systems. In Salvendy G (eds), *Handbook of Human Factors and Ergonomics*, 3rd ed. Hoboken, NJ, Wiley, pp. 761–800.

Refinetti R (2016). *Circadian Physiology*, 3rd ed. Boca Raton, FL, CRC Press.

Silva I, Costa D (2023). Consequences of shift work and night work: a literature review. *Healthcare*, 11(10):1410. https://doi.org/10.3390/healthcare11101410

Sweileh WM (2022). Analysis and mapping of global research publications on shift work (2012–2021). *Journal of Occupational Medicine and Toxicology*, 17:22. https://doi.org/10.1186/s12995-022-00364-0

SECTION V

Applied human engineering

So far in this book, in the first four sections, we have addressed specific categories of knowledge related to the human body and mind, abilities, and limitations. In many cases, the discussions also included the effects of the physical and work environment on human performance.

The fifth section of this book applies this detailed knowledge. Using what has been discussed in the previous chapters, we examine several examples of intricate multipart design, such as homes and workplaces. These are typical cases of human engineering (also called human factors engineering), which require gathering and using information from various specialized knowledge areas. We recognize and address differences in where and how people live and work so that we can design tasks, gear, equipment, and procedures to fit their specific needs. Therefore, the following chapters provide relevant information for ergonomic design:

- Chapter 17, Design of the Home
- Chapter 18, Design of the Office and Office Furnishings
- Chapter 19, Design and Use of Computers
- Chapter 20, Design of the Workplace

DOI: 10.1201/9781032674575-22

- Chapter 21, Load Handling
- Chapter 22, Healthcare for Patients and Providers
- Chapter 23, Emerging Ergonomic Issues with Artificial Intelligence
- Chapter 24, Designing Work to be Efficient and Pleasant

CHAPTER SEVENTEEN

Design of the home

Different homes

A home suitable for living in Norway would not serve its purpose well in Egypt; a community house from Fiji would look strange in Turkey. Different house designs provide shelter from diverse climates, and their occupants may have rather divergent opinions about comfort and privacy. Accordingly, there is not just one type of dwelling that is "right" from an ergonomic point of view; instead, there are many good solutions, depending on climate, availability of materials, needs of the inhabitants, and societal expectations.

This chapter discusses human engineering solutions for homes suitable for moderate climates. Many of these concepts, designed to make living easy and safe, grew from habits originally developed in Europe and therefore reflect "Western" architectural traditions, expectations, and lifestyles. Climates, living conditions, and customs are quite different across the many regions and cultures on earth. Still, many of the ideas presented here have universal appeal.

The main purpose of having a home is to be sheltered from unpleasant or dangerous components of the environment. Once this is achieved, other attributes* come to the forefront: features such as privacy, comfort, convenience, pleasure, and aesthetics. Different inhabitants are likely to have differing opinions about the relative importance of these attributes, but from an ergonomic point of view, the usability of the design of the home is key. For young and healthy persons, rather unusual designs might be quite attractive, such as spiral staircases and lofts with minimal railings. However, such features may be difficult to use, laborious to maintain, potentially dangerous, especially for pregnant women or young children, and impractical for elderly or impaired persons.

DOI: 10.1201/9781032674575-23

17.1 Basic considerations for design of housing

Overall, the design of homes should consider fundamental ergonomic perspectives to ensure they are user-friendly and functional.

- *Anthropometrics*: Understand the dimensions of diverse body types to fit the spaces and furnishings.
- *Adaptability*: Provide flexible designs to address varying preferences and abilities, and consider using materials that are durable, sustainable, and eco-friendly.
- *Accessibility*: Ensure spaces (including passages) can be accessed by all, ensure furnishings are comfortable and functional.
- *Lighting, heating, and cooling*: Optimize climate conditions to suit the use of the space.

To be user-friendly and functional, we must understand who the user is. For example, when a user has a specific physical or mental need, disability, or impairment that needs to be accommodated, we may adopt a *human* (or *user*)-*centered design* approach. We can also use the approach of *universal design*, also called *design-for-all*, which involves creating an environment that is both aesthetically pleasing and usable by everyone to the greatest extent possible, regardless of age, gender, ability, or status in life.

17.2 Designing for households with children

Child-proofing

Most homes serve, usually for many years, as a safe haven for children and their parents. This requires the design of the home's interior to be child-friendly, which primarily involves ensuring safety for young children. Examples of "child-proofing" include eliminating stairs (or at least blocking them off), avoiding protruding hard and sharp objects, and keeping hot items that can burn skin, such as cooking surfaces in the kitchen or scalding hot water in the bath, out of reach. Of course, all these measures also make the home safer for everyone else.

Designing for pregnancy

During pregnancy, many everyday tasks become more difficult for the mother-to-be. Most of these challenges arise from reduced reach and mobility, often related to back pain. Due to the growing size of the abdomen, objects on the ground near the feet can be hard to see. Many women worry about stumbling or falling, especially if their sense of balance is affected. Another common symptom of pregnancy is frequent urination, which necessitates convenient access to a suitable toilet and washroom.

In general, the later stages of pregnancy are accompanied by a reduced ability to perform tasks that require significant exertion, extend over long periods, and/or involve considerable mobility (such as bending low or reaching far). In responding to a survey in which pregnant women listed their most challenging activities, the following were commonly noted as difficult:

- Picking objects up from the floor,
- Reaching high shelves,
- Walking upstairs,
- Household chores such as ironing,
- Getting in and out of bed,
- Using a car (getting in and out, using seat belts, reaching the steering wheel, driving), and
- Using public toilets.

Overall, the responses (excluding those related to car use and public facilities) indicate that a well-considered overall layout of the home, along with specific design measures related to everyday household activities, contributes to creating an environment that supports pregnant women while also serving all inhabitants*.

17.3 Designing for impaired and elderly persons

When a resident is ill or recovering from an injury, the home serves as a temporary care facility, which may need to accommodate a wheelchair, a special bed, or even a gurney. These medical accommodations may also be required on a permanent basis, depending on the needs of the residents. The use of such devices makes careful house layout and design highly important, especially concerning the hallways, bedroom, toilet, and bath. Having wide doorways allows for a wheelchair to pass. Installing handrails and slip-resistant flooring promotes safer passage. Allowing extra room around the toilet facilitates a caregiver's assistance. Having grab bars near toilets, baths, and showers provide more secure options for standing and sitting.

The same carefully incorporated design features also accommodate aging occupants* who are losing some of the physical and mental abilities they had in their younger years. Continuing to live in one's own home has the major advantage of being in a familiar setting with all its physical and emotional implications. These advantages include feeling at home, being comfortable, enjoying desired levels of privacy, maintaining a routine, and having the satisfaction of self-sufficiency and independence; all of these contribute to a better quality of life.

Remodeling

Unless designed with prudent foresight, private homes are commonly ill-suited for aging residents. Often, alterations are necessary in passage areas, kitchens, bathrooms, and bedrooms. Forward-looking considerations (also called "future-proofing") when building or acquiring a dwelling can save much effort and expense since reworking an existing residence can be substantial, difficult, and costly.

17.4 Designing access, walkways, steps, and stairs

In the United States, perhaps the best-known house expressly built for use by a person with restricted mobility is the Top Cottage in Hyde Park, New York, which President Franklin Delano Roosevelt (1882–1945) designed to accommodate himself in his wheelchair. Fortunately, most people can walk about freely and do not consistently require a wheelchair. However, designing a house for a wheelchair user also makes it convenient for every inhabitant, disabled or not, including aging individuals who are no longer as agile and strong as in their youth. Thus, a (real or imaginary) wheelchair is a good tool to assess the suitability of dwellings.

Passageways

Passages to, from, and within a dwelling must be safe and comfortable for even a frail person. The walking surface should be flat, without barriers such as stairs or thresholds, and ideally not sloped. Doors and passageways should be wide enough to allow a wheelchair to pass and turn. Flooring should provide enough friction for safe stepping even when wet and should not have loose rugs or mats. Straight hallways are easier to navigate than those with turns and corners. Passageways must be well illuminated, as should be all other rooms of the dwelling.

No steps

Flights of stairs, steps, and even thresholds often make it impossible for wheelchair users to roll up, down, or over them. They create difficulties for everyone and cause many stumbles and falls, often resulting in serious injuries. Even low sills at doorways or shower stalls can be a nuisance and lead to stumbling and tripping, as can the edges of carpets and loose rugs. Lifts and elevators facilitate movement if a residence spans several stories. Elevators connecting the ground floor to the next one are fairly simple and relatively inexpensive to install, especially when the dwelling is designed to accommodate them and installation is done early during construction or remodeling*.

Doors and windows

Doors and windows must be easy to open and close, even when an additional insect screen or storm door or window is present. They require clear space in front to ensure access. All exterior doors should swing open to the outside to allow for a quick escape in case of a fire. Controls must be accessible and require little strength to operate while still providing security. Push bars and lever handles are easier to use than round knobs, especially for wet or weakened hands. In some cases, a transparent or translucent curtain can replace certain interior doors.

17.5 Designing kitchen and food preparation area

Kitchen

The kitchen is one of the most important and frequently used rooms in the house for many people. Although the kitchen's primary function is to store, prepare, and serve food, it often also serves as a gathering room, a social space, and a communication center. Many kitchen design concepts* follow the idea of a "work triangle." Its corners represent the three primary areas of food-related activity: *storing* (such as the refrigerator and cabinets), *preparing* (such as the stove or cooktop with gas or electric burners, conventional or microwave oven, coffee/tea maker), and *cleaning* (such as the sink, dishwasher, and garbage disposal). Several classic human factors principles, augmented by newer ergonomic findings and evolving alongside modern living, apply to kitchen design:

- Kitchen design and components should facilitate the flow of work at and between the most used activity areas, including serving prepared food and managing used utensils.
- Items should be stored near their point of use.
- Cabinets and other storage facilities should be within comfortable reach, and the contents should be visible; therefore, shelves should be at or below eye height and deep enough that items located in front do not block those at the back.
- Retrieving items should be easy and not require excessive bending, stretching, or twisting. For low storage, the use of rollout shelves is advantageous.

- Using workspaces should be easy and not require excessive bending, stretching, or twisting. The workspace for the hands (at counters, sinks, and cooking areas) should be at about elbow height or slightly below to facilitate manipulation and visual control. Counters and sinks may be lowered for wheelchair-bound individuals.
- The openings of the stove, oven, refrigerator, and dishwasher should be at heights that do not require the user to bend or reach excessively. Special shielding may be needed for cooking that involves very hot oil or surfaces (such as a wok).
- All counters should have a work surface that extends about 10 cm into the room beyond the base or at least have a recessed "toe space" so that the user can step in closer.
- Sufficient illumination needs to be available to provide for safe work.
- Traffic by others should not disrupt the workflow patterns.

17.6 Designing bedroom, bath, and toilet

Bedroom

Most of us spend about one-third of the day in the bedroom, and weak or ill persons spend even more time there. Hence, it is important to pay attention to its ergonomic features. The bed should be at a height that makes lying down and getting up comfortable. In the past, many different mattress properties have been promoted (especially for people with back problems), ranging from hard to pliant to soft, sometimes even adjustable; since no objective criteria seem convincing, users should select the support and design that suit their individual preferences.

The bedroom should be spacious enough to allow maneuvering space for occupants, potential caregivers, and even a wheelchair or specialized bed. It should provide easily reachable shelving and storage for clothing and bedding, as well as direct-access storage for medical supplies. There should be provisions for emergency access and an emergency exit. Very often, the bedroom should also provide privacy and be located near a bathroom or toileting facility.

Bathroom

Bathrooms deserve particular ergonomic attention because they are essential for healthy living. Basic equipment includes washing facilities (such as bathtubs, showers, and washbasins for the body, face, and especially hands) and toileting. Additionally, bathrooms usually contain storage facilities for toiletries, towels, and other supplies. Unfortunately, many traditional bathroom designs in the United States are difficult for elderly people and persons with disabilities (and their caregivers) to use. Major problems include narrow doors and tight spaces, making it difficult for users who need canes, walkers, and especially wheelchairs to move about. Similar to passageways and kitchens, flooring in bathrooms should provide enough friction for safe stepping, even when wet. Loose rugs or mats should be avoided, and proper lighting should be available.

Faucet controls

Using the control handles for hot and cold water faucets can be difficult for impaired persons, particularly when they have to reach across a tub or shower basin to access them. In some setups, both the controls for hot and cold water move in the same direction, while in other designs, the faucets turn in opposite directions; this is another example where standardization of human-engineered solutions would be advantageous. To prevent scalding, thermostatically adjusting the water temperature is a good ergonomic solution that is helpful and safer for all users.

Tub and shower

Bathtubs and showers, the two main areas for cleansing the whole body, are sites of numerous accidents. Most danger stems from the slipperiness of wet surfaces against skin, including bare feet. The most hazardous feature is a bathtub with slanted, slick surfaces. Getting into and out of a tub with high sides is not easy for most people and is particularly difficult for those who have balance and mobility deficiencies. Proper handrails and grab bars within easy reach can be of great help when getting in and out of the tub and shower.

A shower stall is easier to use than a tub because the shower's low enclosure rim at the floor poses less of a hindrance to stepping in and out. A rimless, no-threshold design with a large shower enclosure usually takes up more floor space but is the best solution for a wheelchair user and also allows the addition of a shower chair or bench.

Controls for hot and cold water should be reachable before entering the tub or shower, allowing appropriate water temperatures to be set safely by the user or caregiver.

Washbasin

Proper height of the washbasin is important. The basin may be difficult to use if it requires a far reach; for example, if it is inserted deep into a cabinet so that the user cannot step close to it but must lean forward over it. Large faucets and outlets should not reduce the usable opening area of the washbasin. Hot and cold water controls should be easy to operate.

Toilet

Throughout the world, different customs and installations are typical for designs used for the elimination of alimentary waste while keeping the body clean. For Western-style seated toilets, substantial literature exists about suitable design, sizing, shaping, height, location, and function. Proper handrails and grab bars aid persons with mobility problems, such as those caused by back pain, in sitting down and getting up. Personal hygiene systems, such as a water shower installed atop the toilet bowl or a bidet, can be of great help.

17.7 Designing lighting, heating, and cooling

Windows and lights

Windows are important to many individuals: not only do they provide light (and often ventilation) to the inside of the house, but they also offer visual and emotional connections with the outside world. However, the windows may be challenging to operate if they need to be awkwardly hand-cranked, slid, or locked. If natural light fails, electrical lamps supply illumination. These lamps should be arranged to light up what must be seen, yet not produce deep shadows or glare that can blind the eye.

Automatic lights are advisable in passageways, bathrooms, and bedrooms. Manual switches and all other controls, as well as electrical outlets, should be located at about hip height so that persons both standing and sitting, tall or short, can reach them easily. They should be easy to operate, ideally by a simple push, and not require fine motor skills or digital dexterity.

In North America, many table and floor lamps provide stunning examples of how to make it difficult, instead of easy, to turn lights on and off: they have push–pull or turn switches near the socket of the light bulb, hidden high up inside the lampshade, or somewhere along the electrical cord. Due to their location, they are hard to find by groping and tricky to operate due to the required arm bending and the awkward switch design. This is not an example of good ergonomic design. Using so-called smart light bulbs, controlled by voice or smartphone app, can be an improvement.

Heating and cooling

Houses of the affluent usually have electric or gas-fired heating systems, referred to as air-conditioning or climate control when combined with features for cooling and regulating humidity. Thermostats control the mechanical systems to adjust temperature and humidity; these thermostats may be manual, programmed by the user, or even so-called smart (software-controlled) systems that learn the habits of the inhabitants and optimize the climate settings. Automatic settings for climate control are desirable because they require no judgment, decision, or action by the person. Comfortable temperatures may range from around 20°C–25°C (68°F–77°F), but this will vary widely depending on the individual, clothing, activity, season, and humidity (as discussed in Chapter 8).

17.8 Designing the home office

Working from home

New work technologies bring changes in work practices. A striking example is the recent emergence of the "instant office": wireless and portable computers, lap-mounted or hand-held (even wrist-mounted), and especially mobile smartphones enable us to communicate, write, and record anywhere—including, obviously, our homes. Electronic devices and the internet have enabled many people to work from home, whether part-time or full-time, instead of commuting to an office building. Such a home office can be a curse if it ties us continuously to work, or it can be a blessing when we do not have to spend time working in a tiny cubicle inside a crowded office but can instead find the flexibility to follow our own work schedule while taking care of family and other personal activities.

Setting up a computer workplace

Computer work in the home often starts with occasional use of the kitchen table, a space in the living room, or a spare room; where we talk and video-conference with colleagues and customers, write letters, memoranda, and reports, draw and invent, order supplies, and send bills.

The more frequent and intensive work at home becomes, the more important it is to apply ergonomic thinking to the layout of the home office*. Wherever the office is located, the same human engineering principles apply:

- Use furniture that comfortably supports body movements and postures.
- Provide work tools and equipment that facilitate task execution without overloading human capabilities, particularly regarding repetitive movements.
- Use lighting that is suitable for the task and does not produce direct or indirect glare.
- Incorporate sound-absorbing materials and design to control noise levels, promoting comfort, well-being, and productivity.

- Maintain a suitable thermal climate and acoustic environment.
- Create comfortable and appealing work surroundings that provide appropriate levels of privacy.

See Chapter 18 for more discussion on these issues.

Your choice

When arranging the home office, we can discard the conventions about how to work that govern us in the formal company office. If we work only occasionally from home, the dining table and an odd kitchen chair may suffice; however, cramming an old cabinet with aging equipment—as shown in Figure 17.1—is not a suitable solution. If we begin working in the home office for extended periods, we should be very conscious of the conditions there: in principle, everything discussed in Chapter 18 applies to the home office. So, we should equip the office with carefully selected furniture, ensuring that the components of the workstation fit together well. Most importantly, they should fit us well.

We can sit on a comfortable office chair, as sketched in Figure 17.2 (even if it is expensive), or semi-sit perched on a kneeling chair, ball, or stool, as shown in Figure 17.3, or lounge in an easy chair, as depicted in Figures 17.4 and 17.5. We

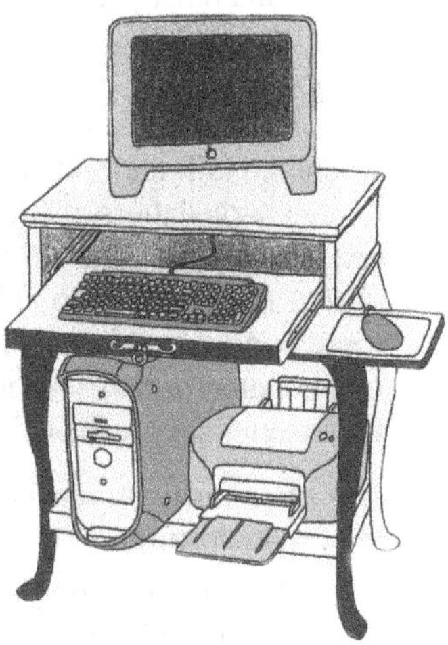

FIGURE 17.1 An improvised and uncomfortable work set-up. The display is too high on the top shelf and faces an awkward direction. The keyboard tray has a sharp edge likely to cut into the user's wrists. The mouse and pad are too far to the side. The overcrowded bottom shelf leaves no room for human legs. Adapted from Kroemer and Kroemer (2017).

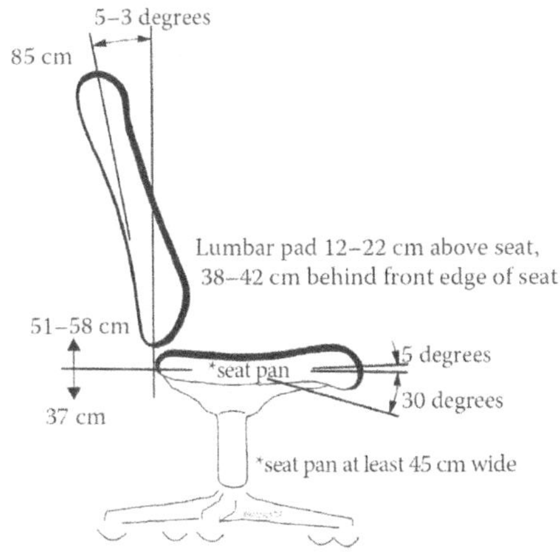

FIGURE 17.2 Dimensions and adjustments of a conventional office chair.

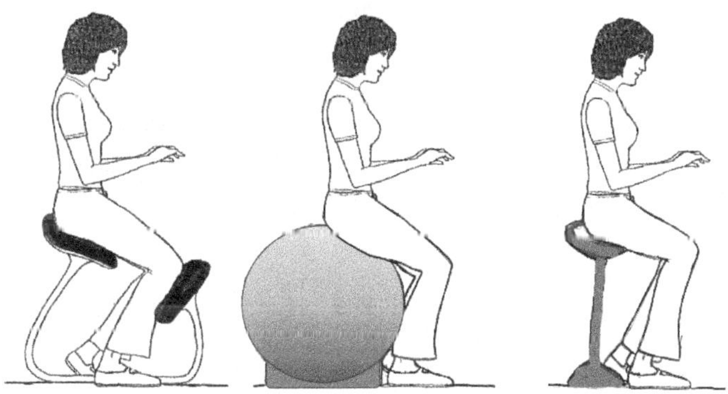

FIGURE 17.3 Seating options other than a traditional office chair: kneeling chair, ball chair, and leaning stool. Adding a backrest is also an option. Adapted and expanded from Kroemer and Kroemer (2017).

should choose whatever feels good to us and what supports the body well over long periods. Alternately, we can work while standing up, at least when taking notes or making phone calls. Some people walk on a treadmill or ride an exercise bike while they work. Since the home office is our own workspace, we should assemble it for our comfort and ease at work. It does not have to match anybody else's setup, nor does it have to be

FIGURE 17.4 Niels Diffrient (1928–2013) executive workstation, circa 1984.

FIGURE 17.5 Home office, circa 2015.

expensive or complex—as even inexpensive furniture can still be well-designed and easy to adjust to suit our preferences.

We should use a quality computer with an up-to-date display and suitable input devices: a keyboard that feels comfortable to us (see Chapter 19), or devices that use voice or other input modalities if they suit us better. The type of computer (desktop, laptop, mini, or other) should be one that works best in all our work locations (office, co-working space, home, or public library). The room should have with good lighting and be private, quiet, and well heated and cooled (Chapters 5, 6, and 8). We may find that we work in the home office more hours than we expected, and our well-being is worth the effort and money we invest in it.

Summary

Laying out and arranging a safe dwelling must be one of the oldest tasks for humankind. We might hope that modern design procedures and building techniques incorporate, at a basic level, knowledge on how to lay out a home to accommodate the inhabitants in terms of ease of use and safety, especially when they are not agile due to pregnancy, age, or disability; however, such human factors design is not as common as it should be.

Numerous publications provide advice on how to design homes that suit pregnant women and children, as well as accommodate disabled and older persons. This chapter summarizes much of this ergonomic information and applies it primarily to the topics of access, walkways, steps, and stairs; lighting, heating, and cooling; and to rooms of special importance, namely the kitchen, bedroom, bath, and toilet.

With computer work increasingly done from home, the layout of the home office becomes important for many people, impacting their well-being and work performance.

Steps to fit the human

Step 1: Determine what the home will be used for, and who will live and work there.

Step 2: Consider the specific needs of the inhabitants as they currently are and what they may become. Where possible, add flexibility to the design of the home to allow for future modifications.

Step 3: If the users or their needs change, consider how these changes can be accommodated. Ask the inhabitants if the dwelling and furnishings suit their needs, and adjust as necessary.

Step 4: Monitor conditions and modify the dwelling, furnishings, or work and iterate as necessary.

Notes

The text contains markers, *, to indicate specific references and comments, which follow.

Purpose of housing: Note the similarity to Abraham Maslow's Hierarchy of Needs theory, discussed in

Chapter 13, which prioritizes existence needs first, followed by relatedness, and then growth.

17.2 Designing for households with children: Kroemer (2006a, 2006b), Kroemer Elbert, Kroemer, and Kroemer (2018), Lueder and Rice (2007).

17.3 Designing for impaired and elderly persons: Kroemer (2006c).

Remodeling: Richard Atcheson describes renovation needs of a home to accommodate older residents in vivid detail in his article "Our Old House" in *Modern Maturity* (November–December 2000, issue 43R: 6, 62–71. Reprinted in a book by Kroemer, 2006c). In the United States, the American Association of Retired Persons (AARP, http://www.aarp.org) provides free brochures on age-friendly buildings and modifications of habitats.

17.5 Designing kitchens: In the 1920s, Lillian Gilbreth and her husband Frank developed the initial idea of the "work triangle" in the kitchen from their time-and-motion studies (Lepore, 2009).

17.8 Designing the home office

Ergonomic layout of the home office: Kroemer and Kroemer (2017).

References and further reading

Kroemer AD, Kroemer KHE (2017). *Office Ergonomics: Ease and Efficiency at Work*, 2nd ed. Boca Raton, FL, CRC Press. https://doi.org/10.1201/9781315368603

Kroemer Elbert K, Kroemer HB, Kroemer Hoffman AD (2018). *Ergonomics: How to Design for Ease and Efficiency*, 3rd ed. London, UK, Elsevier. https://doi.org/10.1016/C2016-0-02516-0

Kroemer KHE (2006a). Designing children's furniture and computers for school and home. *Ergonomics in Design*, 14(3):8–16. https://doi.org/10.1177/106480460601400304

Kroemer KHE (2006b). Designing for older people. *Ergonomics in Design*, 14(4):25–31. https://doi.org/10.1177/106480460601400407

Kroemer KHE (2006c). *"Extra-Ordinary" Ergonomics: How to Accommodate Small and Big Persons, the Disabled and Elderly, Expectant Mothers and Children.* Boca Raton, FL, CRC Press. https://doi.org/10.1201/9780203025246

Lepore J (2009). Not so fast. *The New Yorker*, 85(32):114–122.

Lueder R, Rice VJB (2007). *Ergonomics for Children: Designing Products and Places for Toddler to Teens.* Boca Raton, FL, CRC Press. https://doi.org/10.1201/9780203609163

CHAPTER EIGHTEEN

Design of the office and office furnishings

Different offices

The *Uffici* (offices) in Florence, Italy, is among the best-known classic office buildings, completed in 1581. As seen from above, the structure is shaped like the letter U around an open court: every room has a window to the outside, providing natural light and ventilation in good weather. Heating was provided, if needed, by coal fires and stoves in the rooms, and unpleasant odors abounded when the windows had to be kept closed. Into the 1900s, large office buildings*, and the arrangement of the offices within them, still followed this example. Then, new technology in lighting and air conditioning allowed for radically different office building designs.

The layout, equipment, and decor of our office and office building can affect our quality of work life. Our feelings of well-being, or discomfort, can also appreciably influence our productivity at work. Consequently, there are both ergonomic and economic reasons to design offices that truly fit their human occupants so they can work efficiently and effectively.

Moving into the 21st century, the novel technology of wireless and portable computers profoundly changed many working habits and, consequently, altered the appearance of the office workspace. Large stationary desktop computers are increasingly being displaced by movable laptop and notebook-style computers (and surely, computer design and usage will continue to evolve). Easily portable mobile smartphones, hand-held, and wrist- and head mounted communication devices allow individuals to have "instant offices" wherever they happen to be: in waiting rooms, cafes, on trains and planes, or at home. For remote-eligible workers, whether the resulting continual office

DOI: 10.1201/9781032674575-24

connection is a blessing or a curse can be debated; but it certainly changed many former "cube dwellers" into roaming teleworkers.

18.1 Designing office spaces

Designs of conventional offices can vary from open plans to closed, walled-in individual rooms. A fully open floor plan can include the stereotypical "paperwork factory," comprising a vast room filled with straight rows and columns of desks and chairs. Figure 18.1 depicts an example of such a circa 1990 layout with seas of computers on repeated rows of desks; if this had been 20 or more years earlier, the computers would instead have been typewriters.

Open or private offices

More recent configurations for open office spaces emphasize flexible working environments, featuring a variety of formal and informal groupings of desks and tables intended to allow for collaboration or privacy as needed. Often, low partitions subdivide large spaces into smaller, semi-private cubicles. Despite cubicles being a frequent target of jibes and jokes, when designed appropriately, these partitioned workstation cubicles can indeed provide some privacy and lessen background noise. At the other extreme of office layout are the separate, walled-in offices, each shared by a few employees or used by just one person (often of higher organizational rank). In current practice, many office buildings incorporate various designs: they include areas with open plans, sections with shoulder-high cubicle dividers, and some departments or floors with closed-off, individual offices.

FIGURE 18.1 Computerized paperwork factory, circa 1990.

Office landscape The idea of an interior "office landscape" originated in the 1960s. The basic concept was that a large office space could be made appealing by setting it up like a park or large garden, with irregular arrangements of furniture, office machines, and dividers interspersed with potted plants, bushes, and small trees. This often resulted in rather pleasant office layouts that could readily accommodate changes in interior arrangement. One popular feature was the use of green plants: appreciated in part for their appearance and in part for their (assumed) abilities to prevent noise propagation and absorb air contaminants. Unfortunately, plants do little for noise reduction or air freshening. Still, they can be pleasing to the eye; aesthetics are important for human factors as well.

Pros of open designs Advantages of open plan office designs include ready access to coworkers. There is a heightened feeling of transparency and access since everyone is in the open and not closed off in small individual offices. Furthermore, if workspaces are not permanently assigned, it becomes easy to temporarily co-locate teams to help them interact, collaborate, and communicate. The organization may also be able to accommodate more people in a smaller space (especially if some employees work remotely—outside of the office—some of the time), saving on office costs, and modular furniture is easy to re-arrange.

Cons of open designs Disadvantages of open-plan designs include a lack of privacy, which is an issue when business requires the exchange of confidential or personal information. Another clear drawback of the wide office landscape is disruptive noise from people and equipment, which may reduce performance (see Chapter 6) and job satisfaction. Having many colleagues around can be distracting and interfere with focus on work tasks. In theory, open office spaces are thought to encourage interactions between coworkers; in practice*, however, coworkers in open offices report having fewer interactions with colleagues, likely due to a shared sense of not wanting to disturb their fellow workers and thus overcompensating for the openness of the space.

Office design process The process of designing a new office or remodeling an existing facility involves several steps. The first step is evaluating the needs of the people who will inhabit the office, the tasks they will perform, the work tools and equipment they will use, and their preferences and work styles. Specific statements of these functional requirements guide the design concepts, such as open plans or separate rooms. The next step is identifying a range of design options from which practical solutions can be chosen; this usually involves some compromises. The third step involves evaluating the candidate designs (more on this below). The last step is selecting the final design, followed by implementing it and putting it to use. Employees should be involved in all phases of the process so that their needs are truly met and so that they do not feel manipulated or ignored.

Evaluating different designs

When considering different designs, it helps to have an objective way to make a decision. One method is to use a *Pugh decision matrix*: the proposed designs are assessed against specific criteria, which are then weighted according to their relative importance. As an example, let's evaluate several candidates for a new (or updated) office involving four designs: Layout #1 through Layout #4. These candidate designs are to be evaluated along carefully selected criteria, as listed below:

- Construction cost: what it will cost to build?
- Expandability or ability to expand when needed: how flexible is the design?
- Running costs: what will it cost to maintain and operate?
- Appeal to employees and clients: how pleasing is the new design?
- Expected efficacy, effectiveness, and efficiency of use: efficacy meaning getting things done (is it working?), effectiveness meaning doing the right things (is it actually working well?), and efficiency meaning doing things right (is it working in the most economical way?).
- Availability: when the space will be ready to be occupied?

A panel of experts that includes managers and employees will do the scoring. For this example, we'll assume the raters gave the relative scores (with 1 being worst and 10 being best) listed in Table 18.1.

Rating scores

According to the raw scores, Office Layout #2 is the preferred solution, while Layout #4 is a strong contender. However, we understand that not all criteria are equally important. Therefore, we assign and apply weighting factors to the ratings; this requires us to agree on which factors are relatively more or less important. Applying the selected weighting factors changes the outcome: with weighted criteria, Layout #4 becomes the preferred solution, with Layout #1 not far behind.

Table 18.1 Raw and weight scores of four candidate office layouts

		Layout #1	Layout #2	Layout #3	Layout #4
Raw scores					
Construction cost		3	1	5	10
Expandability		4	10	6	1
Running costs		9	10	1	5
Appeal		7	2	1	10
Efficacy, efficiency, effectiveness		8	10	5	1
Availability (construction time)		1	5	8	10
Raw score total		**32**	**38**	**26**	**37**
Weighted scores					
Construction cost	× 8	24	8	40	80
Expandability	× 3	12	30	18	3
Running costs	× 10	90	100	10	50
Appeal	× 7	49	14	7	70
Efficacy, efficiency, effectiveness	× 5	40	50	25	5
Availability (construction time)	× 1	1	5	8	10
Weighted score total		**216**	**207**	**108**	**218**

Source: Adapted from Kroemer and Kroemer (2017).

A clear approach Setting up specific criteria for judging the candidate designs, including weighting their relative importance, eliminates irrational and highly subjective decisions like the muddled "I like this better." The process outlined above is logical and transparent, and forces all decision-makers to agree on the criteria and their relative importance and to follow the same rules. Furthermore, the selected criteria also consider the human aspect of office design, not only the conventional monetary cost–benefit analysis*.

Flexibility with new technology Current architectural, engineering, and construction technology allows us to design almost any office environment we might wish (depending on our budget). Yesterday's technology tied us to one office, and often to one location within it, because we were bound by wired telephones and computers, cabled machines, and stationary equipment; now, many office workers can be flexible in location, work schedules, and habits.

Looks good, feels good, makes for good work Pleasing office aesthetics convey an appealing image of the organization to the public; they also contribute to attracting new employees and help in retaining them. However, generating true and lasting positive attitudes among workers requires more than just facility design. It necessitates treating individuals as important contributors to the organization's aims; caring about employees' well-being, both on and off the job; and understanding people's concerns about their professional future. In many successful organizations, employees feel engaged with the company and their colleagues, seeing themselves as part of a "family" with extensive interactions among all levels of the corporate hierarchy and a strong prevailing team spirit. They appreciate a supervisor's advice to stay home when feeling sick, to come in late or leave early for important personal reasons, and to switch off their mobile phones and leave their portable computers at work when going on vacation.

Appreciation

Being personally and professionally appreciated is very important and makes employees feel engaged with their employer. People also enjoy job amenities that provide convenience, creating an almost home-like atmosphere in the office. These features go well beyond a typical employee cafeteria and may include free coffee and tea, as well as on-site amenities such as fitness centers, laundry and dry-cleaning facilities, basic healthcare and clinics, and—very important to working parents with young children—on-site childcare. These factors work together to increase job satisfaction, foster emotional ties to the company, including engagement and a sense of belonging, and thus enhance job performance. This is important to the employer because, ultimately, the office is a place to work and perform. (Chapters 13 and 14 expand on these topics.)

18.2 Basics of the physical office environment

The physical environment in the office—specifically in terms of lighting (Chapter 5), sound (Chapter 6), and climate (Chapter 8)—generates conditions that affect workers' health, comfort, and their ability and willingness to perform. Providing the best possible work environment is among the top priorities in office design.

We need light to see

18.2.1 Designing office lighting We can see an object only if it reflects light that strikes the cones and rods located on the retina at the back of our eyes. That light usually comes from a source, such as the sun or a lamp, and is reflected off surfaces of the object.

Photometry

Illumination is visible energy that radiates from the sun, a lamp, or another light source. However, since we should not look directly at the sun and usually don't want to stare at a luminaire, what we "see" is *luminance*—the portion of illumination that objects reflect into our eyes*. Luminance indicates the brightness of light and can be used to quantify the brightness of a display (measured in units of candela per square meter). (Self-luminous electronic displays, such as computer monitors, present surfaces to the eye that have optical properties similar to the luminance coming from a reflecting surface.) Traditional photometry* has commonly used measurements of incoming light (illumination) to describe the lighting conditions of our visual environment, with measurements in units of lux (lx). The primary reason for these measurements seems to be the ease of measuring illumination; however, only luminance is truly relevant for our everyday vision.

Luminance contrast

A strong luminance contrast between the visual target and its background allows us to see details more clearly. A typical example is how we read printed text: the text is easy to read when it features crisp black letters on a white background; if the letters are faded or lack sharp contours, or the background is dirty or gray, deciphering the letters can become very difficult. Similarly, colored objects, such as curves in charts, must show strong luminance contrasts, as similar colors are difficult to distinguish. Computers prior to the 1990s demonstrated this issue: their monochrome monitors displayed lighter green text on a dark green screen, which made reading arduous.

Lighting affects our work

We want an office environment that pleases us in terms of contrast and colors and allows us to see clearly and vividly. Many individuals appreciate the natural light that enters the office through windows because it provides a spacious and organic feel to the workplace. However, to the lighting engineer, daylight has disadvantages: the natural illumination changes throughout the day, and while areas near the windows can be lit glaringly, workplaces further away may be too dark. Well-placed luminaires can provide even lighting throughout the entire room. Task lights, which are lamps at the workstation controlled by the individual worker, allow fine adjustments of local illumination.

Lighting the office

One specific task encountered when setting up office lighting is to prevent glare and annoying bright spots in our field of view. The use of direct lighting, where rays from the source fall directly on the work area, is most efficient in terms of illuminance gain per unit of consumed electrical power. Unfortunately, direct light can produce high glare, poor contrast, and deep shadows. The alternative is to use indirect lighting, where a surface (often the ceiling of the office) reflects the rays from the light source in many different directions so that diffused light reaches the work area. This helps provide even illumination without shadows or glare, but is less efficient in terms of electrical power usage. A compromise is to use a large translucent bowl that encloses the light source and scatters the light before it is emitted from the bowl's surface. This diffused lighting can cause some glare and shadows but is usually more efficient in terms of electrical power usage than indirect lighting. Figure 18.2 shows these kinds of room lighting.

Glare-free lighting

Glare is the experience of intense light that enters the eye and overwhelms the ability of the cones and rods in the retina to distinguish shades of gray and colors (see Chapter 5). An example of direct glare is light from a lamp shining straight into the eyes, whereas indirect glare occurs when the light rays are reflected from a shiny surface and then enter the eye, as shown in Figure 18.3. Strong glare temporarily disables vision, while weak glare acts like a veil in the field of vision, making it difficult to see subtle contrasts.

Avoiding glare

Placing the computer in front of a window can make it difficult to discern details on the screen when strong outside light overwhelms the eyes' ability to see subtle images at your workplace. Figure 18.4 shows direct glare, and Figure 18.5 illustrates indirect glare coming from a window or lamp. Repositioning the

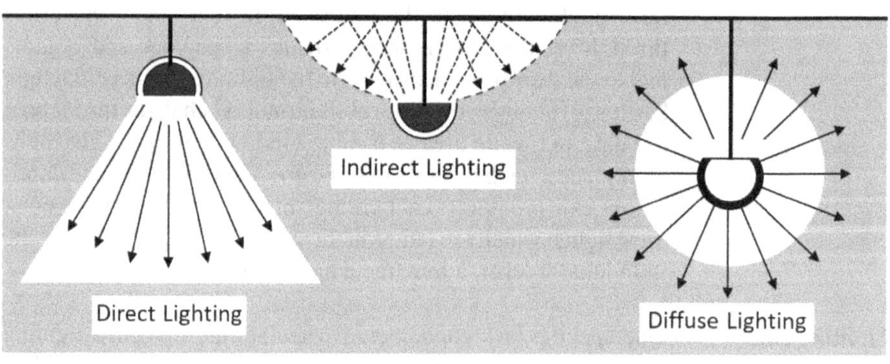

FIGURE 18.2 Direct, indirect, and diffuse lighting. Adapted from Kroemer and Kroemer (2017).

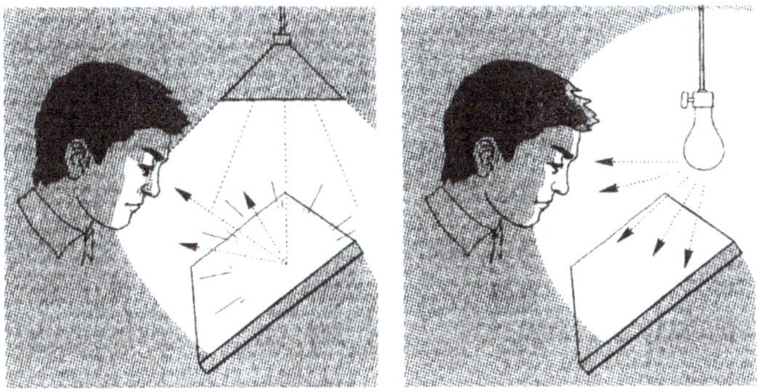

FIGURE 18.3 A lamp providing indirect glare from a shiny surface or direct glare into the eye.

FIGURE 18.4 Direct glare in the office from the sun. Adapted from Kroemer Elbert, Kroemer, and Kroemer (2018).

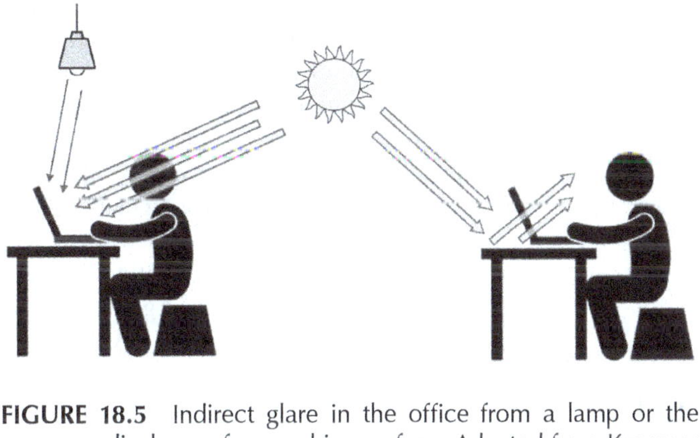

FIGURE 18.5 Indirect glare in the office from a lamp or the sun, on a display, or from a shiny surface. Adapted from Kroemer Elbert, Kroemer, and Kroemer (2018).

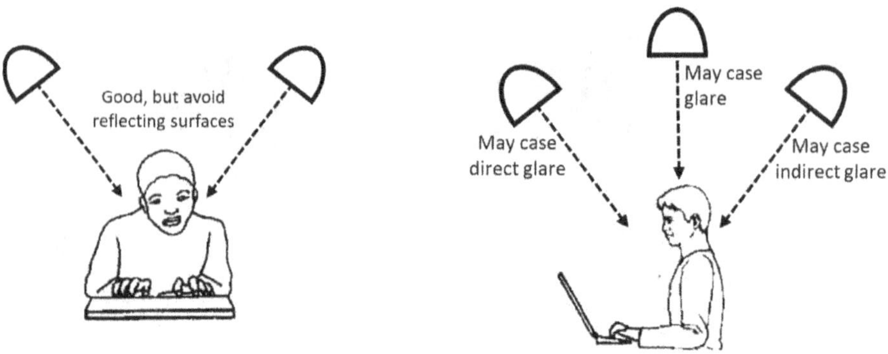

FIGURE 18.6 Placement of lamps. Adapted from Kroemer and Kroemer (2017).

workspace so that light sources are on the worker's left or right side can remedy glare issues, and reducing the intensity of the incoming light (by turning off the lamp or drawing a curtain or screen across the window) would be helpful as well.

Room illumination

Lamps for general office illumination should be arranged to provide glare-free lighting. As the left part of Figure 18.6 shows, locations to the left and right sides of the operator are less likely to cause indirect glare. However, placing a lamp (or a bright window) in front of a person can cause direct glare. To avoid indirect glare, do not position a light source above or behind a person, as the light could reflect off the display surface.

Recommended office illumination

Optimal lighting for good vision depends on many factors, including the condition of the eyes, the task at hand, and the objects to be seen. General office illumination is typically around 500 lx but may be increased to up to 1000 lx if there are many dark (light-absorbing) surfaces in the room. Depending on the type of computer displays (cathode ray, liquid crystal, light emitting, etc.) being used, users may prefer more or less illumination. If the overall space illuminance is a bit dim for specific tasks, such as reading small text on paper, it can be helpful to turn on a task light directed at the visual target; this generates more luminance where needed without significantly raising the overall lighting level in the office. Care must be taken, however, to avoid glare created by shining the light, directly or by reflection, into anyone's eyes.

Easy things to do

The increasing use of electronic screens (such as mobile phones, smartphones, computers, and tablets) has been associated with rising complaints of visual and ocular discomfort, such as eyestrain

Table 18.2 Actions to alleviate vision problems in computer use

Suggestions to reduce risk of vision problems in computer screen use

- *Stop and blink*: When you are working at the computer or reading, pause frequently—say, every 15–20 minutes—to close your eyes, or gaze away from the screen or page, and blink repeatedly. If blinking is not sufficient to address the feeling of dry eyes, consider using sterile eye drops ("artificial tears").
- *Take a break*: Every 30 minutes or so, certainly every hour, get up and move about. A common recommendation that is easy to remember is the "20-20-20 rule": every 20 minutes, take a 20-second break from the screen to look at something at least 20 feet away. Although there are limited clinical data to support this "rule," the practice of regular work breaks is often helpful.
- *Keep your distance*: Position your eyes at the same distance from the screen as you would from a book. If you find that uncomfortable, buy a pair of glasses with a prescription designed for computer work, or use progressive-addition bifocals that have gradually changing optical power from the top to the bottom of the lens.
- *Lower the screen*: Keep the top of the screen below eye level. Gazing upward can strain muscles in the eye and the neck.
- *Hold up the document*: Put a support for the reading matter next to the screen, at the same distance from your eyes.
- *Clean up*: Clean your screen and your glasses. Dust and grime can blur the images.
- *Lighten up*: Age tends to cloud the lens of the eye and shrink the pupil, sharply increasing the need for luminance. So if reading strains your eyes, consider installing brighter lights or at least moving the reading lamp closer to the page. Adjusting the screen's settings for brightness and contrast may also help.
- *Cut the glare*: Position the reading lamp so that light shines from over your shoulder, but make sure it does not reflect into your eyes from the computer monitor. Do not read or do computer work while facing an unshaded window. Wear sunglasses if you are reading outside, but if you feel that you should have shades inside, then the set-up of your visual environment is faulty. The use of an anti-glare screen cover may also help.

If these actions do not reduce the strain from computer work or reading or writing, have an optometrist or an ophthalmologist check whether you need to start wearing glasses or to have your prescription changed.

and headaches. Often, some simple actions, listed in Table 18.2 (and Table 24.2), can help alleviate vision-related annoyances that make us uncomfortable while working at the computer.

Sound or noise

18.2.2 ***Designing office sound*** Sound is everywhere around us. It can excite or annoy us, help us focus or distract us, and at high levels can harm us (as discussed in Chapter 6). Sounds can originate from outside the office (such as from traffic) or inside the office (such as from equipment, activities, and conversations). Such sounds become *noise* if they are unwanted, disruptive, and/or objectionable; these are subjective classifications, and thus, noise is defined by the person hearing the sound.

Excessive sound penetrating from the outside can be reduced by glazed windows, sound-buffering landscaping, and other acoustically insulating materials. Excessive sound generated (or echoing) inside the office space can be reduced with physical means such as noise-absorbing panels, walls, carpeting, and ceilings; acoustical means such as masking sounds; and behavioral means such as how and where workers perform noisy tasks. Office environments often have ambient sounds (such as the sound of air movement from fans or heating and cooling systems) or even background music that masks otherwise intermittent office noises (such as coworkers typing, printing, or conversing, or the rings and pings from electronic device notifications). Some office cultures encourage workers to wear headsets or headphones to answer telephone calls or participate in teleconferences. In this case, coworkers still hear one side of the conversation, which can be even more distracting due to natural conversational pauses. Too little sound can also be frustrating and stressful. Thus, aiming for around 50 dB of sound may be optimal for office well-being. Organizations should design office spaces that consider the acoustical environment: providing zones for louder and quieter (or more private) work tasks, intentionally using sound-dampening and buffering materials, and offering flexibility to accommodate the preferences of workers and tasks (see also Table 24.3).

18.2.3 Designing office climate Controlling the physical climate is important because it influences how the people in the office feel and how they execute their tasks. In particular, air temperature and humidity affect well-being and performance. In our work environment, we also want fresh rather than stale air, but the airflow should not generate a strong draft.

Working when warm or cold

We are not "cold-blooded," neither in the emotional sense nor physically; instead, even while our skin temperatures may change, our body usually succeeds in maintaining a relatively constant temperature of about 37°C in the brain and chest cavity, regardless of the surrounding climate. Achieving a constant core temperature is a complex task for the human body's temperature control system, as discussed in some detail in Chapter 8. The body generates heat energy while simultaneously exchanging energy with the environment. When our office surroundings are cool, we don't want to "feel cold," so we need to prevent excessive heat loss, which we can usually do easily by choosing suitable clothing. When our environment is hot and it threatens to warm up our body core as well, we must shed body heat to prevent overheating. This is rather difficult to achieve because the natural heat flow is always from the warmer to the colder. The most efficient biological way to cool our body in a hot environment is to evaporate bodily water, such as by sweating—but most of us don't want to show much sweat (let alone any resultant odor) in the office. So, we resort to technical engineering approaches; the most comprehensive (and expensive) solution is "climatizing" the office environment by controlling its temperature, humidity, and air movement.

Feeling comfortable

For our well-being and comfort, the temperature difference between exposed skin and the environment is very important, but humidity and airflow also play major roles—as do our attitude and level of acclimatization. In a sheltered office environment, the human thermo-regulatory system has no problem keeping the core close to a constant 37°C. At the skin, in contrast, there are often large temperature differences among various regions. For example, at a given time, the toes may be at 25°C, legs and upper arms at 31°, and the forehead at 34°: for most of us, such a combination feels comfortable.

Exchanging heat

In physics terms, heat energy exchange between our body and the environment takes place via four pathways: convection, conduction, radiation, and evaporation (see Chapter 8 for more details). In each case, heat irreversibly flows from the warmer to the colder medium.

Convection and conduction can cool or warm

Both convection and conduction follow the same thermodynamic rules that govern heat exchange. In each case, skin contact makes for heat transfer: with air (or fluid) in *convection*, and with a solid in *conduction*. The amount of transferred heat depends on the area of human skin involved in the process and on the temperature difference between the skin and the adjacent layer of the outside medium. An air draft, caused by an open window or a fan, can increase convective heat exchange as the air moves quickly along the skin surface, helping to maintain a temperature (and humidity) differential. Wood and some plastics used in office furniture feel warm because their heat conduction coefficients are lower than that of human tissue, allowing them to retain warmth close to the skin. Metal at the same temperature, on the other hand, readily accepts body heat and conducts it away; therefore, it feels colder than wood, even when both are at the same ambient room temperature.

Radiation can cool or warm

Radiation exchanges heat between two opposing surfaces, for example, between a windowpane and a person's skin. Hence, the human body can either lose or gain heat through radiation, depending on whether the skin is warmer (*losing* heat) or colder (*gaining* heat) than the other surface. The amount of radiated heat depends on the temperature difference between the two surfaces and their sizes, but not on the temperature of the air between them.

Evaporation can cool

Heat exchange by evaporation occurs in only one direction: the human body loses heat. (There is never condensation of water on living skin, which would add heat.) Some water evaporates in the respiratory passages, but most evaporation occurs from sweat on the skin. It requires energy to evaporate sweat: about 2430 J (580 calories) per cubic centimeter of water. This energy is primarily taken from the body, reducing the heat content of the body by that amount. Therefore, evaporating sweat is very effective in cooling the body and functions even in a hot environment. Even if we don't notice, there is always some perspiration, and thus sweat evaporation is continually occurring: indeed, our clothes acquire a sweat-related smell when worn too long.

Heat balance

Primarily, the actual amounts of heat exchanged with our surroundings depend, directly or indirectly, on the temperature difference between the participating body surfaces and the environment. Secondarily, the magnitude of exchanged heat depends on the surface areas involved; clothing determines how much skin we expose, greatly affecting how warm we feel. (Clothing also has other heat transfer properties, such as insulation; see Chapter 8.) Humidity plays an appreciable role when it is either very high or very low.

Over time, we feel comfortable in a climate when our body can achieve heat balance, allowing it to maintain its temperature at comfortable levels. A healthy body accomplishes this by responding to a cold environment by cooling its skin surface and to a hot environment by warming the skin. (If that statement seems surprising, recall that heat energy always moves from the warmer to the colder.)

Acclimation

We can help achieve a state of thermal balance by deliberately dressing more lightly in the heat or, conversely, by covering up in insulating layers in the cold. Adjusting how we dress is partly a social action and hence intentional, but it is also part of unconscious acclimatization. We acclimate psychologically by simply expecting and accepting a hotter environment as summer approaches and by preparing mentally for colder conditions

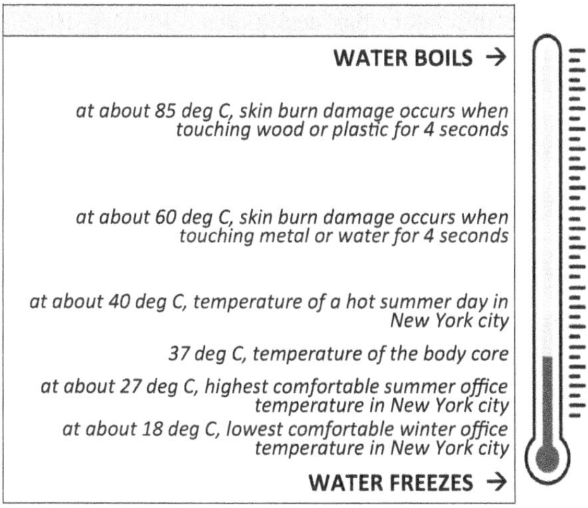

degrees C	degrees F	K
100	212	373.15
90	194	363.15
80	176	353.15
70	158	343.15
60	140	333.15
50	122	323.15
40	104	313.15
30	86	203.15
20	68	293.15
10	50	283.15
0	32	273.15

FIGURE 18.7 Temperature effects between freezing and boiling of water in degree Celsius, degree Fahrenheit, and kelvin.

as fall arrives. We acclimate physiologically through continuous or repeated exposure to hot and (to a lesser extent) cold conditions, which brings about a gradual adjustment of body functions, resulting in better tolerance of climate stresses.

There are no significant differences between females and males regarding their ability to adapt to changing office climates. Any individual tendencies to feel too warm or too cool can be alleviated by using a portable fan or heater; however, simply adjusting clothing habits (and possibly activity) is usually sufficient.

Figure 18.7 provides a summary overview of temperature ranges and their effects on humans. As discussed in more detail in Chapter 8, to convert temperature degree values from one scale to another, we take into account the different settings for freezing and boiling water temperatures as well as the number of degrees between freezing and boiling (so the ratio 100/180 becomes 5/9):

$$\text{Fahrenheit to Celsius}: (°F - 32) \times (5/9) = °C$$

$$\text{Celsius to Fahrenheit}: (°C) \times (9/5) + 32 = °F$$

$$\text{Celsius to kelvin}: (°C) - 273.15 = K$$

Effects of heat on mental performance

It is difficult to evaluate the effects of heat (or cold) on mental or intellectual performance. However, it seems that the mental performance of a non-acclimatized person deteriorates when room temperatures rise above 25°C. That threshold rises to 30°

or even 35° if the individual has acclimatized to heat. Brain functions are particularly vulnerable to heat; keeping the head cool improves tolerance to elevated deep body temperature. A high level of motivation may also counteract some of the detrimental effects of heat. Note that these results stem from laboratory tests; in modern offices, such high temperatures are generally rare.

Good office climate

There are several technical ways to create a thermal environment that meets the physiological needs of people as well as their individual preferences. The primary approach is to adjust the physical conditions of the climate (temperature, humidity, and air movement), usually achieved through automatic "air conditioning" control systems. National regulations and regional recommendations provide guidance, including ASHRAE recommendations in the United States and international standards set by ISO*.

Too hot/too cold: what to do

If there is no way to control the climate in the office automatically, several time-honored and common-sense methods can counteract heat (see also Table 24.4).

- Move air swiftly through the room. (This may require paperweights for loose items.)
- Stay away from warm surfaces, such as windowpanes, that radiate heat.
- Cool the air and hot surfaces by sprinkling water on warm surfaces.
- Rest during the hottest time of the day (e.g., take a siesta).
- Dress lightly and in lighter colors.

If you are cold, do the opposite of what you would do in the heat: stay away from drafts, get close to warm radiating surfaces, sit in the sun, move about (if practical), and dress more warmly.

18.3 Designing office furniture

Sitting at work

The concept of sitting has different meanings depending on local customs. In many regions of the world, sitting implies perching atop a stool or chair, as was the tradition in ancient Egypt and Greece. However, in other cultures, people habitually rest their bodies on a carpet, mat, cushion, or on the bare floor by kneeling, squatting, or sitting with their legs extended or with one or both feet underneath their bodies. Of course, customs can also change: around 1900, subordinates in European offices typically stood while working while their superiors enjoyed sitting comfortably.

Sitting still

It is tiresome to maintain any body position unchanged over extended periods of time. This includes sitting still for hours on end. We would like to move about, but in the modern computer office, we are doubly tied to our equipment: by our hands that must operate the computer keys, and by our eyes that must view the screen, text, and keys. Thus, we keep our hands, eyes, and, accordingly, our whole body fixed at the computer workstation. We may try to ease the effort of maintaining a rigid posture by sitting in a chair as comfortably as possible, shifting and slouching as needed, but our body keeps telling us to get up, walk, and move around for a change.

Erect posture

A century ago, office clerks were men who stood at their desks, using ink to write letters and make entries in ledgers. By the middle of the 20th century, many office employees in subordinate positions were female, and the work posture had changed from standing to sitting, often at desks with typewriters. Now, office workers (both male and female) often sit for long hours at computer workstations.

In the late 1800s, body posture became a significant concern for some physiologists and orthopedists in Europe*. In their opinion, the upright (erect) standing posture was balanced and healthy, whereas curved and bent backs were unhealthy and therefore should be avoided, especially in young people. Consequently, a straight back and neck* became the recommended posture for both standing and sitting, and, logically, seats were designed to promote such an upright body position. However, that cliché applied only to lower-level employees: managers habitually enjoyed ample armchairs with high backs and comfortable upholstery that allowed for various body positions.

Sit as you like

The simplistic concept that upright sitting, with thighs horizontal and lower legs vertical, meant healthy sitting surprisingly lasted for about a hundred years. Finally, around the year 2000, it became generally accepted that people could sit any way they liked (see Figure 18.8) because freely choosing and changing their posture makes them feel comfortable.

FIGURE 18.8 People sit any way they want.

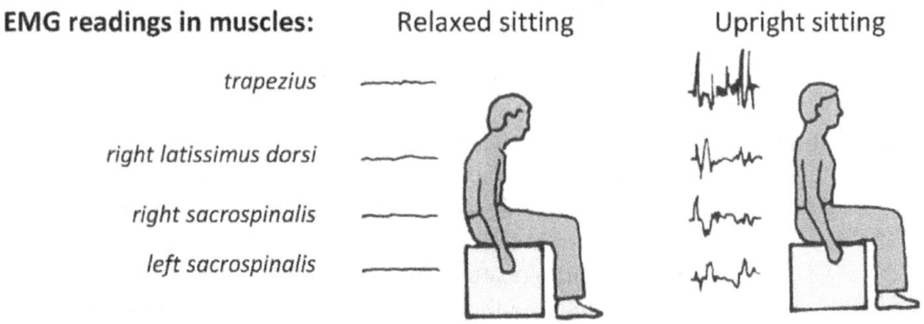

FIGURE 18.9 Activation of back muscles during relaxed and upright sitting without backrest, measured using electromyograms. Adapted from Grandjean (1987) and Lundervold (1951).

Sitting or standing

Sitting, as opposed to standing, is suitable when only a small workspace must be covered by the hands: this is typical for much of today's office work. Sitting keeps the upper body stable, which is helpful for executing finely controlled activities. Sitting supports the body at its mid-point and requires less muscular effort than standing, which is important when posture needs to be maintained over long hours.

The human body is made to move

Of course, there is nothing wrong with sitting or standing upright at one's own will, but maintaining an erect back requires muscle tension, as Figure 18.9 demonstrates for sitting without a backrest. Even when a backrest is present, it does not provide much support unless one presses the back against it, which also requires muscle effort. Slumping in the seat and shifting the body are instinctive actions to reduce strain and tension in muscles that would otherwise be working to maintain prolonged postures. Sitting (or standing) still for extended periods is uncomfortable; it leads to tissue compression, hinders blood circulation, and can result in the accumulation of extracellular fluid in the lower legs. Clearly, the human body is designed for change and movement, as discussed in Chapter 2 of this book.

Comfort and discomfort

Subjective judgments arise from physiological phenomena (such as tissue compression and circulatory events) and incorporate emotional components as well. This becomes evident when we try to assess the feeling of comfort in relation to sitting. One definition of comfort has long been, conveniently yet misleadingly, the absence of discomfort*. However, comfort and discomfort are not at opposite extremes on a single judgment scale. Instead, there appear to be two scales: one for the agreeable feelings related to *comfort* and another for the unpleasant experiences of *discomfort*, which involve sensations such as fatigue, strain, smarting, hurting, and not being at ease. The two scales may overlap but are not parallel to each other. Using the term *annoyance* (instead of discomfort) avoids the misconception of a single scale with comfort and discomfort at opposite ends. Table 18.3 lists rating categories that distinguish chairs by comfort (related to well-being and aesthetics) and annoyance (related to biomechanics and fatigue factors).

Comfortable seats

Feelings of comfort when sitting are associated with such descriptive words as warm, soft, plush, spacious, supported, safe, pleased, relaxed, and restful. However, what feels comfortable depends greatly on the individual and their habits, the environment, the task at hand, and the passage of time. Aesthetics

Table 18.3 Rating of chairs by comfort or annoyance

Comfort statements	Annoyance (discomfort) statements
1. I feel relaxed.	1. I have sore muscles.
2. I feel refreshed.	2. I have heavy legs.
3. The chair feels soft.	3. I feel uneven pressure.
4. The chair is spacious.	4. I feel stiff.
5. The chair looks nice.	5. I feel restless.
6. I like the chair.	6. I feel tired.
I feel comfortable.	I feel annoyed.

Source: Terminology from the study by Helander and Zhang (1997), who used these six specific statements about chair comfort or annoyance, followed by one general statement. Each statement was assessed from "not at all" to "extremely."

play a role: if we like the appearance, color, and ambience, we are more inclined to feel comfortable. Appealing upholstery, for example, can significantly contribute to the feeling of comfort, especially when it is neither too soft nor too stiff but distributes body pressure along the contact area. Mesh fabric, often used on the seat surface and back support instead of traditional cushion upholstery, "breathes" by allowing heat and humidity to escape while the chair supports the body—and it can look fashionable, sleek, and appealing.

Annoying seats

Feelings of annoyance when sitting are associated with descriptors such as stiff, strained, cramped, tingling, numbness, feeling unsupported, fatigue, restlessness, soreness, hurting, and pain. Some of these attributes can be explained in terms of circulatory, metabolic, or mechanical events in the body; others extend beyond physiological and biomechanical phenomena. Users can easily describe design features that lead to feelings of annoyance, such as chairs that are the wrong size, too high or too low, or have hard surfaces or sharp edges. However, avoiding these mistakes does not necessarily make a chair comfortable.

Dynamic design

In the early 2000s, *dynamics* became a key focus in the design of office chairs, in contrast to the statics of stiffly maintained posture. People do move about as they please. Design should encourage and support free-flowing motions, as illustrated in Figure 18.10 (and sitting, in Figure 2.2), with opportunities for transitory postures at the whim of the user.

FIGURE 18.10 People sit or stand any way they want.

18.4 Designing the ergonomic office workstation

Successful ergonomic design of the office workstation* depends on proper consideration of several interrelated aspects: work tasks, work movements, and work activities. All of these must "fit" the person to achieve individual well-being and foster high work output. Of course, job content and demands, control over one's job, and many other social and organizational factors also influence feelings, attitudes, and performance at work, as discussed in Sections III and IV of this book.

Office furniture should accommodate the full range of body sizes, varying body postures, and diverse activities; it should enhance task performance; and it should be appealing and help people feel comfortable in their work environment. At conventional computer workstations, the furniture consists primarily of the seat and a working surface, usually a table (or desk), which serves as support for data entry devices and displays. It is best, though expensive, to have all of these components independently adjustable, as explained below.

Links between person and task

For proper layout of a workstation, it helps to consider three main linkages between a person and the task.

1. Visual interface: One must look at the keyboard, the computer screen(s), and source documents.
2. Manipulation: The hands operate keyboard keys, a mouse, or other input devices and manipulate pen, paper, and telephone.
3. Body support: The seat pan provides support at the undersides of the thighs and buttocks, and the backrest supports the back. Armrests or a wrist rest may serve as additional support links.

Designing for vision

The location of visual targets can greatly affect the body position of the computer operator, as shown in Figure 18.11. Objects upon which we focus our eyesight should be located directly in front, at a convenient distance and height from the eyes. If one is forced to tilt the head up to view the computer screen or to turn it to the side to read a document, "eye strain" is commonly experienced, often accompanied by pain in the neck, shoulders, and back.

As a rule, the display or source document should be at the proper viewing distance for the operator, which is about half a meter from the eyes, the reading distance for which corrective eye lenses are usually designed. A convenient procedure is

FIGURE 18.11 On left, a seated computer worker forced to arch his back and neck to be able to focus on the monitor by poor monitor positioning (too far and too high). On right, a computer worker forced into a twisted upper body posture by a document placed flat and to the side. Adapted from Kroemer and Kroemer (2017).

to place the screen* and source document at slightly less than arm's length, ideally perpendicular to the line of sight (so a document holder may be helpful). Of course, proper lighting is necessary on documents to be read and other objects to be seen distinctly, as discussed earlier in this chapter.

Designing for manipulation

In addition to our eyes, our hands are usually busy with various office tasks: typing, grasping and moving papers, taking notes, and using a computer mouse. When our hands are engaged in different activities, the varied manipulation is likely to keep our arms and upper body moving around in our workspace. Motion is desirable—in contrast to maintaining a fixed posture, such as when typing on a keyboard for extended periods.

When sitting, sweeping our hands at full arm's length provides a fairly large manipulation area. This area is even larger when we move our upper body and larger still when we are standing up. However, we prefer the space in front of our chest and belly for finely controlled hand and forearm movements* (see Figures 2.15 and 11.11). We can also see clearly in this space because objects are at a suitable distance from our eyes, and they are low enough to fit the natural downward direction of gaze, as discussed in Chapter 5.

Designing for motion and support

Our bodies are built to move, not to hold still. It is uncomfortable and tiresome to maintain a body position without change over extended periods. We experience this while driving a car or sitting at workstations with immobile desk-based computers (where the chair is often much less comfortable than our modern car seats). However, unlike when driving, we can get up and move around at will in our office. It is a good practice to get up and stretch our bodies while talking with a colleague, getting a drink, or retrieving some papers or other supplies.

"Active sitting"

Designing for motion means that the office chair should be comfortable for both relaxed and upright sitting, for leaning backward and forward, and for getting in and out. In some chairs, both the seat pan and backrest follow the motions of the sitting person and provide support throughout the range of movement. Other designs start from the premise that the seat must not be a passive device but an active one: the chair as a whole, or its pan or backrest separately, can automatically change configuration slightly over time, perhaps in response to certain sitting postures maintained by the person. The change can involve angles or the stiffness of the material. Seat and back cushions that pulsated were tried in the 1950s to alleviate the strain that military aircrews experienced when they had to fly extended missions, sitting in the same posture for hours on end. Looking ahead to an ergonomic future, should an "intelligent seat" remind us to get up, move about, stretch, or walk after we have sat in a static position for too long?

Designing for diversity

For decades, the prevailing illusion in Western office design was that it was possible to establish strict norms for the dimensions of office furniture that would fit (nearly) everybody. The globalization of trade and the recognition that people are diverse (in terms of size, habits, behavior, and preferences) have effectively put an end to the one-design-fits-all concept. Instead, the diversity of use (and of users) has led to greater diversity in design.

Designing for body sizes

The size of office furniture primarily derives from the body dimensions, work tasks, and working habits of the people in the office. The main vertical anthropometric determinants of height requirements are lower leg (popliteal and knee) heights, thigh thickness, and the heights of the elbow and eye; see Chapter 1 for explanations and data.

Design ranges

As listed in Table 18.4, essential furniture dimensions depend on the body dimensions of the user(s). For example, the height of the seat pan is determined by the user's lower leg length, with the expected shoe heel height added. When designing the adjustment range of the seat, the tallest user's lower leg length

Table 18.4 Dimensions of furniture related to measurements of the user's body

Furniture dimension	Body size of user	Reference numbers in Figure 1.1 and Table 1.3
Display height	Eye height above seat pan	#9
Work surface height	Elbow height above seat pan	#11
Seat pan height	Popliteal height plus shoe heel height	#14
Clearance height under table or desk	Thigh height above seat pan or knee height plus shoe heel height	#12 or #13
Seat pan depth	Buttock–popliteal depth	#25
Seat pan width	Hip breadth	#28

(plus shoe heel height) identifies the highest seat setting, while the lowest adjustment would correspond to the person with the shortest lower leg length. Obviously, these highest and lowest settings, along with the adjustment range between them, would vary for different user populations and may need to be modified for differing sitting habits.

Fitting everybody?

Some relationships between furniture and body dimensions are loose; for example, the location of the display is often not tightly linked to the location of the eyes. Other relationships are definite: the clearance height under a desk or table should not be less than the knee height (plus shoe heel height) of the tallest user. Since that clearance height is usually fixed, even a smaller user is likely to perch on a high-adjusted seat in order to work at the given (high) work surface; such a high seat pan would cause the user's feet to be unable to reach the floor, so a footrest should be available.

Seat slope

Since connective tissue links the bones of the pelvic girdle to the lumbar spine, rotation of the pelvis affects the posture of the lower spinal column. When we sit on a hard flat surface without using a backrest, the lowest protrusions of the pelvic bones (the ischial tuberosities) act as fulcra around which the pelvic girdle rotates under the weight of the upper body. Figure 18.12 illustrates that, if the rear edge of the seat pan is elevated, the resulting slope causes the pelvis to tilt forward, which, in turn, tilts the lower spine into its natural lordosis; a lumbar pad on the backrest can also be helpful. Furthermore, the angles at the hip and knee affect the position of the pelvis and, hence, the curvature of the lumbar spine because leg muscles run from the pelvic area across the hip and knee joints to the lower legs.

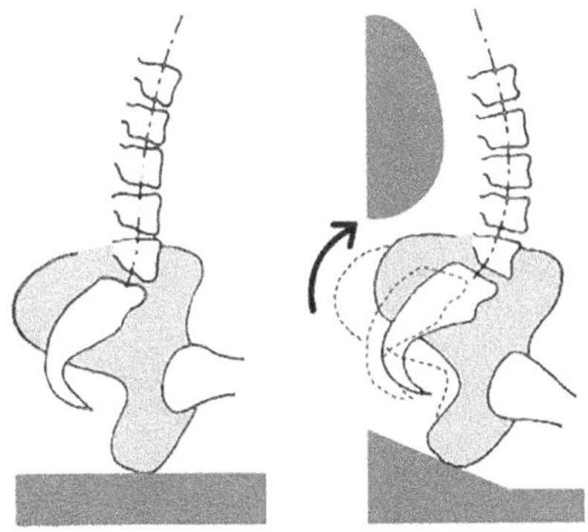

FIGURE 18.12 The tilt (rotation) of the pelvic bone changes when seated on a flat surface or a sloped seat surface with backrest. Adapted from Kroemer and Robinette (1968).

Seat pan variations

The curvature of the spinal column, especially in the lumbar area, has been a major concern for orthopedists and seat designers. Their ideas about proper posture and body support have generated innumerable designs for the seat pan: high and low, hard and cushioned, tilted fore and aft, contoured and flat, saddle-shaped and otherwise curved, all associated with hopes and claims of healthy sitting. However it is shaped, the surface of the seat pan must support the weight of the upper body comfortably and securely. Hard surfaces and edges generate pressure points, which can be avoided by suitable upholstery, cushions, or other surface materials that elastically or plastically adjust to body contours.

Backrest necessary?

Some people think that a backrest is unnecessary when sitting because, without it, back muscles stabilize the trunk, which may provide good muscle exercise. Figure 18.13 depicts one way to semi-sit (on a kneeling chair) without a backrest. However, most people find that a backrest is desirable for several reasons. For one, back support can carry some of the weight of the upper body, which reduces the load that the spinal column must otherwise transmit to the seat pan. A second reason is that a lumbar pad, protruding slightly into the lumbar area of the spine, helps maintain lumbar lordosis, which is believed to be beneficial. A third related reason is that leaning against a suitably shaped backrest allows muscles to relax. Figure 18.14 illustrates the concept of relaxed supported sitting. (Notice in Figure 18.14 that the neck region is not well supported; a better fit could be achieved if the headrest were separately adjustable up/down and forth/back or if an adjustable neck cushion were available.)

FIGURE 18.13 Semi-sitting on a kneeling chair. Adapted from Kroemer and Kroemer (2017).

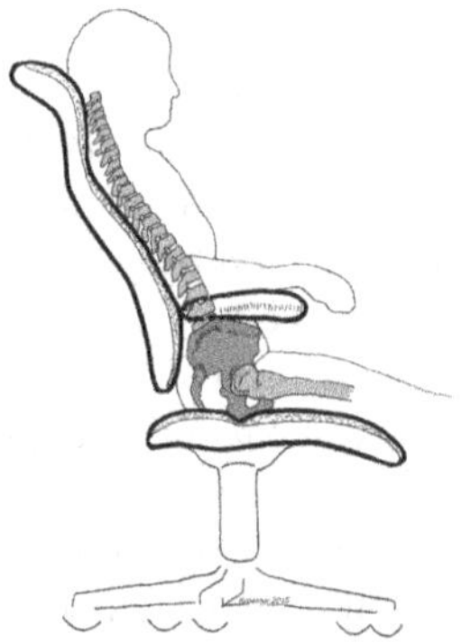

FIGURE 18.14 Relaxed sitting on a supportive office seat.

Chair shapes

Of course, the backrest should be shaped to support the back comfortably. Around 1960, Ridder in the United States and Grandjean in Switzerland found in experiments that their subjects preferred similar backrest shapes, as depicted in Figure 18.16. In essence, these shapes follow the curvature of the rear side of the human body. At the top, the backrest is convex to follow the curve of the cervical lordosis; below, it is nearly straight but tilted backward to support the thoracic area; in the lumbar region, the backrest has a slightly protruding pad accommodating lumbar lordosis; at the bottom, the backrest is concave or open to provide room for the buttocks. The seat pan has some depression (possibly created by a deformable material) to accommodate the ischial tuberosities, and its front is well rounded. Shapes such as these have been used successfully for seats in automobiles, aircraft, passenger trains, and office chairs.

Seat measures and adjustments

The only inherent limitation to the size of the seat for Western-style sitting is that its pan should be short enough that the front edge does not press into the leg's sensitive tissues behind the knees. A seat pan with a well-rounded front edge, between 38 and 42 cm deep and at least 45 cm wide, fits most Western bodies (as shown in Figure 17.2). The height of the seat pan must be widely adjustable, preferably down to about 37 cm and up to 58 cm to accommodate persons with short and long lower legs. It is very important that the person, while seated on the chair, can easily make all adjustments, especially in height. Figure 18.16 illustrates desirable adjustments of the seat pan and backrest, together with other workstation features.

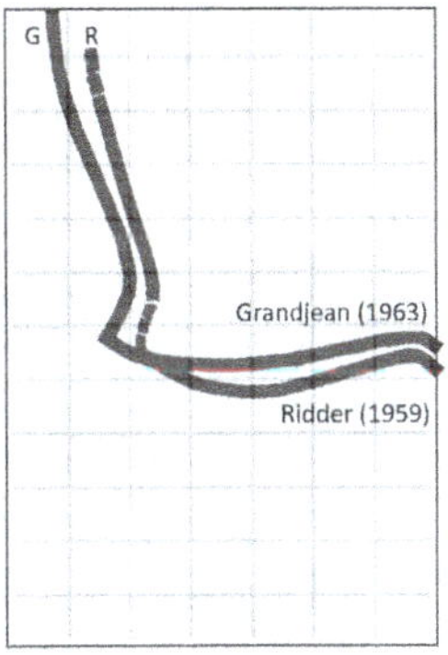

FIGURE 18.15 Preferred contours for seat pan and backrest, as suggested by Ridder in 1959 (R) and Grandjean in 1963 (G). Adapted from Kroemer and Kroemer (2017).

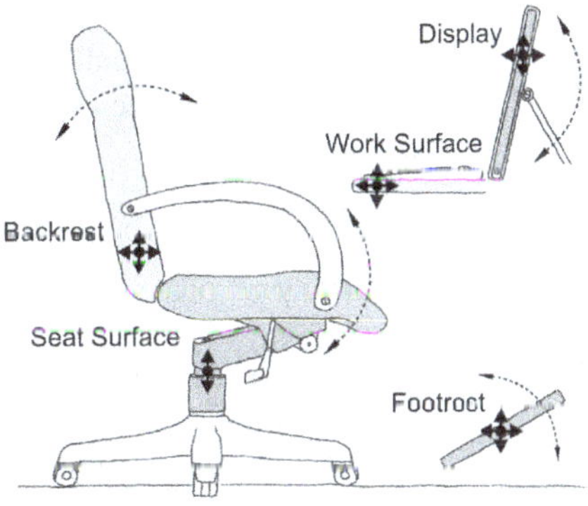

FIGURE 18.16 Adjustments of office furniture: seat, feet, and desk with display.

Work surface and keyboard support

The height of the workstation depends largely on the activities performed with the hands and how well the work must be viewed. Thus, the main reference points for ergonomic workstations are the worker's elbow height and eye level. Both depend on how the workers sit or stand—upright or slumped—and how they alternate among postures. For Western users, the table or other work surface of a sit-down workplace should be adjustable in height between about 53 and 70 cm, even a bit higher for very tall individuals, to permit proper hand, arm, and eye positions. As mentioned, a fixed, high work surface (high so as to provide leg clearance for long-legged individuals) necessitates footrests for shorter individuals perched on a tall seat. A height-adjustable work surface accommodates most users and allows for work while sitting or standing.

Adjustability

Having a fixed height for tables, desks, or seats reduces the price of office furniture but leads to uncomfortable body postures for many users. It is best, though more expensive, to have all dimensions independently adjustable, which allows for an ergonomic fit for all. If technology and working habits continue to favor small portable computers and possibly virtual or head-mounted displays, traditional desks and tables might not be needed anymore—potentially revolutionizing furniture and the appearance of office spaces yet again.

Fitting it all together

The primary responsibility for proper office design lies with management, architects, interior designers, and human factors engineers. They determine the overall layout and other important conditions such as lighting and climate control. However, within the overall boundaries of design, there is considerable freedom for specific workplace arrangements by section managers, supervisors, and especially individual office workers. Table 18.5 summarizes what individuals can do to make office work more agreeable.

Summary

Designs of traditional offices vary from open plans to closed, walled-in individual rooms. Although open designs are popular, disadvantages include a lack of privacy and the propagation of disruptive noise. Many organizations use a mixture of office designs: some areas with wide-open spaces, some areas with shoulder-high cubicle dividers, and some areas with closed-off individual rooms and offices.

Table 18.5 Ergonomic recommendations for your office setup

What to do to feel more comfortable?

- Place all the things you must operate with your hands (keyboard, mouse, trackball, pen, paper, telephone)
  - Directly in front of you,
  - At elbow height,
  - Within easy reach.
- Place the display and the keyboard directly in front of you, at your best viewing and/or reading distance.
- Place the monitor low behind your keyboard. Do not put it on a tall stand.
- Sit on a seat that allows you to change your posture frequently. If long-time sitting is required, then a tall backrest that can recline helps support the back and the head.
- Change your body position often. Changing position helps avoid the continued compression of tissues (especially of the spinal column), facilitates blood circulation, and counteracts muscular fatigue—and it breaks boredom.
- Support your arms and hands by resting them, as often as feasible, on soft arm rests attached to the seat, and/or on padded wrist rests at the keyboard—but avoid hard surfaces and, worse, rigid corners and edges that compress the skin tissues.
- Keep the shoulders relaxed, the upper arms hanging down, the forearms horizontal, and the wrists straight.

What to do if you are not comfortable …

… if your eyes are tired, or hurt, or feel teary or dry

- Place all the things you must clearly see (display, source document, writing pad, template, keyboard) directly in front of you, to the best viewing/reading distance (which is probably shorter than you have now).
- Place your monitor low behind your keyboard.
- Make sure you do not have light (from a window or from lamps) reflected in the display or shining directly into your eyes.
- Talk with your supervisor and, if the condition does not go away, have your eyes checked by an ophthalmologist or optometrist.

… if your back hurts

- Take a break at least every 30 minutes; walk and move your body.
- Make sure that you lean against the backrest of your seat.
- Get a seat that fits your body and accommodates your sitting habits better than your current chair.
- Place all the things you must clearly see (display, source document, writing pad, template, keyboard) directly in front of you, at the best viewing/reading distance (which is probably shorter than you have now).
- Place your monitor low behind your keyboard.
- Place all the things you must operate with your hands (keyboard, mouse, trackball, pen, paper, telephone)
  - Directly in front of you,
  - At elbow height,
  - Within easy reach.
- Talk with your supervisor and get a medical evaluation if the condition does not go away.

(Continued)

Table 18.5 (Continued) Ergonomic recommendations for your office setup

... if your neck or your upper back hurts

- Take a break at least every 30 minutes; walk and move your body.
- Place all the things you must clearly see (display, source document, writing pad, template, keyboard)
  - Directly in front of you, at your best viewing/reading distance (which is probably shorter than you have now),
  - Low behind your keyboard (do not use a tall tilt stand or other equipment under the monitor).
- Place all the things you must operate with your hands (keyboard, mouse, trackball, pen, paper, telephone)
  - Directly in front of you,
  - At elbow height,
  - Within easy reach.
- Talk with your supervisor and get a medical evaluation if the condition does not go away.

... if your shoulder hurts

- Take a break at least every 30 minutes; walk and move your body.
- Put the mouse or trackball next to the keyboard, all at elbow height.
- Operate the mouse or other input device with the other hand. (You can train your other hand to do so well after just a few minutes.)
- Use armrest and wrist support.
- Talk with your supervisor and get a medical evaluation if the condition does not go away.

... if your wrist or hand or arm hurts

- Take a break at least every 30 minutes; walk and move your body.
- Make sure that your wrist remains straight while working keyboard or other input device.
- Strike keys very lightly.
- Use armrest and wrist rest often.
- Put the mouse or trackball next to the keyboard.
- Talk with your supervisor and get a medical evaluation if the condition does not go away.

... if your leg hurts

- Take a break at least every 30 minutes; walk and move your body.
- Make sure that you have ample room at your workstation to position and move your feet freely.
- If the front portion of your seat presses on the underside of your thighs, lower the seat. (Probably, you must also lower the keyboard and the monitor, the table or the desk, or other work surface accordingly.)
- Get another seat that has a soft waterfall shape at its front.
- Use a wide and deep footrest.
- Talk with your supervisor and get a medical evaluation if the condition does not go away.

The specification of suitable office illumination depends on the tasks at hand, the objects to be seen, and the conditions of the eyes that need to be accommodated. General recommendations for office illumination are about 500 lx, but this may decrease or increase depending on the conditions and use. For tasks such as reading fine text on paper, it can be helpful to use a task light; however, care is necessary to avoid glare. Sound levels should be around 50 dB, and the climate should be perceived as comfortable by the workers.

Neither theories nor practical experience endorse the idea of one single proper, healthy, comfortable position, such as erect sitting. Instead, many motions and postures may be subjectively comfortable (healthy, preferred, suitable) for short periods of time, depending on one's body, habits, and work activities. Changing from one posture to another and moving freely among comfortable poses is important. Consequently, furniture should allow for body movements among various postures through easy or automatic adjustments of features, particularly in seat height, seat pan angle, and backrest position. The entire computer workstation should permit easy variations, for example, in the location (especially height) of the input devices and the height, distance, and angle of the display.

Within given overall boundaries, there is usually freedom for specific workplace arrangements that fit the task and suit the individual. Individual arrangements can be essential to achieving ease of work.

Steps to fit the human

Step 1: Determine what the office will be used for, who will work there, and what tasks will be performed in which locations.

Step 2: Consider the specific needs and preferences of the workers, as well as the equipment and tools they use. Involve the workers in the design process. Where possible, add flexibility to the design of the office building and its furnishings to allow for future modifications.

Step 3: If the users or their needs change, consider how this can be accommodated. Ask the inhabitants if the space and furnishings suit them, and adjust as needed.

Step 4: Monitor conditions, modify the office space, furnishings, and work tasks, and iterate as necessary.

Notes

The text contains markers, *, to indicate specific references and comments, which follow.

Office buildings: Detailed discussions by Kroemer and Kroemer (2017), Saval (2014).

18.1 Designing office spaces

Theory versus practice: As the saying goes, in theory, there is no difference between theory and practice, while in practice, there is.

A clear approach, cost–benefit analysis: Mishan and Quah (2007).

18.2.1 Designing office lighting

Luminance and lluminance: Luminance is light reflected or emitted from an object, whereas illuminance is light falling upon an object. One way to remember these terms is to note that (**L**)uminance is the light (**L**) eaving an object, whereas (**I**)lluminance refers to the (**I**)ncident light (see also Chapter 5).

Photometry in office illumination: In traditional photometry, the apparent reason for measuring illuminance was the ease of doing so. Instruments to measure luminance, which determines how well we can see objects and distinguish their details, are more complex and expensive and harder to use (see Rea (2005)). Illuminance is measured in units of lux (lx) or lumen per square meter. A moonless night sky may have illuminance of 0.001 lux, an office or a sunrise 400 lx, and full daylight more than 10,000 lx.

Traditionally, the relationship between lighting and human vision has been concerned primarily with two elements: the object being viewed (e.g., print on paper) and its illumination (e.g., by daylight or electrical lamps). However, the widespread use of self-luminous sources, such as television screens and computer monitors, has changed the relationship because visual characteristics of these light-emitting objects are not dependent on illuminating light (see Howarth (2005)).

18.2.3 Designing office climate: Further information for the built environment, such as in offices in Europe and North America, can be found in the ANSI-ASHRAE

Standard 55 and ISO standards, as listed by Parsons (2003, 2005, 2014).

18.3. Designing office furniture

Late 1800s, concern for body posture: In 1884, orthopedist Franz Staffel had publicized theories about "hygienic" sitting and standing postures, recommending to hold trunk, neck, and head erect. In 1889, it was reported that farmers and laborers often had back curvatures, which were either too flat or overly bent. Staffel and his colleagues were also concerned about the postural health of children and recommended that all be exhorted to maintain an erect posture of the back, neck, and head. Starting in the late 1880s, a great number of "hygienic and healthy" designs for school furniture were proposed. These ideas were also applied to office furniture with the intent to make office personnel sit "correctly upright" (Kroemer Elbert, Kroemer, and Kroemer Hoffman, 2018; Kroemer and Kroemer, 2017).

Straight back and neck: Seen from the side, a healthy spinal column is not "straight" but slightly bent forward in the neck and lower back regions and bent backward in between (see Chapter 2).

Comfort and discomfort: Corlett (2005), Helander (2003).

18.4 Designing the ergonomic office workstation

Ergonomic design of the office workstation: Discussion of the ergonomics of office furniture is not new. Recommendations by Kroemer and Robinette (1968) are still applicable today, even if typewriters have largely been replaced by computers. "Researchers generally agreed that the trunk, including the neck and shoulders, should be in a natural, upright, but relaxed position. The seated person must be able to select and change body position from time to time. Weight transfer through the thighs should be avoided; body weight should be transmitted mainly through the buttocks. The height of the seat should be adjustable so that the individual's feet can be placed firmly on the floor while the thighs remain horizontal. A back rest should be provided on all chairs to allow for temporary relief of weight form the spinal column. Chair and desk (or table) should be treated as a unit and the height of the working surface should be derived from the height of the seat surface."

Place the screen: Rempel, Willms, Anshel et al. (2007).

Space ... for finely controlled hand and forearm motions: Kroemer (1965).

References and further reading

Corlett EN (2005). The evaluation of industrial seating. Chapter 27 in Wilson JR, Corlett N (eds), *Evaluation of Human Work*, 3rd ed. London, UK, Taylor & Francis.

Grandjean E (1987). *Ergonomics in Computerized Offices*. London, UK, Taylor & Francis. https://doi.org/10.1201/b12564

Helander M (2003). Forget about ergonomics in chair design? Focus on aesthetics and comfort! *Ergonomics*, 46(13–14):1306–1319. https://doi.org/10.1080/00140130310001610847

Helander MG, Zhang L (1997). Field studies of comfort and discomfort in sitting. *Ergonomics*, 40(9):895–915. https://doi.org/10.1080/001401397187739

Howarth PA (2015). Assessment of the visual environment. Chapter 25 in Wilson JR, Sharples S(eds), *Evaluation of Human Work*, 4th ed. Boca Raton, FL, CRC Press. https://doi.org/10.1201/b18362

Kroemer AD, Kroemer KHE (2017). *Office Ergonomics: Ease and Efficiency at Work*, 2nd ed. Boca Raton, FL, CRC Press. https://doi.org/10.1201/9781315368603

Kroemer Elbert K, Kroemer HB, Kroemer Hoffman AD (2018). *Ergonomics: How to Design for Ease and Efficiency*, 3rd ed. London, UK, Elsevier. https://doi.org/10.1016/C2016-0-02516-0

Kroemer KHE (1965). *Ergonomic Aspects of Control Operation* (in German). PhD dissertation, Technical University Hannover.

Kroemer KHE, Robinette JC (1968). *Ergonomics in the design of office furniture: a review of European literature* (AMRL-TR 68-90). Wright-Patterson Air Force Base, OH: Aerospace Medical Research Laboratories. (Also published with shortlist of references in 1969). *International Journal of Industrial Medicine and Surgery*, 38(4):115–125.

Lundervold AJ (1951). Electromyographic investigations of position and manner of working in typewriting. *Acta physiologica Scandinavica Supplementum*, 24(84):1–171.

Mishan EJ, Quah E (2007). *Cost-Benefit Analysis*. London, UK, Routledge.

Parsons KC (2003). *Human Thermal Environments: The Effects of Hot, Moderate, and Cold Environments on Human Health, Comfort and Performance*. London, UK, Taylor & Francis. https://doi.org/10.1201/9781420025248

Parsons KC (2005). Ergonomic assessments of thermal environments. Chapter 23 in Wilson JR, Corlett N (eds), *Evaluation of Human Work*, 3rd ed. London, UK, Taylor & Francis.

Parsons KC (2014). *Human Thermal Environments*, 3rd ed. Boca Raton, FL, CRC Press. https://doi.org/10.1201/b16750

Rea MS (2005). Photometric characterization of the luminous environment. Chapter 68 in Stanton N, Hedge A, Brookhuis K, Salas E, Hendrick HW (eds), *Handbook of Human Factors and Ergonomics Methods*. Boca Raton, FL, CRC Press. https://doi.org/10.1201/9780203489925

Rempel D, Willms K, Anshel J, Jaschinski W, Sheedy J (2007). The effects of visual display distance on eye accommodation, head posture, and vision and neck symptoms. *Human Factors*, 49(5):830–838. https://doi.org/10.1518/001872007X230208

Saval N (2014). *Cubed: A Secret History of the Workplace*. New York, Doubleday.

CHAPTER NINETEEN

Design and use of computers

In just a few decades, computers have become essential work tools, widely used leisure devices, and entertaining gadgets. Computer technology has facilitated long-distance interactions. Much business communication that was once conducted via mail and telephone is now wireless. Computers open the world to many people, especially those who are sick or elderly, who would otherwise feel isolated; electronics enable them to communicate directly with others, shop and bank, and access news. The internet provides an abundance of information at the fingertips of anyone with just a few keystrokes.

QWERTY keyboards

However, even current computer keyboards are not well-designed from a human factors perspective. They still adhere to the 1878 typewriter layout, often in a miniaturized form. The QWERTY arrangement complicates typing unnecessarily, making it time-consuming and potentially leading to repetitive strain injuries in the hands and arms of users. For more than a century, inventors have proposed various alternatives to replace the outdated keys and keyboards; yet, widely adopted solutions have so far failed to emerge.

This chapter explores ergonomic aspects of computer design and operation, deriving suggestions for better, more user-friendly designs. This review illustrates a typical process for solving a technical problem: starting with a workable prototype, incrementally improving it, and ultimately seeking a fundamentally different, effective solution that meets users' needs and accommodates human ergonomics.

DOI: 10.1201/9781032674575-25

19.1 Sholes's "typewriting machine" with its QWERTY keyboard

Throughout the 1800s, numerous inventors proposed a variety of typographical devices that, through manipulation of an input device (usually keys of some sort), generated imprints of letters on paper. Given the technology available at that time, the inner workings of these "typewriting machines" relied on a complex set-up of mechanical levers. The overriding technical challenge was to find workable mechanisms. Thus, the inventors of these machines paid little attention to the usability of the input side: the design and arrangement of the keys themselves.

Sholes's Typewriting Machines

One of these inventors, Christopher Latham Sholes (1819–1890), was granted eight US patents for various designs of a type-writing machine from 1868 on. Two patents were issued in 1868, both with Carlos Glidden and Samuel Soulé, followed by one patent in 1876 with Matthias Schwalbach. In 1878, he obtained five US patents: 199,382; 200,351 (with Glidden); 207,557; 207,558; and finally 207,559.

Sholes's keyboards

The first two patents, both in 1868, featured rectilinear rows of keys. The first patent (79,265) shows 21 unmarked keys, described as being "similar to the keyboard of a piano," with 10 shorter keys atop 11 longer keys, akin to the white and black clavier keys. The 36 keys in patent 79,868 are also described as "similar to the keys of a piano or melodeon," but they all lie in one plane, side by side, alternating in length. The keys on the left side, as the operator sees them, display numbers in increasing order, while the keys in the middle and on the right side carry alphabetic letters (possibly in alphabetical order). Neither patent provides an explanation for the choice of key layout beyond the quoted statements.

The 1876 patent (182,511) and the first four patents of 1878 exhibit arrays of three straight rows of button-like keys affixed to lever-type bars: 32 keys in the 1876 patent and 21 keys in the 1878 patents. No labels with letters or numbers are present on the buttons, and the texts of the patents provide no further explanation or description.

QWERTY keyboard

Sholes's final patent (207,559), contains fourteen specific technical claims, but none refer to the key layout. One drawing in this patent shows a frontal view of four straight and horizontal rows of key tops. The rows are staggered in height so that the row closest to the

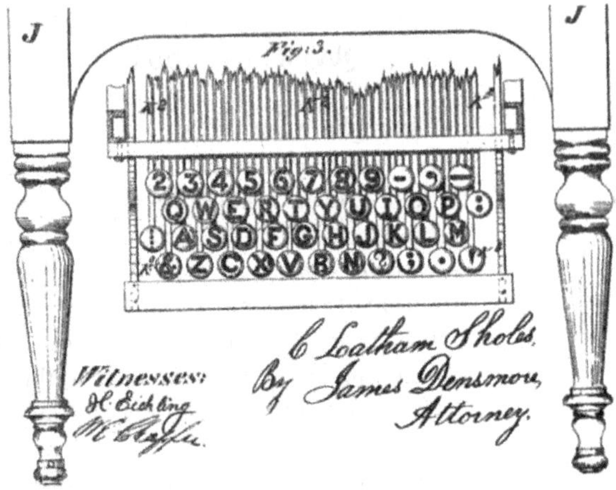

FIGURE 19.1 QWERTY keyboard in Sholes's 1878 US patent 207,559.

operator is the lowest and the farthest row is the highest. Another drawing depicts a top view of the four straight rows, each with 11 round keys, for a total of 44 keys—see Figure 19.1. The top surfaces of the keys carry inscribed numerals, letters, and signs.

Why QWERTY?

We know of no statements by Sholes or his co-inventors or contemporaries, either in their patents or reported elsewhere, that explain why the specific keyboard layouts were chosen in the patents. However, the layout in the final 1878 patent (207,559) shows remnants of an alphabetic arrangement. Sholes was a printer by trade, so we may surmise parallels to the arrangement of printer's type cases, where pieces were presumably organized by convenience of use rather than strictly alphabetically. Another possible reason for the arrangement of the characters and keys may have been the intent to avoid collisions or sticking together of type bars in the then-used mechanical lever mechanisms when activated in quick succession. This could have led to a separation of certain bars and, consequently, the keys on the keyboard. Letters that frequently pair may have been placed relatively far apart to allow for faster typing since different hands then type the letter pairs. Alternatively, the key layout might have been designed for the convenience of telegraph operators who were transcribing Morse code. However, there is no contemporaneous evidence to support any of these speculations. Sholes himself continued to change the key arrangement, as seen in his final patent 568,630 (awarded posthumously), where the six leftmost keys are not QWERTY but XPMCHR.

19.2 From typewriter to computer keyboard

Sholes's 1878 invention became a global success, and his puzzling layout of the keyboard is still in general use, with only the positions of X and C exchanged and the M moved to the first row in English-language versions. The six leftmost keys on the third row, counted from the operator, carry the letters *QWERTY.* Even on today's computers, the term "QWERTY keyboard" serves as a shorthand for any arrangement in which the letter keys essentially follow Sholes's keyboard layout (in countries that use the Latin alphabet). Some countries have QZERTY, AZERTY, QWERTZ, and other national variants, accommodating local variations in letters and punctuation. Countries that do not use a Latin alphabet often still utilize the QWERTY keyboard layout with additional software to convert into, for example, Indian scripts or Chinese characters.

20th-century typewriters

None of the numerous patents and proposals for alternate key and keyboard designs* have been commercially successful. Changes within the Sholes layout were technically difficult as long as the mechanism relied on mechanical levers. The first significant change to typewriter designs was the introduction of electric motors in the 1950s, which supplied auxiliary energy to strike the type bar (later, a type-ball on some machines) against the inked ribbon to make the imprint on paper; until then, the typist's hands had provided that energy. In the 1960s, fingertip-operated switches controlling electric and then electronic circuitry began to replace the lever mechanisms of the keys. This drastically reduced both the finger force needed to actuate the key and the displacement of the key, diminishing the dynamic work (the integral of force and distance traveled) required from the operator's fingers for every keystroke.

Typewriters and computers

Starting in the 1960s, electronics replaced the mechanical innards: typewriters transformed into personal computers. The new technology would have easily allowed the keys to be relocated and the keyboard to be redesigned. However, instead of creating novel solutions, keys were simply added to the original keyboard. These were mostly so-called function keys, placed around the perimeter of the QWERTY set. With an extra numeric keypad and an additional cursor control keypad (both commonly on the right-hand side), by the 1980s, the total number of keys was customarily just over one hundred, and in some cases about 125—more than twice as many keys as on the old Sholes typewriter. The use of so many keys required larger keyboard space, which then necessitated larger finger and hand movements. Mouse, trackball, touch pad, and other accessories created new tasks and required new body motions for the keyboard user.

The 1980s saw a wave of computers entering modern offices, sweeping out the remaining mechanical and even most of the newer electric typewriters. For example, IBM introduced its first personal computer in 1981; only 10 years later, it stopped producing typewriters. Two decades after that, IBM sold off its entire PC business to a Chinese firm.

From around 2000 on, wireless mobile smart telephones became immensely popular worldwide. They are often used for texting or sending instant messages typed on miniature keyboards, which still largely follow Sholes's key layout.

19.3 Considering human factors for keyboarding

Obviously, the keyboards on early typographic devices, including Sholes's 1878 invention, were not "human-engineered." Nevertheless, his QWERTY layout became by default the most commonly manufactured keyboard when his type-writing machine became a global success.

Body posture and effort

Current desktop computer technology imposes two postural requirements on the user that are similar to what typists had to do: focusing the eyes on the display while keeping the hands on the keyboard. Such prescribed locations of the eyes and hands fix the overall positions of the head and upper body, thus allowing little variation in body posture. Even portable laptop computers impose similar posture requirements on their users.

Overloading typists

The spatial arrangement of the keyboard and its keys (as per Sholes's design) forced typists' arms into a strong inward twist (pronation) and the hands into a lateral bend (ulnar deviation) at the wrist—similar to playing on a piano keyboard—and required complex motions between the misaligned keys. The unfortunate combination of forced posture and strenuous effort overloaded many typists' hands, wrists, arms, shoulders, and necks.

1920s typists

EA Klockenberg provided an illustration in 1926, reproduced as Figure 19.2, of a typical typist. In a moving narrative, Klockenberg described how young women, after just a few years as typists, found themselves with painful hands, unable

FIGURE 19.2 Klockenberg's illustration of the posture of a 1920s typist.

to type or perform everyday manual tasks, and even incapable of lifting their small children. The typists' musculoskeletal injuries were the result of the heavy repetitive work required to operate their typewriters in unsuitable body postures. Today, the posture of a keyboarder is usually much less contorted, but still constrained by the need to keep the fingertips on the keys and the eyes directed at the display.

Heidner's 1915 keyboard designs

Among the early proposals for improved keyboards, one stands out: In 1915, Fritz Heidner was awarded US patent 1,138,474 for his novel keyboard layouts. In his patent, he wrote that his keyboard designs "enable the operator to obtain a better view of the keys and write with greater ease, in a less cramped position ... and there being thus much less strain ... writing is rendered considerably less fatiguing" and that his design "facilitates operation of the keys by the fingers in their natural position ... in accordance with the natural form of the hand." Figure 19.3 illustrates how Heidner divided the keyboard into left and right halves; on them, he arranged the keys in various layouts that predate many ergonomic recommendations proposed by others long after 1915. Surprisingly, his patent was apparently forgotten or overlooked until the 1980s.

Repositioning keys

Keying performance* has been an issue since the early years of keyboarding. Measures of performance commonly focus on the total number of keystrokes completed during a certain time, as well as on the ratio between correct and incorrect strokes. Starting in the early 1900s, several patents for new key arrangements appeared that were designed to improve typing

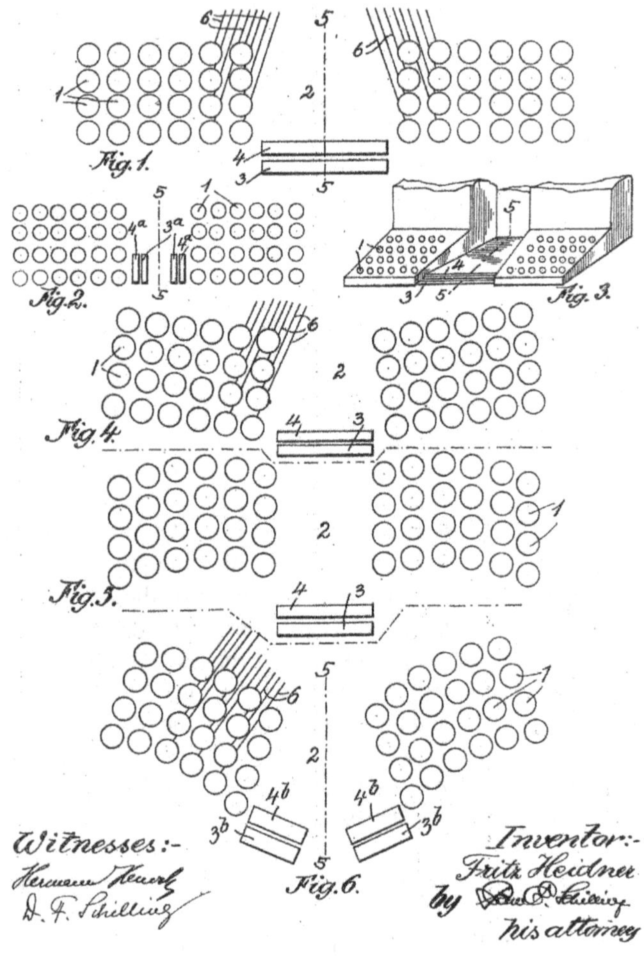

FIGURE 19.3 Alternative keyboard designs in Heidner's 1915 US patent 1,138,474.

performance by addressing the shortcomings of the QWERTY layout. These new designs usually relocated keys* but retained Sholes's original layout of offset columns and straight rows of keys. The "Simplified Keyboard" proposed by August Dvorak (1894–1975) is probably the best known and most long-lived of all these plans, introduced in 1936 and 1943. However, a comparative test in 1956 showed that it took a long time to retrain typists to become proficient on Dvorak's layout; they still made more errors on it than on their customary QWERTY keyboard and gained less improvement in typing rate with ongoing training. These findings apparently discouraged further attempts to reposition certain keys on the regular keypad. Even several well-reasoned proposals to use alphabetic layouts* have not found wide acceptance.

Repetitive injuries during the 1800s

The physician George Vivian Poore (1843–1904) stated in 1872 and 1887 that he and other authors, following the 1713 observations from Ramazzini, had traced writer's cramp to "muscular impotence" and "spasms" caused by the often repeated use of the same muscles. Poore stated that this health problem was not limited to writers but also occurred in tailors, cobblers, fencing masters, and musicians. He wrote that one might speak of pianist's cramp because this overuse disorder occurred so frequently among pianists. In 1892, physician William Osler (1849–1919) also associated these spasms "with continuous and excessive use of muscles in performing a certain movement" in writers and musicians, as well as in operators of the Morse key apparatus who suffered from a disabling injury then called *telegraphist's wrist.**

Keying "myalgia"

Thus, we can see that since the second half of the 1800s, the associations between repetitive strain disorders and repetitive muscle overuse in certain occupations have been well established. It was common knowledge that players of keyboard instruments are at risk of developing chronic musculoskeletal disorders. After Sholes's typewriter became widely used, the same "myalgia" (i.e., muscle aches and pains) appeared in typists as well.

Muscles used in typewriting

In 1951, AJ Lundervold published the first report of his groundbreaking studies to gain knowledge about the use of individual muscles while typewriting. He measured electromyographic (EMG) signals from 47 healthy typists and 88 patients, most of whom were suffering from "occupational myalgia" in muscles that had been overstrained during repetitive typewriting. Lundervold demonstrated that changes in the EMG records reflected the actual muscle activity and the status of muscle fatigue. His experiments provided support for the long-held belief that repetitive typing can lead to a cumulative overexertion injury, even when single keystrokes are not harmful by themselves.

Occupational overuse disorders

In Europe, ailments such as tendinitis, tenosynovitis, and tendovaginitis of the upper extremities became recognized as occupational repetition-related musculoskeletal diseases* of typists in the late 1950s. Researching the causes of cumulative trauma disorders (especially carpal tunnel syndrome, CTS) associated with repetitive keyboarding activities became a topic of engineering and medical concern. In the late 1990s, the US National Research Council convened a workshop on work-related musculoskeletal disorders. In the summary of its findings, the NRC (1999, p. 59) stated that musculoskeletal disorders are multifactorial since individual, social, and organizational factors can contribute to their occurrence; however, the biomechanical demands of work constitute the most important risk factors.

Overuse pathomechanics

A slow keyboard operator, tapping just 20 words per minute (with five letters per word), performs 12,000 keystrokes (20 × 5 × 60 × 2) during a 2-hour work session. A fast keyer, typing 100 words per minute over 6 hours, performs 180,000 keystrokes. Each stroke requires digit flexion followed by digit extension. These many thousands of motions can create pathomechanical conditions*.

The biomechanical stresses that generate cumulative trauma disorders, especially CTS, became well researched in the 1980s and 1990s. The anatomy of the human hand and wrist, along with its kinematics* while keyboarding, explains the overuse conditions that particularly afflict the flexor tendons* of the digits in the carpal tunnel: see Figure 2.6 in Chapter 2. The forces in the flexor tendons are often more than three times stronger than the impulses that the fingertips transmit to the tops of the keys. The longitudinal travel of the tendons can be up to 2 cm, and wrist position strongly affects the ease of their gliding within synovial sheaths. Inflammation of the tendons and their sheaths causes swelling of the tissues within the limited space of the carpal tunnel. The resulting pressure can damage the median nerve, leading to the CTS that affects hand function.

Anthromechanical issues

Sholes's keyboard concept, with its mechanical keys, required efforts from the typists that often exceeded their musculoskeletal capacities. The underlying biomechanical problems arose from several categories of work demands: the physical energy required to accelerate the masses of the typewriting machine mechanisms; the postures of the hands, arms, upper body, and neck; the extensive use of primarily the flexor muscles and tendons in the typists' hands; and the rate and repetitiveness of muscle contractions and tendon motions.

Human factors design recommendations

Attempts to lighten the typist's load resulted in many proposals and patents from the early 1900s onward. The mechanical nature of the machinery available at the time severely hindered new design solutions. In the 1960s, the terms "ergonomics" and "human (factors) engineering" emerged. Consequently, the importance of fitting tasks and equipment to the human became widely recognized; reconfiguring the keyboard to suit the size and mobility of the hand became an important challenge. In 1969, Remington and Rogers compiled more than 300 publications on keyboard entry devices. In the same year, Kinkead and Gonzalez formulated "human factors design recommendations for touch-operated keyboards." No new keyboard designs were introduced, even with the emerging use of electric and then electronic circuitry in typewriters; standards in the USA* in 1968 and 1988 still incorporated the QWERTY layout while adding more keys on the sides of the keyboards. Reviews of keyboard design issues and novel solutions for operator-centered designs of keys, keyboards, and keyboarding workstations, as well as proper ergonomic practices for keying work, appeared in the international literature*.

19.4 Designing other possible solutions

Customary computer keyboards

Neither the original QWERTY keyboard nor its derivatives have been truly "human re-engineered" in spite of Heidner's 1915 proposals and the many other designs that followed. Even on today's customary computer keyboards, all keys are arranged side by side in straight rows (while our fingertips are not). The columns of keys follow two different design rules: on the QWERTY set, the columns zigzag slanted to the left (as seen by the operator); on other key sets, however, the columns are usually straight. That irrational design has largely been transferred to portable computers, then miniaturized and applied to mobile smartphones, where the small size of the keys seems to intensify the old QWERTY layout problem, even if the columns of keys are stratified.

Designing for "big" changes

Design issues intermingle with existing technology, user expectations and practices, and marketing and fashion. In spite of these interactions, for clarity's sake in the following discussion, improvement ideas will be assigned to certain categories. Obviously, solutions can combine aspects from two or more categories, such as the design of keys, keyboards, and alternate input devices.

A major argument against changing the basic QWERTY design of the Latin-alphabet keyboard is the expectation that altering the layout would require retraining the operators and hence would slow keying performance, at least in the short term. This reasoning is probably valid for most minor changes, such as exchanging the letter designations of certain keys (as in Dvorak's proposal). In these cases, the new skills are so close to the old habits that confusion is indeed likely. However, if there is a basic change in design, such as relocating large sets of keys or employing keys that have a clearly different operational mode (e.g., ternary keys or joysticks instead of binary keys), then the new procedure is distinctly dissimilar from the old ways, and there may be little or no interference from carryover of previous practices. Humans are remarkably fast at acquiring novel keyboarding skills, as demonstrated by research* and by the phenomenally rapid spread, across the globe, of texting on miniature keyboards, which, according to experiments and design principles, should be practically unusable*.

Redesigning the key

Changing the basic key design can have major consequences for the user's efforts and performance. The traditional mechanical binary key has two activation states: "off" when up, "on" when pressed down. However, electrical switch and electronics technology make it easy to replace tap-down binary keys with devices that can establish or break connections through different actions. Just a few examples are as follows: we can use a ternary device (having three states); or a device that responds to movements in many directions other than just down (like a joystick); or a force controller that does not require appreciable displacement for activation; or a multifunction gadget that can react to combinations of positions, displacements, and contact modes; or a sensor that responds to user body motion or to sound.

Redesigning the keyboard

Heidner's 1915 patent included several significantly improved designs. Numerous later proposals, in essence, repeated Heidner's ideas to rearrange the angles of slant, slope, and tilt of key sets, to change key spacing, and to split* the keyboard into left and right sections in order to make keying less strenuous and to improve typing performance—see Figure 19.4. In 1962, US patent 3,022,878, assigned to IBM, positioned the keys like a glove around the hands*. Several keyboards, commercially available since the 1900s, have similar features, in some cases with adjustable arrangements.

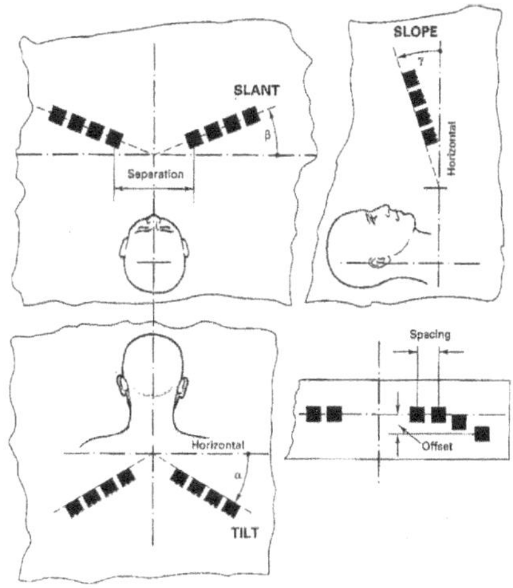

FIGURE 19.4 Terms describing keyboard angles and key positions: separation, slant, slope, tilt, spacing, and offset. Adapted from Kroemer (2001).

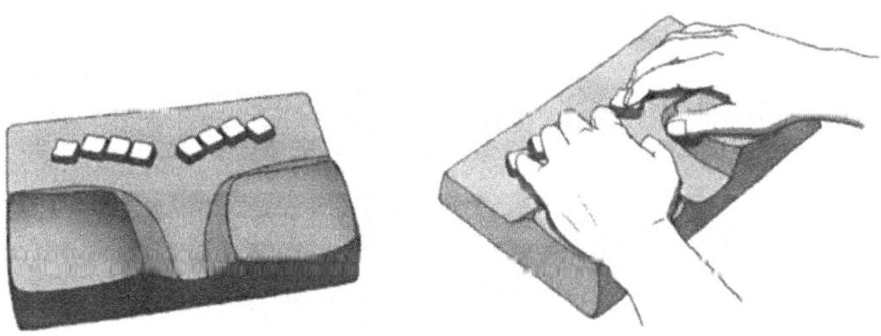

FIGURE 19.5 A ternary keyboard with four keys for each hand, as described in Langley's 1988 US patent 4,775,255.

Ternary keys

Keyboard design decisively depends on the keys used. The choice of ternary instead of binary keys in Lawrence Langley's 1988 US patent 4,775,255 provides a good example. Here, keys are not tapped (binary *on/off*) but are toggled. Each toggle key can assume three different positions: *off* when in its center position, *first on* when pulled from there, and *second on* when pushed. Langley used *chorded* activation: simultaneous activation of two or more keys (similar to a musical chord) generates the desired letter or other character. Accordingly, his keyboard needs only eight keys, four for each hand, as illustrated in Figure 19.5. The hands can rest on built-in support pads. The keys support the fingers, which do not move from one key to another but stay on their assigned key. We can see that choosing other key types can result in different keyboards.

Repositioning the keyboard

The traditional piano keyboard, which served as a model key set for many early typographic machines (including those in Sholes's early patents), can be divided into several sections that can be arranged side by side or elevated at different levels, as seen in organs and modern electronic music keyboards. In his final 1889 patent, without explaining why he did so, Sholes used a keyboard with the keys arranged in a 4 × 11 pattern: 4 straight horizontal rows of keys and 11 columns of zigzagging and rising keys. This layout required tiring arm positions and complex finger movements, as well as significant effort from the typist to exert on the keys, leading to postural problems and repetitive musculoskeletal trauma. In 1915, Heidner was apparently the first patent holder to attempt to address these postural problems by dividing the keyboard into sections and repositioning them at angles and locations that are more suitable for the operator. Many subsequent proposals and patents presented similar ideas*, some with adjustment features. So far, however, the more radical designs have not conquered the market.

Resizing keys and keyboard

Recent electronic, easy-touch, lightweight, and wireless keyboards allow users to place them in various locations on a support surface, on their laps, or elsewhere as desired. Being able to relocate the keyboard can help alleviate many of the past postural problems associated with stationary (fixed-location) keyboards. However, one new problem has emerged: current mobile smartphones have such extremely small keypads that many users are tempted to operate them with one or two thumbs while cradling the phone in their hand. This is resulting in a new overuse disorder, commonly and descriptively called *texter's thumb.*

19.5 Designing alternatives to keyboards

There is no overwhelming reason to use the principle of "one keystroke for every bit (letter, numeral, sign) of input." That idea follows the tradition of Western writing; it is exact but also inefficient and can cause repetitive strain injuries to the user's hands and arms. Even most virtual keyboards on touch-sensitive screens only alleviate the force used to operate physical keys; they still require hand positioning and posture. Using fewer keys, moveable keys, virtual keys, or no keys at all would alleviate, or even completely avoid, past and current overexertion problems and facilitate body movements instead of the fixed posture caused by "hands on keys, eyes on display." Ways to communicate with the computer without manipulating keys are of great interest for everyday use, especially for persons with physical disabilities and impairments. Several approaches are already in use.

Alternative keys and keyboards

Chords, traditionally used by pianists and other musicians, combine single tones to convey complex sounds by simultaneously triggering several keys instead of activating them sequentially. Chording through the simultaneous activation of two or more keys has been repeatedly proposed for typewriter and computer operation to replace the sequential "one key for one character" procedure. Stenographers also use chords on stenotype machines to allow for faster transcription of human speech. Langley's 1988 US patent, already mentioned, is a recent example of chording design, which can simplify keyboards, cut the number of keys, reduce hand movements from key to key, and altogether, make key input faster while relieving hand and arm strain.

Alternative input devices

With advances in technologies, sensors, and activation devices, non-keyboard input devices are ever-increasing. Today, there are many available input devices to move a cursor on the computer monitor to select and manipulate items, open and close files and applications, and even sketch letters and drawings:

- *Mouse or puck*: usually a palm-sized, hand-contoured block with one or more finger-operated button(s), often slid over a surface (mouse pad).
- *Trackball*: a modification of a mouse that includes a ball mounted in an enclosure whose spinning surface moves the cursor; it may also be incorporated into a hand-held unit with other controls.
- *Joystick controller*: a short lever that pivots on a base to respond to angle or direction, usually with buttons for fingertips.
- *Stylus or light pen*: a pencil-shaped, hand-held device used on a contact-sensitive surface (often preferred for drawing).
- *Touch- or contact-sensitive screen or trackpad*: used with a stylus or fingertips.

Speech and sound recognition

There are means to generate inputs to the computer that require no keys of any kind. We already communicate by voice, using speech-to-text recognition programs that work well (with training) within confined topic areas, such as computer control commands, medical evaluations, or music composition. For general text, both speed and recognition accuracy still have opportunities for improvement (careful proofreading is highly recommended). It is likely that as improvements in speech recognition are made, conversing with the computer to request information or actions will continue to proliferate. Advances in artificial intelligence (including machine learning and natural language processing and understanding) allow *digital assistants* to act as an interface to the computer—more on this in Chapter 23. Of course, many other sounds could transfer information to the computer: whistles, grunts, claps, and so forth, depending on the user's needs and capabilities

Non-verbal communication

Another set of solutions employs the recognition of body movements*. Some examples are listed below:

- The eyes for tracking;
- The mouth for lip movement, use of the tongue, or a blow/suck tube;
- The face for grimaces and other expressions;
- The head for positioning;
- The hands and fingers for sign language and gestures in addition to activating control devices;
- The arms for gestures, making signs, and actuating control devices;
- The torso for positioning and operating actions;
- The legs for gestures and for moving or pressing devices; and
- The feet for motions and gestures, for moving and actuating devices.

These design solutions require sensors that respond to the positions, movements, and forces of the operator's body: cameras, force platforms, motion sensors, position detectors, instrumented gloves, and similar devices are among the long-established technologies in the larger field of virtual and augmented reality. Other approaches use surface EMG associated with muscle actions*; at the moment, employing neural signals such as electroencephalograms to interface with brain activities is still under development. Certainly, there are many as-yet unexploited possibilities, including *nanoergonomics**, which may make it possible to access the human central nervous system and integrate important features of human cognition into computers to allow for direct interaction.

19.6 Designing for new syntax and diction

Changing language

Klemmer (1958) and Lockhead and Klemmer (1959) recognized how changes in syntax and diction—new abbreviations, words, and phrases—can interact with changes in key and keyboard technology. Stenographers and court reporters have long demonstrated that using shorthand code for words and parts of sentences, chunking, batch processing, and related techniques save time and effort. The telegram style in the middle of the last century created a new shortened mode of written expression; in recent times, email, text messages, and instant messaging have changed communication methods, word-writing modes, spelling, abbreviations, and punctuation: *thru* replaces *through*; *BRB* quickly communicates that the other person will *be right back*; *thx* and *rgds* demonstrate that most text is decipherable without vowels. Emoticons and emojis can convey complex messages and add emotional context.

Language changes constantly. New technologies bring shifts in communication; conversely, transformations in our language can also provide designers with opportunities to create novel devices for communicating with computers.

19.7 Designing with "smart" software

Software

Cleverly designed software facilitates work with computers by reducing the demands on manipulative skills and short-term memory*. Software can use context to distinguish between *mail* and *male*, or *then* and *than*; it can complete words and phrases after the input of only a few initial letters; it can provide a set of stock responses to certain situations; it can anticipate the use of specific words; and it can correct sentence structure and insert punctuation. With further advances in artificial intelligence and augmented reality, there is much more to come that will require varying degrees of human intervention and resulting ergonomic considerations (see also Chapter 23).

19.8 Combining solutions

Today's electronics and tomorrow's novel technologies provide fundamentally new alternatives to Sholes's typewriting machine design. We may sort these into certain categories, but it is the combination of solutions that is most likely to succeed: for example, the blending of traditional keys with speech recognition and smart software. However, this approach still essentially employs the long-outdated QWERTY key arrangement, which makes its use difficult for persons with disabilities and, in fact, everyone else, because it requires repetitive manipulation and constrains the operator's body posture via "hands on keys, eyes on monitor."

One essential design decision is to move away from the principle of "one binary keystroke for every bit (letter, numeral, sign) of input" and instead use some sort of code or shortcut supported by smart software. The general acceptance of innovative mobile smartphones and tablets, and of gaming devices, for example, shows that the public is willing to accept, even embrace, fundamentally new techniques in human–computer interaction. These are likely to come together with technology-related neologisms in our language. Ergonomic innovations will address both input and display to improve the loop of feed-forward and feedback between computer and human.

Summary

The current challenge for applying human engineering to the computer workplace is specific to existing computer technology. Prescribed locations of the eyes and hands fixate the overall position of the head and upper body, which does not allow much variation of body position over time. New human–device interface technology can free the user from these constraints and allow for more freedom in choosing and changing body posture.

Since the nature of repetitive trauma disorders (and thus the means to avoid them) was well documented even in the 1960s, we might have expected that key and keyboard designs for typewriters and then for computers would have been improved accordingly. However, this did not happen. Even current electronic devices essentially follow established paths; while they created novel uses, they perpetuated the old key-related overuse problems. New technologies, both those currently in use and those still emerging, allow designers to establish innovative solutions for human–computer interfaces that make communication easier and more efficient.

One argument against changing the basic QWERTY design of the Latin-alphabet keyboard has been that altering the layout would require operator retraining and would initially slow their typing performance. This reasoning is probably valid for most minor changes, such as swapping the letter designations of keys or slight repositioning. In such cases, carryover confusion is indeed likely because the new uses are close to old habits. However, fundamental changes in key and keyboard design, or even abandoning these input devices altogether, create distinctly novel use procedures: when new processes are very different from the old ways, there is little interference from crossover or confusion based on previous practices.

We humans are remarkably willing to quickly acquire novel keyboarding and other input skills. A striking recent demonstration is the phenomenally fast spread, across the globe, of texting and typing via miniature keys and keyboards on mobile smartphones, along with associated new use procedures.

Steps to fit the human

Step 1: Determine what tasks will be performed with the computerized device, where it will be located, and who will operate it.

Step 2: Consider the specific needs and preferences of the computer operator, as well as the other equipment and tools that will be used. Involve the users in the design process. Where possible, add flexibility to the design of the computer system, its input devices, and any associated workstation or furniture to enable user adjustments.

Step 3: If the users or their needs change, consider how this can be accommodated. Ask the users if the computer set-up and furnishings suit them, and make adjustments as needed.

Step 4: Monitor conditions, modify the computer set-up, workstation, or work tasks and iterate as necessary.

Notes

The text contains markers, *, to indicate specific references and comments, which follow.

19.2 From typewriter to computer keyboard

Patents and proposals for alternate key and keyboard designs: Compiled by Kroemer (2001, 2010), Kroemer and Kroemer (2017), Kroemer Elbert, Kroemer, and Kroemer Hoffman (2018).

19.3 Human factor considerations for keyboarding

Keying performance: Pereira, Hsieh, Laroche et al. (2014).

Relocated keys: Dvorak's "Simplified Keyboard" of 1936 and 1943 and Strong's 1956 report.

Proposals to use alphabetic layouts: Kroemer (2001, 2010).

Repetitive injuries in the 1800s and earlier: Osler (1892), Poore (1872, 1887), Ramazzini (1713).

Occupational repetition-related diseases: Lundervold (1951), Arndt and Putz-Anderson (2014), Garg and Marras (2014), Kroemer (2001, 2010), National Research Council (1999).

Pathomechanical conditions: Armstrong (2006), Arndt and Putz-Anderson (2014), Kumar (2001), Marras and Karwowski (2006), Putz-Anderson (1988).

Anatomy of the human hand and wrist and its kinematics: Baker, Cham, Cidboy et al. (2007), Freivalds (2006), Hughes and An (2007), Lee and Healy (2005), Marras and Radwin (2006).

Overexertion conditions that especially afflict flexor tendons: Ettema, Zhao, Amadio et al. (2007), Goodman and Choueka (2005), Dennerlein (2005, 2006), Freivalds (2006, 2011), Ugbolue, Hsu, Glitz et al. (2005), Zhao, Ettema, Osamura et al. (2007).

19.4 Input-related anthromechanical issues

US Standards: 1968, Proposed USA Standard for a General-purpose Keyboard; 1988, ANSI/HFS 100-1988 Standard.

Reviews... in the international literature: Marklin and Simoneau (2004), Kinkead and Gonzalez (1969), Klemmer (1958), Kroemer (2001, 2010), Noyes (1983a, 1983b), Remington and Rogers (1969).

19.5 Possible design solutions

Acquiring novel keyboarding skills as demonstrated by research: Anderson, Mirka, Joines et al. (2009), McMulkin and Kroemer (1994), Kroemer (2001), Marklin and Simoneau (2004).

Miniature keyboards ... should be practically unusable: Pereira, Hsieh, Laroche et al. (2014).

19.6 Design alternatives for keyboards

Split the keypad: Rempel (2008).

Placing the keys around the hands: Noyes (1983b) and Rempel (2008) compiled information and commented on new designs.

Subsequent proposals and patents presented similar ideas: Dennerlein (2006), Kroemer (2001, 2010), Noyes (1983b), Rempel (2008).

Recognition of body movements: Kroemer (2001), McMillan (2001), McMillan and Calhoun (2001), Pereira, Wachs, Park et al. (2015).

Use of EMGs: Reddy and Gupta (2007).

Nanoergonomics: Karwowski (2006)

19.8 Designing "smart" software

Reducing the demands on manipulative skills and short-term memory: Czaja and Lee (2001, 2002); Fisk, Rogers, Charness et al. (2009).

References and further reading

Anderson AM, Mirka GA, Joines SM, Kaber DB (2009). Analysis of alternative keyboards using learning curves. *Human Factors*, 51(1):35–45. https://doi.org/10.1177/0018720808329844

ANSI/HFES 100 1998 (2007). Human Factors Engineering of Computer Workstations (2024 revision expected).

Armstrong TJ (2006). The ACGIH TLV for hand activity work. Chapter 41 in Marras WS, Karwowski W (eds), *The Occupational Ergonomics Handbook: Fundamentals and Assessment Tools for Occupational Ergonomics*, 2nd ed. Boca Raton, FL, CRC Press, pp. 421–442. https://doi.org/10.1201/9781420003635

Arndt R, Putz-Anderson V (2014). *Cumulative Trauma Disorders*. Boca Raton, FL, CRC Press.

Baker NA, Cham R, Cidboy EH, Cook J, Redfern MS (2007). Kinematics of the fingers and hands during computer keyboard use. *Clinical Biomechanics*, 22(1):34–43. https://doi.org/10.1016/j.clinbiomech.2006.08.008

Czaja SJ, Lee CC (2001). The internet and older adults: design challenges and opportunities. In Charness N, Parks DC, Sabel BA (eds), *Communication, Technology, and Aging: Opportunities and Challenges for the Future*. New York, Springer, pp. 60–80.

Czaja SJ, Lee CC (2002). Designing computer systems for older adults. Chapter 21 in Jacko J, Sears A (eds), *The Human-Computer Interaction Handbook: Fundamentals, Evolving Technologies and Emerging Applications*. New York, Lawrence Erlbaum and Associates (IEA), pp. 1384.

Dennerlein J (2006) The computer keyboard system design. Chapter 39 in Marras WS, Karwowski W (eds), *The Occupational Ergonomics Handbook: Interventions, Controls, and Applications in Occupational Ergonomics*, 2nd ed. Boca Raton, FL, CRC Press, pp. 848. https://doi.org/10.1201/9781420003642

Dennerlein JT (2005). Finger flexor tendon forces are a complex function of finger joint motions and fingertip forces. *Journal of Hand Therapy*, 18(2):120–127. https://doi.org/10.1197/j.jht.2005.01.011

Dvorak A (1943). There is a better typewriter keyboard. *National Business Education Quarterly*, 12:51–58.

Ettema AM, Zhao C, Amadio PC, O'Byrne MM, An KN (2007). Gliding characteristics of flexor tendon and tenosynovium in carpal tunnel syndrome: a pilot study. *Clinical Anatomy*, 20(3):292–299. https://doi.org/10.1002/ca.20379

Fisk AD, Rogers WA, Charness N, Czaja SJ, Sharit J (2009). *Designing for Older Adults: Principles and Creative Human Factors Approaches*. Boca Raton, FL, CRC Press. https://doi.org/10.1201/9781420080681

Freivalds A (2006). Upper extremity analysis of the wrist. Chapter 45 in Marras WS, Karwowski W (eds), *The Occupational Ergonomics Handbook: Fundamentals and Assessment Tools for Occupational Ergonomics*, 2nd ed. Boca Raton, FL, CRC Press, pp. 361–404. https://doi.org/10.1201/9781420003635

Freivalds A (2011). *Biomechanics of the Upper Limbs: Mechanics, Modelling and Musculoskeletal Injuries*, 2nd ed. Boca Raton, FL, CRC Press, pp. 564. https://doi.org/10.1201/b11547

Garg A, Marras WS (2014). Epidemiological studies of workplace musculoskeletal disorders. *Special Issue of Human Factors*, 56(1):5.

Goodman HJ, Choueka J (2005). Biomechanics of the flexor tendons. *Hand Clinics*, 21(2):129–149. https://doi.org/10.1016/j.hcl.2004.11.002

Hughes RE, An KN (2007). Biomechanical models of the hand, wrist, and elbow in ergonomics. Chapter 14 in Kumar S (ed), *Biomechanics in Ergonomics*, 2nd ed. Boca Raton, FL, CRC Press, pp. 20.

Karwowski W (2006a). From past to future: building a collective vision for HFES 2020. *HFES Bulletin*, 49(11):1–3.

Kinkead RD, Gonzalez BK (1969). *Human Factors Design Recommendations for Touch-Operated Keyboards – Final Report* (Document No. 12091-FR). Minneapolis, MN, Honeywell, Inc.

Klemmer ET (1958). *A Ten-Key Typewriter* (Research Memorandum RC-65). Yorktown Heights, NY, IBM Research Center.

Klockenberg EA (1926). *Rationalisierung der Schreibmaschine und ihrer Bedienung* [Rationalization of the typewriter and its operation]. Berlin, Springer.

Kroemer AD, Kroemer KHE (2017). *Office Ergonomics: Ease and Efficiency at Work*, 2nd ed. Boca Raton, FL, CRC Press. https://doi.org/10.1201/9781315368603

Kroemer Elbert K, Kroemer HB, Kroemer Hoffman AD (2018). *Ergonomics: How to Design for Ease and Efficiency*, 3rd ed. London, UK, Elsevier. https://doi.org/10.1016/C2016-0-02516-0

Kroemer KHE (2001). Keyboards and keying. An annotated bibliography of the literature from 1878 to 1999. *Universal Access in the Information Society UAIS*, 1(2):99–160. https://doi.org/10.1007/s102090100012

Kroemer KHE (2010). 40 Years of human engineering the keyboard. *Proceedings of the 54th annual meeting of the Human Factors and Ergonomics Society*, 54(15):1134–1138. https://doi.org/10.1177/154193121005401509

Kumar S (2001). Theories of musculoskeletal injury causation. *Ergonomics*, 44(1):17–47. https://doi.org/10.1080/00140130120716

Lee JC, Healy JC (2005). Normal sonographic anatomy of the wrist and hand. *Radiographics*, 25(6):1577–1590. https://doi.org/10.1148/rg.256055028

Lockhead GR, Klemmer ET (1959). *An Evaluation of an Eight-Key Word-Writing Typewriter* (Research Report RC-180). Yorktown Heights, NY, IBM Research Center.

Lohse C (1968). Proposed USA standard: general-purpose alphanumeric keyboard arrangement for information interchange. *Communications of the ACM*, 11:126–129.

Lundervold AJ (1951). Electromyographic investigations of position and manner of working in typewriting. *Acta physiologica Scandinavica Supplementum*, 24(84):1–171.

Marklin RW, Simoneau GG (2004). Design features of alternative computer keyboards: a review of experimental data. *Journal of Orthopaedic & Sports Physical Therapy*, 34(10):638–649. https://doi.org/10.2519/jospt.2004.34.10.638

Marras WS, Karwowski W (2006). *The Occupational Ergonomics Handbook* (2 volumes), 2nd ed. Boca Raton, FL, CRC Press. https://doi.org/10.1201/9781420003635 and https://doi.org/10.1201/9781420003642

Marras WS, Radwin RG (2006). Biomechanical modelling. Chapter 1 in Nickerson RS (ed), *Reviews of Human Factors and Ergonomics*. Santa Monica, CA, Human Factors and Ergonomics Society, pp. 588–589.

McMillan GR (2001). Brain and muscle signal-based control. In Karwowski W (ed), *International Encyclopaedia of Ergonomics and Human Factors*. London, UK, Taylor & Francis, pp. 379–381.

McMillan GR, Calhoun GL (2001). Gesture-based control. In Karwowski W (ed), *International Encyclopedia of Ergonomics and Human Factors*. London, UK, Taylor & Francis, pp. 237–239.

McMulkin ML, Kroemer KHE (1994). Usability of a one-hand ternary chord keyboard. *Applied Ergonomics*, 25(3):177–181. https://doi.org/10.1016/0003 6870(94)90016-7

National Research Council (US) Steering Committee for the Workshop on Work-Related Musculoskeletal Injuries: The Research Base (1999). Work-Related Musculoskeletal Disorders: Report, Workshop Summary, and Workshop Papers. Washington, DC, National Academies Press. https://doi.org/10.17226/6431

Noyes J (1983a) Chord keyboards. *Applied Ergonomics*, 14(1):55–59. https://doi.org/10.1016/0003-6870(83)90221-1

Noyes J (1983b) The QWERTY keyboard: a review. *International Journal of Man-Machine Studies*, 18(3):265–281. https://doi.org/10.1016/S0020-7373(83)80010-8

Osler W (1892). *The Principles and Practice of Medicine, IX. Professional Spasms: Occupation Neuroses*. New York, Appleton.

Pereira A, Hsieh CM, Laroche C, Rempel D (2014). The effect of keyboard key spacing on typing speed, error, usability, and biomechanics, Part 2: vertical spacing. *Human Factors,* 56(4):752–759. https://doi.org/10.1177/0018720813502524

Pereira A, Wachs JP, Park K, Rempel D (2015). A user-developed 3-D hand gesture set for human-computer interaction. *Human Factors*, 57(4):607–621. https://doi.org/10.1177/0018720814559307

Poore GV (1872). "Writer's cramp": its pathology and treatment. *Practitioner*, August issue:341–350.

Poore GV (1887). Clinical lecture on certain conditions of the hand and arm which interfere with the performance of professional acts, especially piano-playing. *The British Medical Journal*, February 26 issue, 441–444.

Putz-Anderson V (1988). *Cumulative Trauma Disorders.* Boca Raton, FL, CRC Press. https://doi.org/10.1201/9781315140704

Ramazzini B (1713). *De morbis artificum diatriba* [diseases of workers]. Translated in *American Journal of Public Health*, 2001, 91(9):1380–1382. https://doi.org/10.2105/AJPH.91.9.1380

Reddy NP, Gupta V (2007). Toward direct biocontrol using surface EMG signals: control of finger and wrist joint models. *Medical Engineering & Physics*, 29(3):398–403. https://doi.org/10.1016/j.medengphy.2005.10.016

Remington RJ, Rogers M (1969) Keyboard Literature Survey, Phase 1: Bibliography, TR 29.0042. Research Triangle Park, NC, IBM Systems Development Division.

Rempel D (2008). The split keyboard: an ergonomics success story. *Human Factors*, 50(3):385–392. https://doi.org/10.1518/001872008X312215

Strong EP (1956). *A Comparative Experiment and Simplified Keyboard Retraining and Standard Keyboard Supplementary Training.* Washington, DC, (US) General Services Administration.

Ugbolue UC, Hsu WH, Goitz RJ, Li ZM (2005). Tendon and nerve displacement at the wrist during finger movements. *Clinical Biomechanics*, 20(1):50–56. https://doi.org/10.1016/j.clinbiomech.2004.08.006

Zhao C, Ettema AM, Osamura N, Berglund LJ, An KN, Amadio PC (2007). Gliding characteristics between flexor tendons and surrounding tissues in the carpal tunnel: a biomechanical cadaver study. *Journal of Orthopaedic Research*, 25(2):185–190. https://doi.org/10.1002/jor.20321

CHAPTER TWENTY

Design of the workplace

Workspace design constraints

Design of the workplace must always balance competing objectives: the need to accommodate equipment, tasks, and the operator against constraints on space, budget, and time. Consequently, a workspace will not be perfect for everyone but will include both poor and good ergonomic features. The workspace for a driver in a modern motor vehicle provides examples of such poor and good designs within applicable constraints. Overall, the driver's workspace is poorly designed ergonomically; it forces the driver to maintain the same body position for long periods to keep the feet on the pedals, hands on the steering wheel, and eyes focused on the road ahead. Good ergonomic design typically involves a supportive, adjustable, and comfortable seat, suitable designs and arrangements of displays, and the addition of electronic driver-assist features such as adaptive cruise control, parking assistance, cameras, and other sensors to enhance visibility around the car, along with safety features such as enhanced braking and blind spot warnings. Most arrangements of hand controls are good examples; in contrast, the usual arrangement of foot-operated controls illustrates one of the worst designs imaginable. The driver cannot see the pedals and must move the foot quickly and accurately between the accelerator and brake (and the clutch pedal, if present). These pedals are both operated in a similar forward push direction, although they cause opposite effects: one accelerates the vehicle while the other slows it down. This illustrates that we need to understand how we perform within constraints in the workplace to fit task to the human.

DOI: 10.1201/9781032674575-26

20.1 Sizing the workplace to fit the body

Design to fit the body

Consider the design of 1960s-era space capsules intended to reach the moon: they were small and needed to be lightweight for technical reasons. This required that the astronauts be fairly small to be able to fit. The workspace around them, with its controls and gauges, was carefully arranged for reach and visibility. The shell and seat had to support very high accelerations. Given technical and other constraints, comfort was not on the human engineering to-do list; the priority was to get the astronauts up, around, and down from space and keep them safe.

Another classic example of workplace design concerns equipment developed without proper attention to the human operator, as shown in Figure 20.1. This image depicts a lathe, which, after a long technological development, is one of the machines that functions reliably well—but it requires an operator who would need to be much shorter and wider than an actual human to easily operate the machine. Of course, it should be the other way around: all the controls that need to be manipulated and the cutting tools that must be observed should be arranged in size and location to fit the real human body.

Body dimensions

Obviously, body dimensions are important for the design of large pieces of equipment and workspaces. Generally, the most crucial anthropometric data* are those that describe overall size (e.g., stature) and identify eye location as well as hand height and location because they indicate where objects should be placed that need to be seen and manipulated.

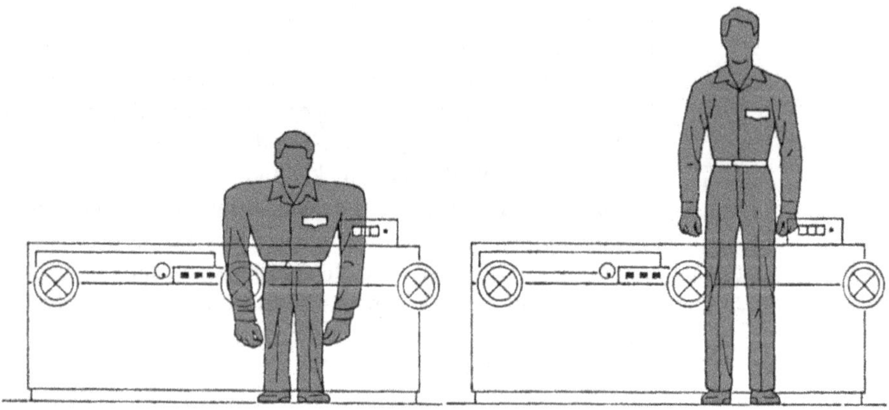

FIGURE 20.1 A lathe designed for an imaged operator (at left) and shown with a real operator (at right). Adapted from Eastman Kodak Company (1983).

Workspace height The worker's body size plays a major role in determining the working height of a workbench, for example. The work task is another major determinant of the proper work height: Figure 20.2 (same as 2.16) illustrates that different types of work—precision, light, or heavy—are best done at distinct heights, with elbow height serving as a suitable anthropometric reference. If constraints on a workplace result in equipment that cannot be adjusted for operator height, then simple ergonomic adjustments, such as a platform, can lift a shorter worker into position, as shown in Figure 20.3; however, note that an elevated platform may become a tripping hazard.

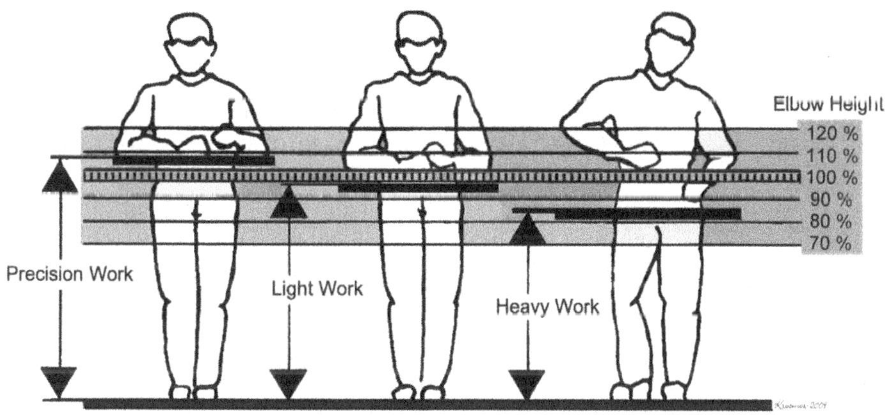

FIGURE 20.2 Elbow height is a suitable reference to determine the heights of working surfaces for different work tasks.

FIGURE 20.3 Using a platform to stand on can help accommodate operators with different heights to a fixed machine. Adapted from ILO (1979).

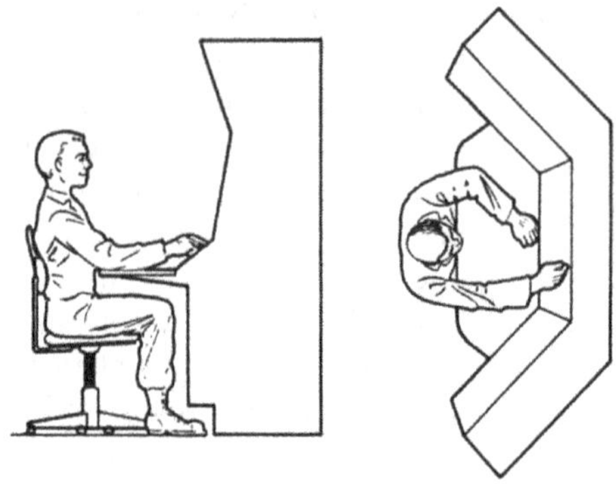

FIGURE 20.4 A console arranged around the operator. Adapted from Kroemer Elbert, Kroemer and Kroemer (2018).

Console

Another example of simultaneously considering several human engineering aspects is the design of consoles, where instruments and hand controls are arranged around the upper body of the operator, as shown in Figure 20.4. Such design allows for both quick reaches to controls and good viewing of instruments, all at about an extended arm's length from the operator.

20.2 Designing for walking, standing, or sitting down

Walk, stand, or sit

There are tasks that require walking about, for example, to pick up material, take it to the workbench, and, after the work is done, return it to another spot. Moving around like this can be quite healthy for a fit person. In this case, the workbench is probably best positioned a bit below elbow height, as shown in Figure 20.5. There may also be periods in which the worker performs best while sitting: for this, a tall seat is appropriate, as also shown in Figure 20.5 (of course, the ability to adjust the seat or workbench height, or the addition of a platform to stand on, is also appreciated). In either case, the work surface should extend some distance beyond the vertical front panel of the workbench so that the worker can step up close to it or extend their legs under it when sitting. If such legroom is lacking, as shown in Figure 20.6, sitting becomes very difficult and awkward, limiting the work that can be done.

Shop seats

On the shop floor, elaborate seats are usually not suitable. Instead, simple designs that are robust and easy to clean are

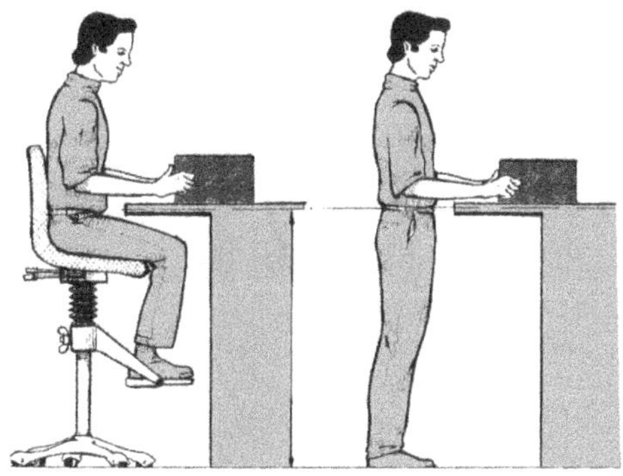

FIGURE 20.5 Workplace suitable for sitting (with seat height adjustments and foot support) and standing. Adapted from ILO (1979).

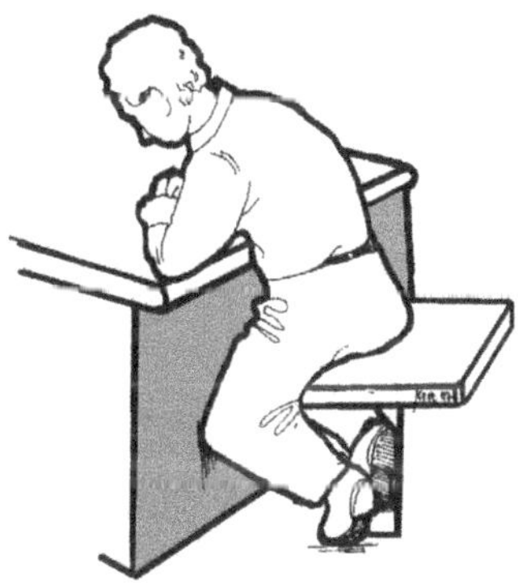

FIGURE 20.6 A work surface without legroom makes sitting awkward and uncomfortable.

more useful. Various kinds of stools and lean-ons, such as those shown in Figure 20.7, can take some load off the operator's feet, at least temporarily.

One footed standing

As mentioned, requiring a person to maintain the same position for long hours is not a good human factors solution: we humans need to move our bodies, not maintain the same posture. Keeping the same posture is particularly tiring if it involves standing on one foot, as shown in Figure 20.8 (same as 2.18). In this case, to operate a pedal with the right foot, the operator has to support their whole body weight on their left foot since a

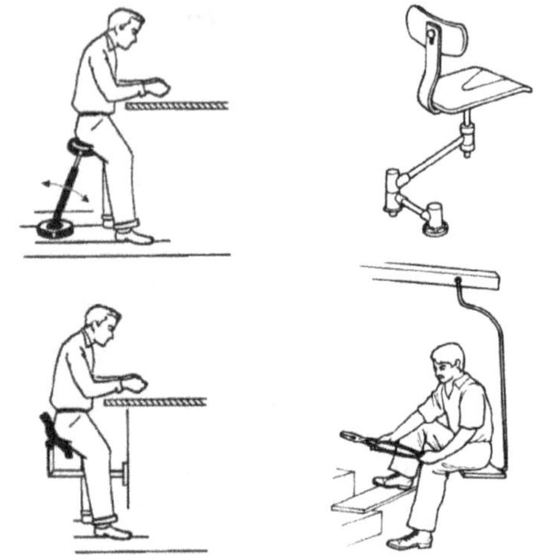

FIGURE 20.7 Examples of stools and lean-ons to allow sitting at a worksurface. Adapted from Kroemer Elbert, Kroemer, and Kroemer (2018).

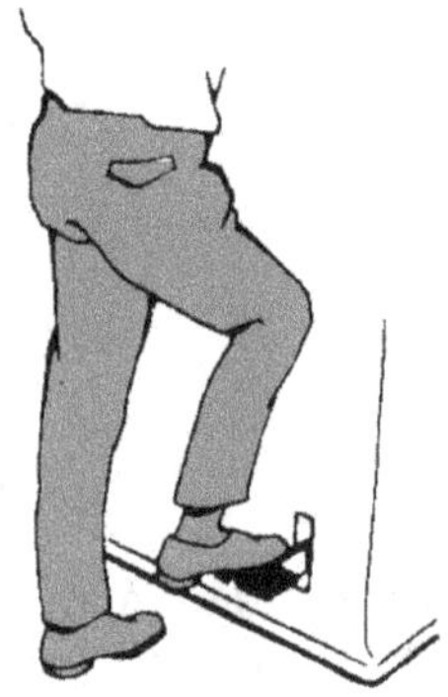

FIGURE 20.8 Operator is forced to stand on one foot in order to operate a pedal with the other foot.

standing person can only use one foot at a time. Birds can stand on one leg for a long time, but people find this very tiring.

Seated reaches

Even when we have a chair to sit on, we should not sit still. However, this does not mean that we should perform awkward and possibly harmful motions, such as shown in Figure 20.9: for the worker on the left, the workplace is so poorly set up that the operator has to awkwardly bend over to the right side to place items into a container. The worker on the right also has an awkward twist to read a document next to the keyboard.

FIGURE 20.9 Operator forced to awkwardly bend the body to the right: to pack items into boxes or to read a document for typing on a keyboard.

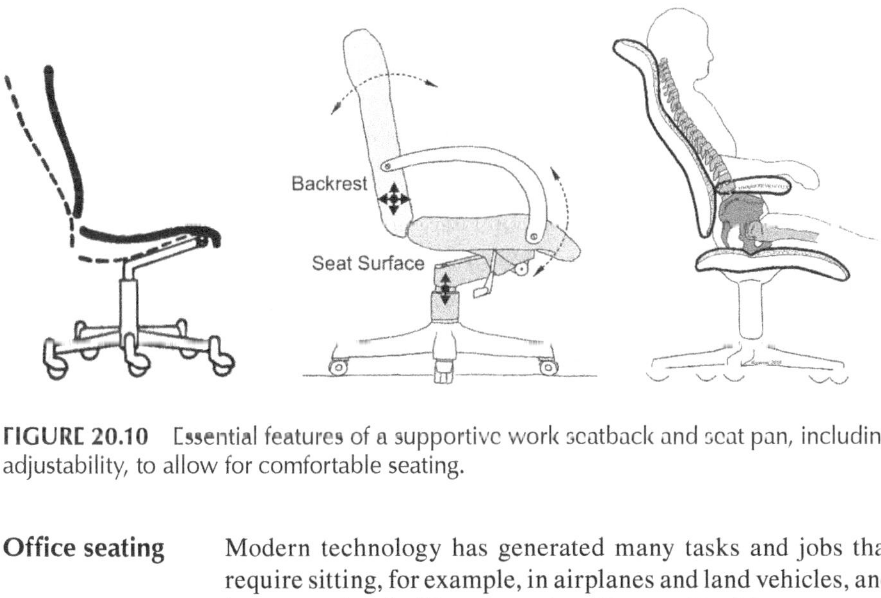

FIGURE 20.10 Essential features of a supportive work seatback and seat pan, including adjustability, to allow for comfortable seating.

Office seating

Modern technology has generated many tasks and jobs that require sitting, for example, in airplanes and land vehicles, and in conventional offices (see also Chapter 18). Many automobile seats are quite comfortable and supportive, well designed to support the driver or passenger. Many of the same design concepts are applied to chairs for use in the office. The times of the hard-surfaced and primitive task chair are largely over; now, office chairs are available that are well-designed yet relatively inexpensive. They provide such features* as sketched in Figure 20.10 (also including features from Figures 18.14 and 18.16): a contoured and padded seat pan with a rounded contour at the front that avoids pressure behind the knee; a high backrest that conforms to the shape of the back and even provides a headrest. Both seat pan and backrest should incline and decline, independently or together, to allow for humane seating.

20.3 Designing for manipulating, reaching, and grasping

Openings

The work task, together with the worker's body size, determines the necessary size of openings, such as doors and hatches*, that must allow passage, often while carrying equipment and wearing gear that makes the body more bulky. The bulk caused by gloves and clothing, along with the need to maneuver tools, determines the size of openings in machinery or enclosures to provide access for the hand to perform repairs and adjustments. Figure 20.11 shows examples.

Hand workspace

Quite a few work tasks require finely controlled handling of delicate objects and instruments, such as in repair and assembly work. Exact manipulations are easiest at about elbow height, in front of the chest, and close to the body, allowing for steady and secure arm and hand motions; see Figure 20.12 (see also Figures 2.15 and 11.11). Additionally, this location facilitates good visual control because it is at a close viewing distance with a well-declined angle of the line of sight*.

Repeated reaches, such as in assembly work, are easy if the supply bins are carefully positioned along the periphery of hand–arm movements, as shown in Figure 20.13. In addition to the bins, tools needed for the job may be hung nearby, as Figure 20.14 illustrates, so that the worker can readily pull them in and then simply release them when no longer needed.

20.4 Designing displays and controls

Simple and complex controls

Many of us have learned, through trial and error, how to ride a bicycle to our desired destination. We can simply control its direction by turning the handlebar to point the front wheel where we want to go; we can readily control its speed by pedaling faster or more slowly; and we can easily see the path ahead of us. The relationships between controls (inputs) and resulting actions are more complicated in most technical systems. In an automobile, for example, there are several controls to determine its movement, and the results of our control actions are displayed on gauges and through the windows. It takes much longer to learn how to drive a car than to ride a bicycle,

		Approximate dimensions (cm)	
		A	B
For screwdriver use so that hand can turn up to 180 degrees	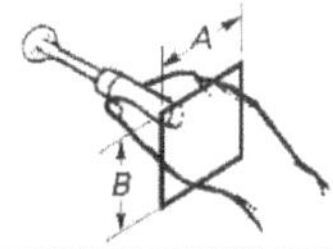	11	12
For use of pliers and similar slender tools	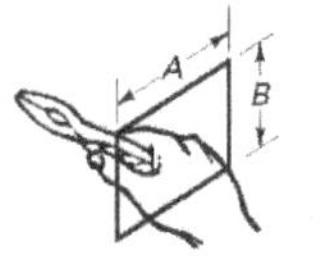	13	12
For turning T-handle wrench up to 180 degrees	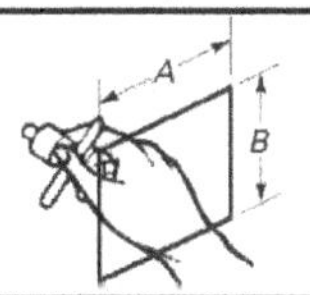	14	16
For turning open-end wrench up to 60 degrees	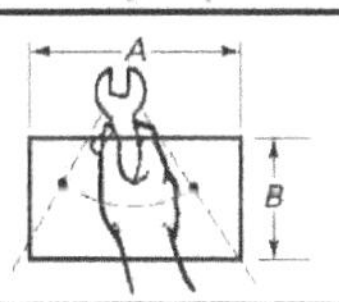	27	20
For turning Allen-type wrench up to 60 degrees	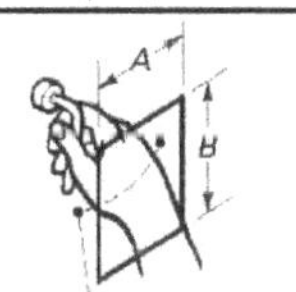	12	16
For using test probe and similar slender devices	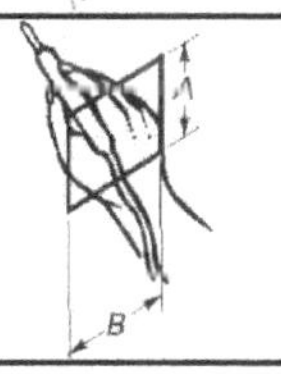	9	9

FIGURE 20.11 Minimal opening sizes (in centimeters) that allow one hand to pass when holding a tool. Adapted from Kroemer Elbert, Kroemer, and Kroemer (2018) and from MIL-HDBK 759.

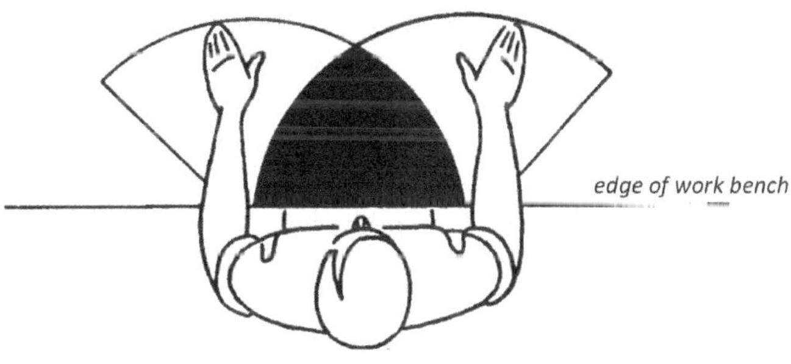

FIGURE 20.12 Normal work area suitable for finely controlled manipulation. Adapted from ILO (1992).

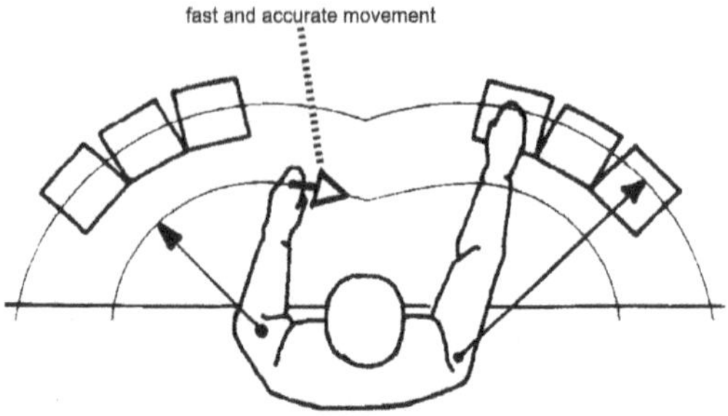

FIGURE 20.13 Supply bins placed along the semi-circular area of easy reach. Adapted from ILO (1992).

FIGURE 20.14 Workplace arrangement of tools and bins. Adapted from ILO (1992).

and the consequences of faulty control actions or misreading a display can be severe. Other human–machine systems are even more intricate, such as piloting a large ship into a tight harbor. A power plant is operated from the control room in abstract ways by pushing buttons and turning knobs, with its functioning solely displayed via gauges, as Figure 20.15 illustrates in a simplified way. Aircraft cockpits are among the most complex human factors design tasks since they involve many controls and displays. Figure 20.16 sketches the multitude of controls and displays in a large airplane; not surprisingly, the pilots have found ingeniously simple ways to mark objects of specific concern.

Coding controls Controls can be coded* in many ways to help identify them, point out how to operate them, indicate the effects of their

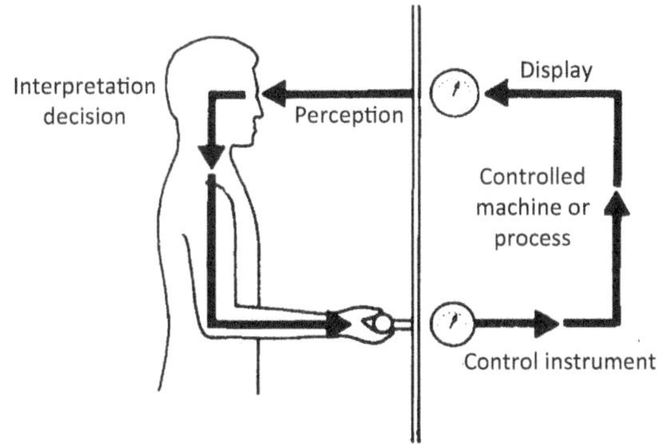

FIGURE 20.15 Operating a system by activating a control and reading a display.

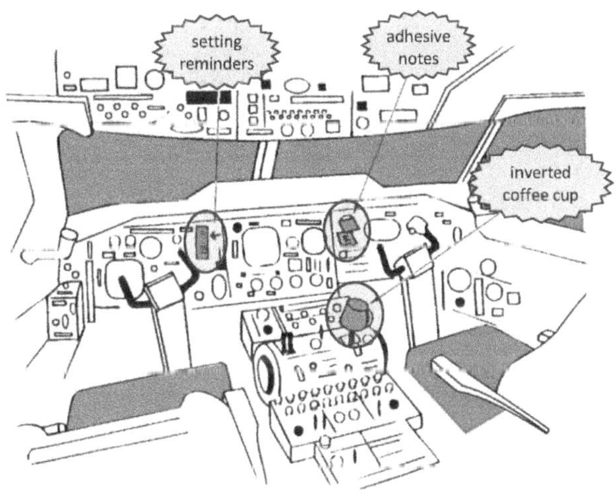

FIGURE 20.16 Aircraft cockpit with improvised markers as cognitive aids. Adapted from Norman (1991), Kroemer Elbert, Kroemer, and Kroemer (2018).

activation, and show their status. The major coding practices employ:

- *Shape*: appeals to both vision and touch (see Figure 20.17).
- *Location*: indicates importance and sequence of operation.
- *Size*: makes operation fast and easy.
- *Mode of operation*: such as pushing, turning, or sliding.
- *Color*: effective when lit to distinguish among them.
- *Labeling*: effective when readable by all.

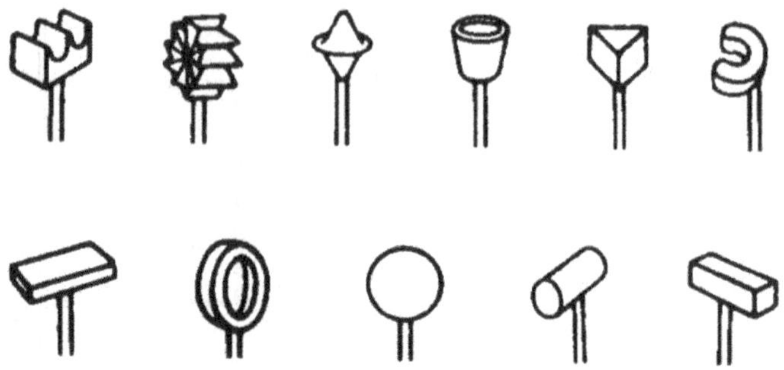

FIGURE 20.17 Easily distinguished control shapes. Adapted from Jenkins (1953).

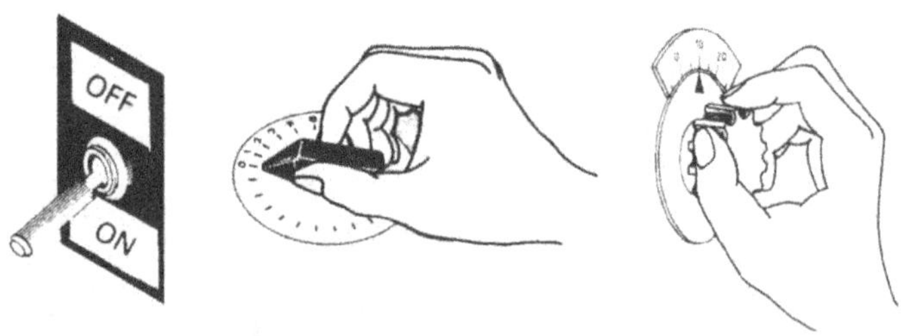

FIGURE 20.18 Examples of a toggle switch, knob bar, and rotary knob.

Coding is related to the concept of affordances and signifiers, popularized by cognitive engineer Donald Norman*. Affordances are the possible interactions between a person and an object, such as pushing or pulling a door. Signifiers signal what action to take or where to perform the action, such as a push plate or a handle on a door. Buttons invite being pushed, and knobs invite being turned.

Often, several of these coding techniques are used together, creating redundancy. In a car familiar to the driver, he or she relies on locating controls by their position and feel (produced by shape, size, and type). If the driver has to look for the controls, then he or she is wasting attention that should instead be focused on the road and traffic.

Knobs, switches, buttons

Recommendations for knobs, switches, and buttons are readily available. Some examples are shown in Figure 20.18: a toggle switch is easy to grasp and operate, but it has only two or three

FIGURE 20.19 Aircraft altimeter, circa 1960s; note the confusing controls and pointer that covers counter windows.

possible settings. A bar knob is also easy to grasp and operate (if of proper dimensions) and, by virtue of its shape, indicates its setting. A rotating knob is handy as well, but its setting is not obvious unless a special indicator highlights it; it can have more settings than a toggle or bar control, and the settings can be continuous or in steps. Push buttons are fast-operated controls, but they usually have only two settings, which may be difficult to distinguish.

Altimeters, instruments in aircraft that indicate flying height, have been notorious for being difficult to read, easy to misread, and for causing numerous emergencies and crashes. Figure 20.19 shows a 1960s model, already improved from previous designs, but still complex, and the pointer can still cover the counter windows. Current designs are not much different.

Indicators

Figure 20.20 lists the characteristics of three indicator gauges: a moving pointer over a fixed scale, a fixed pointer (or marker) over a moving scale, and a digital counter. The main advantage of the moving pointer is that its motion catches the eye and conveys information about both the magnitude and direction of change. The advantage of a fixed marker over a moving scale is that a large range can be covered, often with only the currently used section appearing in the window. The digital counter is particularly good for exact readings, whereas pointer displays can have errors of over 35%, depending on the relative locations of numbers, scale marks, and the pointer itself.

Often, exact information, such as that presented by a numerical dial or counter, is not needed: all the observer has to know is whether a condition is normal or either below or above normal. Figure 20.21 presents such a simple but effective gauge. The display may be made even more effective with bright and distinct signal colors, such as green for normal, blue for cold,

Type of display:	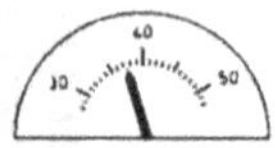 *Moving pointer*	 *Fixed pointer, moving scale*	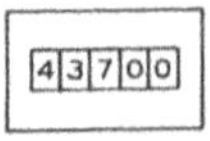 *Counter*
Ease of accurate reading	Acceptable	Acceptable	Very good
Detection of change	Very good	Acceptable to poor	Poor
Setting to a value to control a process	Very good	Acceptable	Acceptable
Tracking the value to detect change	Very good	Acceptable	Poor

FIGURE 20.20 Relative advantages of pointer and counter displays.

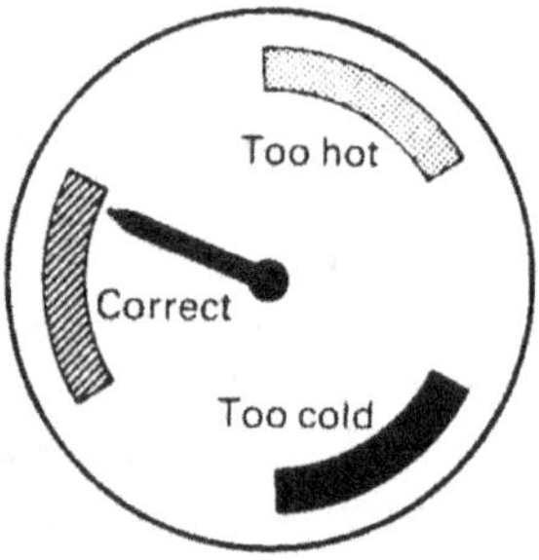

FIGURE 20.21 Status display: too hot, correct, or too cold.

and red for hot. The designer must keep in mind, however, that the meanings of colors may be different for individuals from various cultural backgrounds. Humans do differ: this is what makes ergonomic research exciting and human factors engineering interesting.

20.5 Designing layout of the workplace

So far in this chapter, we have been discussing ergonomic considerations for the individual's specific workplace. However, most of us work in teams and within larger organizations. Thus, it is ergonomically important to consider how the overall workspace is organized and how workers and their activities interact. This is easiest to demonstrate when considering the manufacturing of physical goods, but the principles apply to all organizations.

Optimizing layout Generally, businesses want to optimize the layout of the workplace not just for each worker, but also in the interactions between workers and their specific workstations and activities. This may include streamlining the work tasks themselves and the interactions between tasks (and locations) for efficiency and safety. To do this, businesses must first understand the overall workflow and processes: the flow of materials, the placement of equipment, and the movement of workers. Analysis tools can be as simple as a flowchart sketched on paper or as complex as the increasingly sophisticated computer-based simulation models available.

Lean manufacturing Optimizing the manufacturing process and method is often called *lean manufacturing*. This approach aims to reduce production times: both within the manufacturing process itself and in the response times from suppliers and customers. It can also involve *just-in-time* concepts, where production and demand are matched so that the storage of raw materials, interim work-in-process, and final products is minimized. To implement a lean process, we focus on eliminating unnecessary activities and workflow. It is important to design for capacity and flexibility; for example, if new products or processes are anticipated, or as technology and automation change how work is performed.

Laying out workflow The workflow can be organized by process or by product. When organizing the workflow by process, all equipment or work of the same type is grouped together, such as when one room contains all the molding equipment, another the heat-treating ovens, and another the final packaging. This requires more dedicated space but allows for focused control of special conditions, such as clean room filtration. When organizing the workflow by product, all the equipment and activities needed to make a particular product are grouped together. This requires less floor space and shorter travel distances for the product itself, but a malfunction in any single machine or process may halt the entire production line.

Figure 20.22 contains an example of a workflow diagram: the initial layout shows long travel distances and waste of workers, floor space, and production time. After reconfiguration, one workbench combines unpacking, inspection, and inventory to reduce transportation and delays, and adding conveyors improves material handling.

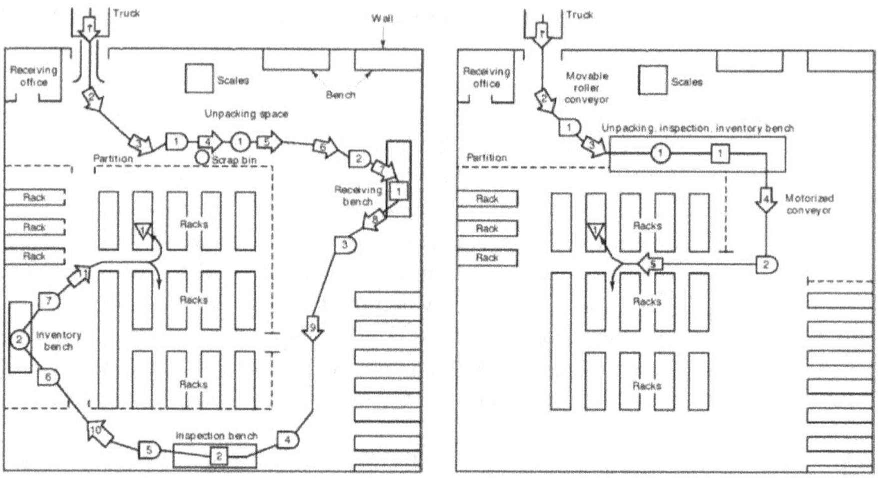

FIGURE 20.22 Diagram of work flow through a processing plant, from unpacking, receiving, inspection, inventory, and storing: before (left) and after (right) assessment and reconfiguration. Adapted from ILO (1992).

20.6 Continuous improvements in the workplace

Kaizen continuous improvement

Most workplaces seek to improve work efficiency and enhance worker engagement continuously. Involving workers in discussions about individual or organizational work is important since they know the processes best and are most directly impacted. One way to formally approach workplace improvements is loosely termed Kaizen (a combination of two Japanese words that can translate as "good change," based on the work of W. Edwards Deming and Masaaki Imai), which encourages small but ongoing changes to foster broader positive improvements. This technique includes ten broad principles (stated slightly differently in various sources):

1. Never stop improving.
2. Question current practices and do not accept the status quo.
3. Rely on data, not opinions, and do not assume all new methods will work.
4. Be proactive about solving problems.
5. Correct mistakes immediately and look for solutions.
6. Let go of perfectionism and use creativity to identify small, low-cost improvements.
7. Create an environment in which everyone feels empowered to contribute.

8. Seek information and knowledge from multiple people.
9. Ask "Why" five times to uncover the root cause.
10. Continuously and consistently improve.

5S framework

Kaizen also includes a *5S framework* regarding the set-up and maintenance of an idealized workplace.

- *Seiri/Sort* (organize). Separate necessary workplace items from unnecessary ones and remove unnecessary items.
- *Seiton/Straighten* (set in order). Arrange needed items to allow for easy access in a manner that makes the most sense for work.
- *Seiso/Shine* (cleanliness). Keep the workspace clean and tidy.
- *Seiketsu/Standardize* (standardized clean-up). Systematize workplace clean-up to maintain best practices.
- *Shitsuke/Sustain* (discipline). Maintain the habit of these established procedures.

Even if a workplace does not formally adopt these principles and approaches, we may find that incorporating them into the work within our control yields benefits.

20.7 Addressing safety and risks at work

Keeping workers safe

Any discussion of workplace design should ensure that safety is also addressed. It is understood that keeping workers safe is in everyone's best interest: after all, ergonomics seeks to optimize human well-being, which clearly includes safety. Additionally, it's also good business sense, with fewer injuries and illnesses, less need for protective equipment and training, and more engaged and productive employees—all of which serve to optimize overall human–system performance.

Hazard and harm

When we talk about risk, we are referring to the likelihood and severity of harm. *Harm* is a physical injury or incident that impacts health. A *hazard* is a source of potential harm. Hazards can fall into many categories:

- Thermal, such as temperatures higher than 65°C (150°F) or below –10°C (14°F).
- Mechanical, such as from a pinch, crush, shear, cut, entanglement, stab, or abrasion.

- Chemical, especially if inhaled, absorbed, ingested, or injected.
- Electrical, such as burns, shock, arc blast, fire, or electrocution.
- Vibration, depending on the severity and duration of exposure (e.g., low risk if less than 2–3 hours per workday, high risk if more than 4 hours per workday).
- Laser beam exposure, which can cause eye and skin damage.
- Noise, especially if greater than 85 dBA for an 8-hour time-weighted average base, or an impact noise greater than 112 dBA.
- Radiation, both ionizing and non-ionizing.
- Biological, including allergens and infectious agents.
- Physical ergonomic, especially regarding cumulative trauma disorders.

Ranking hazards

Taking a risk-based approach means that we need to be most concerned with hazards that occur more often or cause more severe harm. Therefore, hazards can be ranked using a formula that multiplies together the critical factors:

- The *likelihood* of occurrence or probability that injury will occur,
- The *frequency* of exposure to the hazard,
- The *severity* of potential injury or degree of potential harm, and
- The *number* of people exposed to the hazard.

Hierarchy of controls for safety

Preventing (or reducing the risk of) injury is best accomplished through the design of the work task and workplace. There is a hierarchy of controls for workplace safety* to identify and rank safeguards. They are arranged from most to least effective and are often used in *combination.* The primary way to address safety risk is through *elimination*: physically removing the hazard. If this cannot be done, the next step should be *substitution*: exchanging or replacing the hazard with a less risky task. If this cannot be done, then we can use *engineering controls*: these are ways to isolate people from the hazard. Next, we can also use *management (administrative) controls*, which include training, work certifications and permits, and appropriate labeling to change the way people work. Another mitigation strategy is to use *personal protective equipment* (PPE) to physically protect the worker.

All of these together are intended to make the work and workplace safer for all involved; this is the ergonomic goal of improving human well-being and overall system performance.

Summary

Human-centered design of the workplace follows these guidelines*:

Consider …	in order to …
Human body size	Establish the dimensions of workspace, equipment, and tools to accommodate the human body.
Human strength	Facilitate exertion of strength (work, power) by object location and orientation.
Human speed	Place items so that they can be reached and manipulated quickly.
Human effort	Arrange work so that it can be performed with least effort.
Human accuracy	Select and position objects so that they can be manipulated and seen with ease.
Importance	Place the most important items in the most accessible locations.
Frequency of use	Place the most frequently used items in the most accessible locations.
Function	Group items with similar functions together.
Sequence of use	Lay out items that are regularly used one after the other in that sequence.
Floor plan and layout	Organize the workflow between operators and workstations to be efficient, effective, and safe.
Human safety and comfort	Arrange all details to assure that the human will not be injured, will not suffer health consequences, and instead will be secure and feel comfortable.

Steps to fit the human

Step 1: Determine what tasks will be performed, with what equipment, by which persons, and what information must be conveyed and acted upon via displays and controls. Consider safety and address hazards that can lead to harms.

Step 2: Consider the specific needs and preferences of the worker, and what other tools will be used. Involve the workers in the design process. Where possible, add flexibility to the design of the work equipment and furnishings, the displays, and controls, to allow user adjustments.

Step 3: If the users or their needs or the tasks change, consider how this can be accommodated. Ask the workers if the workplace set-up and furnishings suit them, and adjust as needed.

Step 4: Seek to continuously improve. Monitor conditions, modify the work place or work tasks, and iterate as necessary.

Notes

The text contains markers, *, to indicate specific references and comments, which follow.

20.1 Sizing the workplace to fit the body

Anthropometric data: See Chapter 1 in this book.

20.2. Designing for walking, standing, or sitting down

Seat features: see Chapter 18.

20.3 Designing for manipulating, reaching, and grasping

Doors, hatches, and passageways: Kroemer Elbert, Kroemer, and Kroemer Hoffman (2018).

Line of sight: Discussed in Chapters 5 and 11.

20.4 Designing displays and controls

Coding of controls: Freivalds (1999), Kroemer, Kroemer, and Kroemer (2018).

Signifiers and affordances: Norman (2013).

20.7 Addressing safety and risks

Hierarchy of controls: Many governments and regulations prescribe workplace safety, so be sure that they are (at a minimum) understood and complied with (see also ISO 45001).

Summary: *Human-centered design, guidelines*: More conceptual information and guidelines can be found in international standards and various handbooks, including

- Human Factors and Ergonomics Society (HFES) (2004)
- Karwowski, Soares, and Stanton (2011)
- Marras and Karwowski (2006)

- Kroemer (2006)
- Kroemer Elbert, Kroemer, and Kroemer Hoffman (2018)
- Nagamachi (2011)
- NASA (2023)
- Pew and Mavor (2007)
- Salvendy (2012)
- Stanton, Hedge, Brookhuis, Salas, and Hendrick (2004)
- Wilson and Sharples (2015).

Ergonomics books with a practical orientation are also listed at the end of Chapter 24.

References and further reading

Eastman Kodak Company (1983). *Ergonomic Design for People at Work*. New York, Van Nostrand Reinhold.

Human Factors and Ergonomics Society (HFES) 300 Committee (2004). *Guidelines for Using Anthropometric Data in Product Design*. Santa Monica, CA, Human Factors and Ergonomics Society.

ILO (1979). *Introduction to Work Study*, 3rd ed. Geneva, CH, International Labour Office.

ILO (1992). *Introduction to Work Study*, 4th ed. Geneva, CH, International Labour Office.

Jenkins WL (1953). *Design Factors in Knobs and Levers for Making Settings on Scales and Scopes, WADC-TR-53-2*. Wright-Patterson AFB, OH, Aero Medical Laboratory.

Karwowski W, Soares MM, Stanton NA (2011). *Human Factors and Ergonomics in Consumer Product Design: Uses and Applications*. Boca Raton, FL, CRC Press. https://doi.org/10.1201/b10949

Kroemer AD, Kroemer KHE (2017). *Office Ergonomics: Ease and Efficiency at Work*, 2nd ed. Boca Raton, FL, CRC Press. https://doi.org/10.1201/9781315368603

Kroemer Elbert K, Kroemer HB, Kroemer Hoffman AD (2018). *Ergonomics: How to Design for Ease and Efficiency*, 3rd ed. London, UK, Elsevier. https://doi.org/10.1016/C2016-0-02516-0

Kroemer KHE (2006c). *"Extra-Ordinary" Ergonomics: How to Accommodate Small and Big Persons, the Disabled and Elderly, Expectant Mothers and Children*. Boca Raton, FL, CRC Press. https://doi.org/10.1201/9780203025246

Marras WS, Karwowski W (2006). *The Occupational Ergonomics Handbook: Fundamentals and Assessment Tools for Occupational Ergonomics*, 2nd ed. Boca Raton, FL, CRC Press. https://doi.org/10.1201/9781420003635

MIL-HDBK 759 (2018). *Human Factors Engineering Design for Army Material (Metric)*. Philadelphia, PA, DLA Document Services.

Nagamachi M (2011). *Kansei Engineering*, 2 volumes. Boca Raton, FL, CRC Press. https://doi.org/10.1201/b16799

NASA/SP-2010-3407/REV1 (2014). Human Integration Design Handbook (HIDH). https://www.nasa.gov/wp-content/uploads/2023/03/human-integration-design-handbook-revision-1.pdf?emrc=664e5339d33ee

NASA-STD-3001 (2023). Space Flight Human-System Standard Volume 1 (Crew Health); Volume 2 (Human Factors, Habitability and Environmental Health) https://standards.nasa.gov/standard/NASA/NASA-STD-3001-VOL-1 and https://standards.nasa.gov/standard/NASA/NASA-STD-3001-VOL-2

Norman DA (1991). Cognitive science in the cockpit. *CSERIAC Gateway*, 2:1–6.

Norman DA (2013). *The Design of Everyday Things*. Cambridge, MA, The MIT Press.

Pew PW, Mavor AS (2007). *Human-System Integration in the System Development Process*. Washington, DC, The National Academies.

Salvendy G (2012). *Handbook of Human Factors and Ergonomics*, 4th ed. New York, Wiley. https://doi.org/10.1002/9781118131350

Stanton N, Hedge A, Brookhuis K, Salas E, Hendrick HW (2004). *Handbook of Human Factors and Ergonomics Methods*. Boca Raton, FL, CRC Press. https://doi.org/10.1201/9780203489925

Wilson JR, Sharples S (2015). *Evaluation of Human Work*, 4th ed. Boca Raton, FL, CRC Press. https://doi.org/10.1201/b18362

CHAPTER TWENTY-ONE

Load handling

Handling loads We all "handle" loads daily. As part of our everyday lives, we lift, lower, hold, carry, push, pull, and drag loads while packing, moving, and storing them. The load may be soft or solid, bulky or small; it may come as bags, boxes, or containers, with or without handles. We handle objects occasionally or repeatedly, both during our leisure activities and as part of our occupational work. On the job, ergonomic design of loads, containers, and workstations can help us avoid overexertion and injuries, especially if we are also provided instructions and training on how to "lift properly." For some jobs, it may be appropriate to consider selecting people who are especially physically capable of strenuous material handling. We need to understand how to handle loads appropriately in order to fit the task to the human.

When handling loads is part of the job, the activities are often labeled *manual material(s) handling* (an example of a tautology because the Latin *manus* means hand).

21.1 Basics of material handling

Material handling is among the most frequent and often severe causes of workplace injury all over the world*, with strains in the low back being the most common. The direct and indirect costs are enormous, and the human suffering associated with back and other injuries is immeasurable.

Stress by material handling Bending, stretching, reaching, or otherwise moving our own bodies stresses our muscles and loads our joints. Manipulation of even lightweight and small objects adds strain, especially if done frequently. Heavy and bulky loads pose additional strain

DOI: 10.1201/9781032674575-27

on the body because moving more mass requires more force from us (Newton's second law*). This is even more demanding if we must do so in awkward body positions.

External and internal forces

When we exert force and energy to move an object with our hands, we also strain our hands, arms, shoulders, trunk, and (often) legs. Lifting or lowering, pushing or pulling, holding or carrying loads all basically stress the same parts of the musculoskeletal system; however, the directions and magnitudes of the external and internal force vectors vary depending on the specific task.

Pains related to lifting

The primary area of physiological and biomechanical concern is the back, particularly the lumbar spine of the lower back. Any contraction of the longitudinal trunk muscles compresses the spinal column, which is the only actual structure in the trunk capable of bearing weight (see Chapter 2). This loads the spine, primarily the intervertebral discs (also spelled disks) and the facet joints of the vertebrae; however, all the connective tissues of the spine (ligaments and cartilage, as well as muscles with their tendons) may experience injuries, sprains, or trauma. The result is often low back pain (LBP). Hence, many ergonomists (and workers and employers) are concerned about the association between LBP and lifting.

Strainful activities

Lifting loads at work is not the only activity that can cause injury. For example, the sport of weightlifting also generates such musculoskeletal strain, as do other sports and many leisure activities. Most of the time, we exert energy intentionally onto an object to move it, but our bodies may be subjected to external energy unexpectedly, such as when we catch an object or brace ourselves if we are about to slip or fall. The strain can be static, such as when we hold a load and keep our bodies still, or dynamic, with fast or slow onset, and of short or long duration. The strain may be a single event or consist of multiple events. If the same or similar strains recur, that repetition may lead to cumulative trauma disorders, as discussed in Chapter 11.

21.2 Designing for body capabilities with load handling

"Strength" is individual and situational

Material handling, such as lifting and lowering, pushing and pulling, requires our body to exert energy. The energy needed to perform these tasks must be generated within the body by its musculature and then exerted in terms of force or torque over time to the external object. Such strength capabilities depend on our individual fitness and on situational conditions*, such as body stance, as discussed in Chapters 3 and 4. Figures 3.7 and 3.8 in Chapter 3 show examples of static push and pull forces measured under experimental conditions set up so that the subjects kept their bodies still while they exerted one-time maximal efforts. In reality, however, people usually move their bodies and external loads dynamically, often repeatedly.

Overexerting the back

Exertions of muscle strength while lifting, lowering, pushing, pulling, carrying, and dragging loads are dynamic efforts, often in the form of heavy work (see Chapter 10). A major concern is the associated wear and tear on the back, which can overexert spinal structures, especially the lumbar intervertebral discs. LBP* reduces one's mobility and vitality; it often leads to long absences from work and appears as one of the main causes of early disability in North American and European statistics. LBP is common even in younger age groups and in certain occupations, such as nursing, manual labor, farming, baggage handling, and warehouse work, where individuals are frequently afflicted. Overexertion injuries, especially in the lower back, account for about one quarter of all reported occupational disabling cases in the USA; some industries report that more than half of all injuries are due to overexertion. Accident and health statistics in the United Kingdom and Germany, for example, show similar figures. Low back strain and the associated LBP are among the most common causes of injury and disability in many industrial populations.

Loading the spine

Many LBP victims cannot pinpoint when their back problems started. Often, there was not a specific moment or action when pain appeared; rather, it developed slowly until it became strong enough to be troublesome, even disabling. When observing a person standing freely, a healthy spinal column has a slim S shape when viewed from the side, as shown earlier in Chapter 2. The spinal facet joints provide bone-on-bone mobility and sturdy load transmission; the intervertebral discs provide both elasticity and mobility. As we consider the loads on the spine, the weight that must be supported increases from the neck downward; it is greatest near the base of the spine, in the lumbar area. Just keeping the upper body in balance requires tensioning the muscles in the trunk, which further compresses the spine. Weights being carried, along with hand/arm/shoulder/trunk forces and motions, intensify the compression, bending, and twisting strains on the spinal column.

"Slipped" disc

The spinal discs separate the vertebrae and provide shock absorption and flexibility. Disc degeneration is typical with aging but also occurs from repeated motions: a sudden overload can easily lead to an acute injury, especially when aging and wear come together. Discs degenerate primarily at their outer layers; additional fluid loss can make that fibrous ring brittle and fragile. Initially, the degenerative changes cause the disc to flatten, which reduces shock absorption and mobility of the spine. When this happens, even small actions such as lifting one's own body, carrying a light load, or experiencing a slight stumble may injure the disc and cause severe back pain. Progressive degeneration of the disc and/or a sudden load can lead to disc herniation, where a rupture of its fibrous ring allows some of the gel-like nucleus to protrude. Such damage further narrows the space between vertebrae, generating additional tension in the muscles and ligaments of the spine. The deformed portion of the disc can also press on spinal nerves. These occurrences can lead to a variety of discomforts and aches and may result in disabling health problems such as lumbago and sciatica.

Biomechanics of disc loading

Figure 21.1 illustrates, in simplified terms, a biomechanical approach* to approximate the conditions of disc pressure. The external load **L** pulls downward. The (upper) body weight **W**

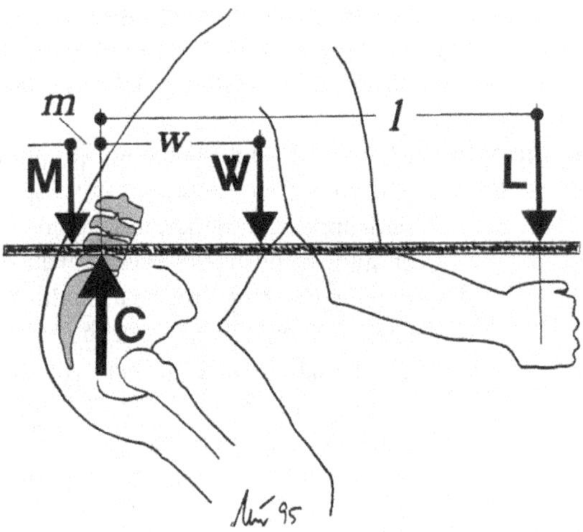

FIGURE 21.1 Biomechanical model for calculating musculoskeletal back loading due to lifting.

acts downward as well. The distances l and w are the lever arms of load and weight to the point of support (the fulcrum of the system) at a disc of the spinal column. The forces **L** and **W** exert the moments $l \times \mathbf{L}$ and $w \times \mathbf{W}$. Counteraction comes mostly from muscles that attach to the spinous processes of vertebrae; together they produce the force **M**. This force uses its lever arm m to the fulcrum to generate the moment $-m \times \mathbf{M}$ where the minus sign indicates torque in the direction opposite to the torques due to **W** and **L**. The upright force **C** at the spinal column keeps the system in place.

The system is in balance when these equations apply:

Sum of moments: $m \times \mathbf{M} = l \times \mathbf{L} + w \times \mathbf{W}$; therefore,

$$\mathbf{M} = (l \times \mathbf{L} + w \times \mathbf{W})/m$$

Sum of forces: $\mathbf{C} = \mathbf{L} + \mathbf{W} + \mathbf{M}$

Assume the external load $\mathbf{L} = 100\,\text{N}$, its lever arm $l = 25\,\text{cm}$; the upper body $\mathbf{W} = 500\,\text{N}$ with a lever arm $w = 15\,\text{cm}$; the lever arm m of the muscle force **M** is 5 cm; then the muscle force M = 2000 N, and the column force C = 2600 N

In this example, the relatively large muscle force **M** (20 times the external load **L**) and the significant column compression force **C** (26 times the external load) primarily result from the disadvantageously short lever arm m of the muscle **M**. That lever is anatomically determined, so we cannot change it—but the external load **L** is under our control, both in magnitude and in location (l) relative to the body.

Both back muscle force **M** and column compression force **C** diminish as **L** and l get smaller; hence, the recommendation to *reduce the external load and keep it close*. **M** and **C** decrease even further if we *maintain a more upright posture* (which reduces w and l) and if we decrease upper body weight **W**—so, *slim down that belly*.

These simple biomechanical (or anthromechanical)* considerations demonstrate that the following recommendations can help decrease the risks to spinal connective tissues and the spinal column:

- Keep the external load small,
- Keep the external load close to the body,
- Keep the upper body upright, and
- Get rid of excess upper body weight, including belly protrusion*.

21.3 Assessing load handling capabilities

Everyday experience and formal epidemiological studies have long demonstrated the expected association: the level and frequency of strain caused by load handling determine the incidence and severity of injuries. Naturally, the ability to tolerate load handling strain differs between individuals. Physiology, biomechanics, and psychophysiology provide scientific disciplinary approaches that can be used, often in combination*, for assessing individual and population capabilities.

Intervertebral disc loading

Around 1970, researchers were able to measure the pressure inside intervertebral discs as affected by body postures and holding loads. Figure 21.2, based on one of the early studies, illustrates the relative changes in strain on the disc between the third and fourth lumbar vertebrae in various postures. Compared to no-load upright standing (A), standing straight with 10 kg weights in each hand (B) naturally increases the disc pressure; holding the weights while standing with bent knees and a bend in the back (C) further increases disc loading; and the largest pressure occurs when 20 kg is held with locked legs and a rounded back (D). Other researchers repeated these studies and found essentially the same result: a bent or twisted back causes higher and possibly dangerous pressures on the intervertebral discs* compared to a straight back. Bending the back leads to large compression pressure and also shear on the front edge of the disc, as Figure 21.3 illustrates. Straightening the

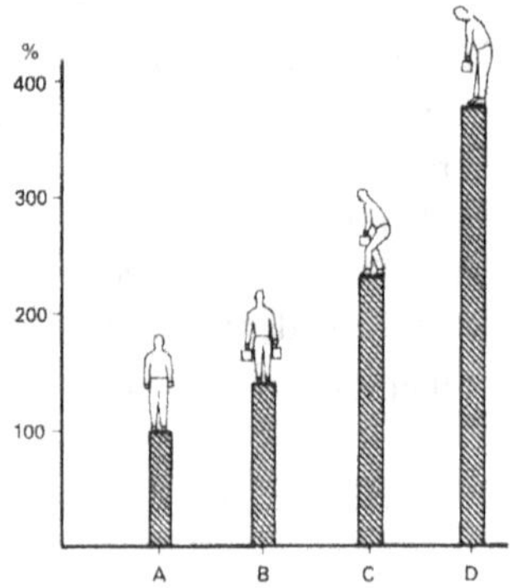

FIGURE 21.2 Relative strain on the disc between the third and the fourth lumbar vertebrae with the body in various postures without weights (A) and while holding weights of 20 kg (upright B, crouching C, bending D). Adapted from Nachemson and Elfström G (1970).

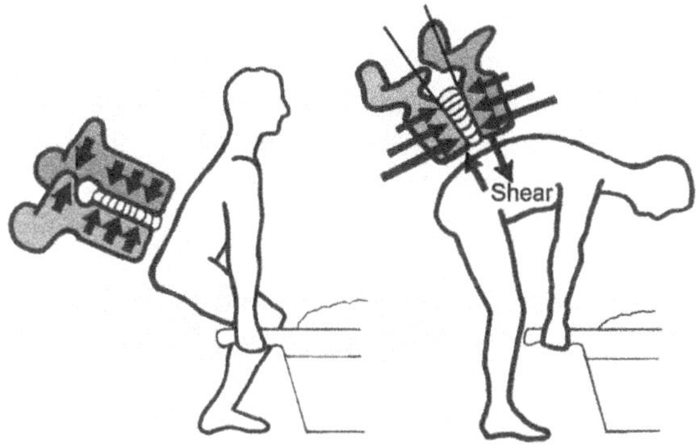

FIGURE 21.3 Diagram of typical loading of the vertebrae of the spine when lifting with straight and bent back (longer arrows indicate relatively higher forces in compression and shear).

back eliminates shear, causes the facet joints to carry some of the compressive load, and distributes the compression forces more evenly over the disc surface, all of which reduce the risk of damage.

Assessing lift strength

While measuring a person's lift strength, we want the person to fully cooperate, but we cannot subject them to the risk of injury during the test. As discussed in Chapter 3, the maximum voluntary contraction relies on subjects being motivated to give their best effort without injuring themselves. This depends on various physiological, psychological, and environmental conditions that exist at the time of measurement. Chapter 4 explains that strength capabilities are both individual and situational. In addition, until about the middle of the 20th century, information about human lifting capabilities relied largely on measurements of static (isometric) "upward" body strength. Until the 1980s, such static "lift assessments" were the basis for rather simplistic and sometimes surprising statements about the loads that men, women, and even children (!) could supposedly lift safely*.

Compared to static tests, designing and controlling experiments with realistic dynamic strength exertions is much more complicated*. In these tests, the subject decides how much effort to exert during the dynamic test; this includes the complex task of judging* (often unconsciously) how strenuous the test is perceived to be. This judgment relies partly on experience and mostly on integrating internal feedback about various physiological (muscular, circulatory, metabolic, etc.) body functions involved in the effort. Based on this assessment, the subject decides how much strength or energy they are willing to exert over a given time, say 10 minutes or 8 hours, without endangering or exhausting themselves. Such "psychophysical" experiments have been conducted to assess tasks performed particularly in industrial material handling* and by firefighters, soldiers, and nurses.

21.4 Designing with NIOSH guidelines for lifting and lowering

Starting in the late 1970s, the US National Institute for Occupational Safety and Health (NIOSH)* began developing guidelines for lifting and lowering. NIOSH combined disciplinary approaches from biomechanics, physiology, and psychophysiology. The limiting criteria used by NIOSH are:

- *Biomechanical* for high loads and low handling frequency,
- *Physiological* for low loads and high handling frequency, and
- *Psychophysiological* for moderate loads and low to moderate handling frequency.

Biomechanics

Biomechanical (anthromechanical) criteria are based on the physical mechanics of the biological system: muscle strength and tissue capabilities. These criteria include assessments of load-bearing capabilities of spinal joints (see Chapter 2) and of static and dynamic strength (see Chapters 3 and 4). Prominent among the physical criteria used to develop guidelines was the highest acceptable disc compression in the lumbar spine (see the biomechanical case discussed above). NIOSH uses 3.4 kN as the maximally allowable disc compression force.

Physiology

NIOSH used *physiological* criteria primarily to prevent whole-body strain and fatigue. Important task characteristics include the location of the external load and the duration of the work, especially the frequency of work and the work/rest pattern. NIOSH uses the following limits for energy expenditure, where VO_2max is the measure of the worker's maximum aerobic capacity (see Chapter 10): 50% of VO_2max for up to 1 hour of continuous work; 40% of VO_2max for 1–2 hours of continuous work; 33% of VO_2max for 2–8 hours of continuous work. To put this in practical terms, energy expenditure for continuous work over an 8-hour shift is limited to 33% of the worker's maximum capacity.

Psychophysiology Human psychological perception appears able to integrate anthropomechanical and physiological demands. Based on *psychophysiological* studies conducted in the United States, NIOSH recommends that the job should be acceptable to most of the exposed population.

NIOSH weight limits The NIOSH guidelines apply to:

- Smooth lift (no jerking),
- Two-handed lift,
- Load width below 75 cm,
- Unrestricted posture,
- Good foot traction, and
- A suitable environment with low humidity, appropriate temperature, and good lighting.

The latest NIOSH guidelines contain recommended weight limits (RWL) that most US industrial workers, male or female, may lift or lower. The maximal weight to be lifted with two hands is 23 kg (51 lb) under optimal conditions. However, optimal conditions are rare; therefore, the load should usually be lower, as determined by several factors, which include the following:

- The start and end points of the paths of the hands,
- Whether or not the action is in front of the body,
- The frequency of the lifting and lowering actions, and
- The quality of coupling between hand and load.

These and other factors become multipliers in an equation by which we calculate the RWL that applies under the given conditions*.

21.5 Designing with Liberty Mutual guidelines for material handling

At about the same time that NIOSH developed its guidelines, Stover Snook and Vincent Ciriello at Liberty Mutual Insurance Company conducted psychophysical tests with US workers to determine the efforts these workers were willing to exert in pushing, pulling, carrying, lifting, and lowering loads.

Acceptable forces and weights The research primarily utilized psychophysical methodology but also included measurements of oxygen consumption, heart rate, and anthropometric characteristics. The subjects controlled either the weight of the objects or the force they applied; all other task variables such as frequency, size, height, and distance were determined by the experimenter. During the experiments, the subjects monitored their own feelings of exertion or fatigue. Accordingly, they adjusted the weight or force so that they could work as hard as possible without straining themselves, becoming unusually tired, weakened, overheated, or out of breath. The results of the tests were compiled into tables of maximal weights and forces considered acceptable to 10%–90% of the male and female US population*.

Women or men handling loads

The Liberty Mutual tables on lifting, lowering, pushing, pulling, and carrying provide data about the capabilities and limitations of US workers, grouped by gender. Specific findings are noted below:

- *Bending*: Any task that begins or ends with the hands below knuckle height presents some risk. The deeper the bending motion, the greater the physical stress on the lower back. Regardless of weight, frequent bending is not recommended.
- *Twisting*: This motion puts uneven forces on the back, generating additional physical stress. The greater the twist, the more physically stressful the task.
- *Reaching*: The distance from the body at which a load is held significantly affects the forces on the back, shoulders, and arms. The farther the reach, the more physically stressful the task.

The Liberty Mutual tables cannot be used to evaluate one-handed lifting tasks. By nature, these tasks place uneven loads on the back and present greater physical stress than two-handed lifts. The tables also cannot be used to assess catching or throwing items since these tasks are physically demanding. In both cases, such work tasks are good candidates for ergonomic redesign.

Comparing NIOSH and Liberty Mutual guidelines

Altogether, there is good overlap and fair agreement between the recommendations for lifting and lowering from NIOSH and the Liberty Mutual guidelines. If there is a difference, a prudent choice is to use the lower value. The NIOSH values are unisex, while the researchers at Liberty Mutual considered female and male workers separately. Both sets of recommendations indicate, as expected, that high loads, far reaches, repetition, and deep bending or twisting of the body reduce acceptable efforts. Both sets of recommendations also suggest that the coupling between hand and load influences how much people are willing to exert. Objects that lack handles, have sharp edges, or are so wide that they are difficult to grasp* reduce the acceptable load values.

The recommendations compiled by NIOSH and Liberty Mutual Insurance Company apply to North Americans (United States and Canada). Other populations may have different body builds and strength characteristics; their work tasks, customs and procedures, and working conditions may be quite different. Hence, regional circumstances and local, regional, or national codes, standards, or regulations should be considered when designing load handling tasks.

21.6 Designing for easy load handling

As already noted, the spinal column is the only solid structure in the upright human torso that keeps the rib cage from falling into the pelvis. Even without an external load, a bent back generates shear stresses in the discs and heavy pressure on their front edges, creating a risk of damage to and rupture of a disc. Furthermore, bending and twisting the body increase the strain on the spinal column and its connective tissues. "Straightening" the back to its natural curves (shown in Figure 2.8) eliminates shear, makes the facet joints carry some of the compressive load, and distributes the compression forces more evenly over the disc surface, all of which reduce the risk of injury.

Intra-abdominal pressure and lift belts

Lifting, lowering, or carrying commonly produce considerable increases in pressure within the abdominal cavity, naturally accompanied by contraction of the abdominal muscles. The resulting intra-abdominal pressure* may help stabilize the trunk and reduce the load on the spinal column and its supporting structures. This association between increased intra-abdominal pressure and reduced spinal loads has led to the idea of wearing a supportive, wide belt around the trunk. This "back belt" is intended to help those lifting weights (whether materials handlers or competitive weight lifters) avoid back problems. However, studies to date have not shown that back belts reduce spinal loads or injuries. Instead, they are often ineffective*: wearing a belt may generate a false sense of security, and even competitive weight lifters who wear belts often suffer injuries. Support belts are not a substitute for proper ergonomic design of work tasks and equipment.

Making load handling easy

Machines do not suffer from back pain; therefore, load handling, especially lifting and lowering, is better performed by machinery than by people. Machines and robots of varying sophistication can move loads without human assistance, and exoskeletons can help humans move loads if they are properly designed and worn. If humans must physically do the job, the most effective way to avoid lifting and lowering is to convert those activities into carrying, or even better, into pushing and pulling (see Figure 21.4). Carrying is best done with machinery or on both shoulders with a yoke, as depicted in Figure 21.5. Dollies, carts, conveyors, and rollers, as shown in Figure 21.6, can convert a carrying job into the least risky load-handling category of pushing and pulling.

Training to lift safely?

It seems that one should be able to learn not to lift objects awkwardly from the floor but rather to lift them more safely between one's legs (as shown in Figure 21.7) or, better still, to pick them up from an elevated storage place (as in Figure 21.8). Likewise, it should be possible to learn to carry, lift, lower, and handle other loads in safe ways. For example, we should learn not to exert strong force with the body twisted or the trunk severely bent, or to make sudden movements. It should be natural to follow such simple advice, but apparently it is also human nature not always to adhere to the recommended methods. Numerous agencies and commercial outfits have developed systematic and often rather sophisticated courses for instructing nurses, warehouse workers, miners, and masons (to mention a few) to perform material handling safely. Unfortunately, many surveys and systematic evaluations of the outcomes of such instruction have shown that a large number of injuries occur even after proper load handling techniques are taught. Instructions and training have only limited success*. With continuing improvements in sensors and other related technologies, ergonomic wearable devices* show promise in helping remind workers to use safer postures and motions. However, even with the various interventions available, the best way to eliminate the risks of overexertion and injury is proper task engineering*. Therefore, *removing the hands from load handling* is good ergonomic advice.

FIGURE 21.4 Convert lifting to carrying or, better, pushing or pulling loads with equipment or machinery makes handling tasks easier and safer.

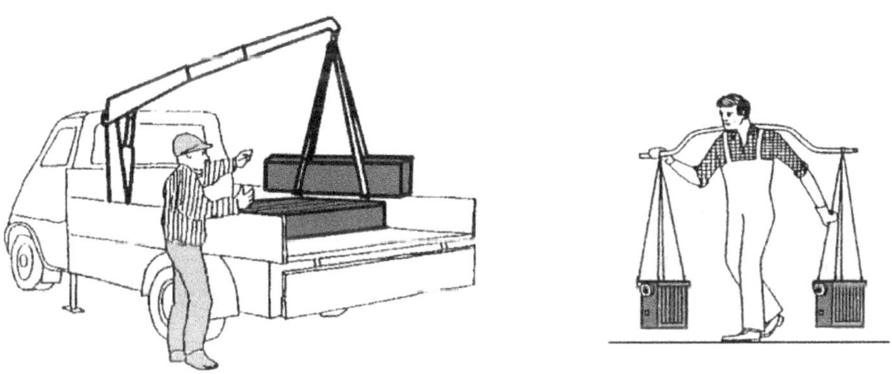

FIGURE 21.5 Suspending loads from machines or even a yoke makes moving loads easier and safer.

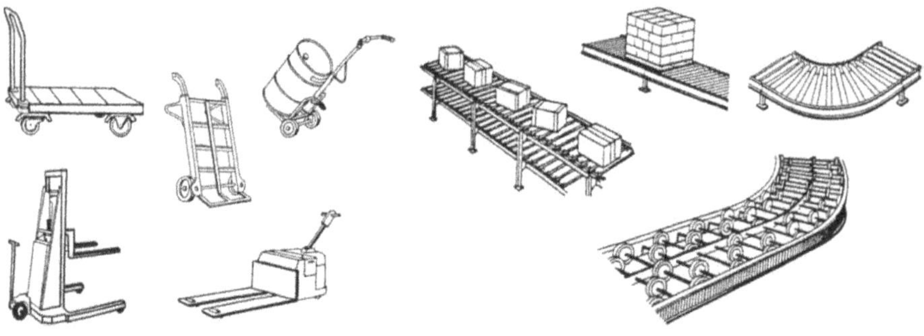

FIGURE 21.6 Using dollies and carts, rollers, and conveyors instead of carrying loads makes moving loads easier, faster, and safer.

FIGURE 21.7 Rather than attempting to lift a load while awkwardly bending, instead try to lift with the load between your legs. Adapted from ILO (1998).

FIGURE 21.8 Rather than taking loads from the floor, take them from an elevated location. Adapted from ILO (1998).

Summary

The human body is vulnerable to overexertion and injury due to load handling. Damage often occurs to the musculoskeletal elements of the lower back. Efforts to educate load handlers in proper and safe procedures are commonly less effective than desired. The only truly successful way to reduce the risk of human injury is to "design out" the load handling risk.

Lifting and lowering, usually the most risky activities, should ideally be eliminated altogether. If that is impossible, mechanization of the task avoids human involvement; mechanization might become economically attractive if the load is increased, for example, by packaging several heavy items together. The opposite action, reducing the load, is appropriate if human involvement cannot be avoided.

Carrying is also likely to lead to injuries. It is best to eliminate the task entirely; if that is not possible, mechanization of the transporting might be feasible (possibly also with bundled loads). If humans must do the carrying, load and distance should be kept minimal.

Pushing and pulling are better techniques than lifting, lowering, and carrying loads; yet, even pushes and pulls can cause overexertions and therefore should be eliminated or at least performed by machinery. If humans must push and pull, the weight of the loads should be kept low.

Steps to fit the human

Step 1: Determine what the material handling tasks are and whether it is necessary for handling to be performed by humans. Can a machine do the work? Can the loads be bundled for a machine, or made smaller and lighter for human handling?

Step 2: Consider redesigning the workflow to reduce or eliminate awkward loads and transits.

Step 3: Try to eliminate the need to lift or lower by hand by using equipment or machines like hoists, lift tables, or cranes. Try to eliminate the need for loads to be carried by using equipment or machines like carts, rollers, and sliding surfaces. If loads must be carried, give them handles or other ergonomic interfaces to the human, keep them light and close to the body, and consider using two or more people if needed.

Step 4: Ensure handling tasks are well below recommended guidelines and provide frequent rest breaks.

Step 5: Monitor conditions, modify the conditions or tasks, and iterate as necessary.

Notes

The text contains markers, *, to indicate specific references and comments, which follow.

21.1 Material handling strains the body

Workplace injuries to lower back: Garg and Marras (2014), Marras and Karwowski (2006), Snook (2000). Leading causes of workplace injuries, Liberty Mutual Insurance Company (2024) Workplace Safety Index https://business.libertymutual.com/insights/2024-workplace-safety-index/

Newton's second law: force = mass × acceleration.

21.2 Body capabilities related to load handling

Individual and situational strength factors: Daams (1993, 2001), Kumar (2004, 2008).

Low back pain: Deyo and Weinstein (2001), Marras (2008), Snook (2000), Violante, Armstrong, and Kilbom (2000).

Biomechanical approach: Chaffin, Andersson, and Martin (2006), Kumar (2004), Marras and Karwowski (2006a), Oezkaya, Nordin, Goldsheyder et al. (2012), Zhang and Chaffin (2006).

Anthromechanical considerations: Kroemer (2008).

Belly protusion: Marras (2008).

21.3 Assessing load handling capabilities

Combinations of measurement methods: Dempsey (1998, 2006), Kumar (2004), Stanton, Hedge, Brookhuis et al. (2004).

Disc pressure: Chaffin, Andersson, Martin (2006), Kumar (2004), Marras and Karwowski (2006b).

Loads which, supposedly, men, women, and even children could lift safely: ILO (1988).

Dynamic tests are more complicated than static tests: Astrand, Rodahl, Dahl et al. (2004), Kroemer (1974, 1999), Marras and Karwowski (2006a), Winter (2009).

Judging the strenuousness of the task: Dempsey (1998, 2006), Snook (2004), Snook and Ciriello (1991).

Industrial material handling: Kroemer (1985).

21.4 NIOSH lifting and lowering guidelines

NIOSH: Colombini, Occhipinti, Alvarez-Casado et al. (2012), Lu (2012), NIOSH (1981, 1994), Waters (2006), Waters and Putz-Anderson (1999).

RWL for the given conditions: Since the guidelines are subject to updating, check the latest NIOSH and OSHA information.

21.5 Liberty Mutual load handling guidelines

Acceptable to the population: Ciriello (2001, 2007); Ciriello, McGorry, Martin et al. (1999); Liberty Mutual guidelines are regularly updated at https://libertymmhtables.libertymutual.com/; Snook (2004), Snook and Ciriello (1991).

Objects that are difficult to grasp: Ciriello (2001, 2007), Kroemer (1997). Special considerations for moving healthcare patients are discussed in Chapter 22.

21.6 Designing for easy load handling

Intra-abdominal pressure: Easily measured by swallowing a capsule that contains a pressure-sensitive element.

Back belts are often ineffective: McGill (2006), US NIOSH website reviewed July 2024.

Instructions and training: Black, Shah, Busch et al. (2011), Kroemer (1997), Kroemer Elbert, Kroemer, and Kroemer (2018), Kurowski, Buchholz, and Punnett (2014), Lavender (2006), Martimo, Verbeek, Karppinen et al. (2008).

Wearable devices: Stefana, Marciano, Rossi et al. (2021), Figueira, Silva, Costa et al. (2024).

Proper task engineering: Ramsey, Davis, Kotowski et al. (2014).

References and further reading

Astrand PO, Rodahl K, Dahl HA, Stromme SB (2004). *Textbook of Work Physiology: Physiological Bases of Exercise*, 4th ed. Champaign, IL, Human Kinetics.

Black TR, Shah SM, Busch AJ, Metcalfe J, Lim HJ (2011). Effect of transfer, lifting, and repositioning (TLR) injury prevention program on musculoskeletal injury among direct care workers. *Journal of Occupational and Environmental Hygiene*, 8(4):226–235. https://doi.org/10.1080/15459624.2011.564110

Chaffin DB, Andersson GBJ, Martin BJ (2006). *Occupational Biomechanics*, 4th ed. New York, Wiley.

Ciriello VM (2001). The effects of box size, vertical distance, and height on lowering tasks. *International Journal of Industrial Ergonomics*, 28(2):61–67. https://doi.org/10.1016/S0169-8141(01)00012-9

Ciriello VM (2007). The effects of container size, frequency and extended horizontal reach on maximum acceptable weights of lifting for female industrial workers. *Applied Ergonomics*, 38(1):1–5. https://doi.org/10.1016/j.apergo.2006.02.001

Ciriello VM, McGorry RW, Martin SE, Bezverkhny IB (1999). Maximum acceptable forces of dynamic pushing: comparison of two techniques. *Ergonomics*, 42(1):32–29. https://doi.org/10.1080/001401399185784

Colombini D, Occhipinti E, Alvarez-Casado E, Waters T (2012). *Manual Lifting: A Guide to the Study of Simple and Complex Lifting Tasks*. Boca Raton, FL, CRC Press.

Dempsey PG (1998). A critical review of biomechanical, epidemiological, physiological and psychophysical criteria for designing manual materials handling tasks. *Ergonomics*, 41(1):73–88. https://doi.org/10.1080/001401398187332

Dempsey PG (2006). Psychophysical approach to task analysis. Chapter 47 in Marras WS, Karwowski W (eds), *The Occupational Ergonomics Handbook: Fundamentals and Assessment Tools for Occupational Ergonomics*, 2nd ed. Boca Raton, FL, CRC Press, pp. 421–442. https://doi.org/10.1201/9781420003635

Deyo RA, Weinstein JN (2001). Low back pain. *New England Journal of Medicine*, 344:363–370. https://doi.org/10.1056/NEJM200102013440508

Daams BJ (1993). Static force exertion in postures with different degrees of freedom. *Ergonomics*, 36:397–406.

Daams BJ (2001). Push and pull data and torque data. In Karwowski W (ed), *International Encyclopaedia of Ergonomics and Human Factors*. London, UK, Taylor & Francis, pp. 299–316, pp. 334–342.

Figueira V, Silva S, Costa I, Campos B, Salgado J, Pinho L, Freitas M, Carvalho P, Marques J, Pinho F (2024). Wearable's for monitoring and postural feedback in the work context: a scoping review. *Sensors (Basel)*, 24(4):1341. https://doi.org/10.3390/s24041341

Garg A, Marras WS (2014). Epidemiological studies of workplace musculoskeletal disorders. *Special Issue of Human Factors*, 56(1):5.

ILO (1988). *Maximum Weights in Load Lifting and Carrying, Occupational Safety and Health Series #59*. Geneva, CH, International Labour Office.

Kroemer Elbert K, Kroemer HB, Kroemer Hoffman AD (2018). *Ergonomics: How to Design for Ease and Efficiency*, 3rd ed. London, UK, Elsevier. https://doi.org/10.1016/C2016-0-02516-0

Kroemer KHE (1974). Designing for Muscular Strength of Various Populations. AMRL-Technical Report 72-46, Wright-Patterson AFB, OH, Aerospace Medical Research Laboratory.

Kroemer KHE (1985). Testing individual capability to lift material: repeatability of a dynamic test compared with static testing. *Journal of Safety Research*, 16(1):1–7. https://doi.org/10.1016/0022-4375(85)90032-5

Kroemer KHE (1997). *Ergonomic Design for Material Handling Systems*. Boca Raton, FL, CRC Press. https://doi.org/10.1201/9780203756683

Kroemer KHE (1999). Assessment of human muscle strength for engineering purposes: a review of the basics. *Ergonomics*, 42(1):74–93. https://doi.org/10.1080/001401399185810

Kroemer KHE (2008). Anthropometry and biomechanics: anthromechanics. Chapter 2 in Kumar S (ed), *Biomechanics in Ergonomics*, 2nd ed. Boca Raton, FL, CRC Press, pp. 48. https://doi.org/10.1201/9780849379093

Kumar S (2001). Theories of musculoskeletal injury causation. *Ergonomics*, 44(1):17–47. https://doi.org/10.1080/00140130120716

Kumar S (2004). *Muscle Strength*. Boca Raton, FL, CRC Press. https://doi.org/10.1201/9780203503591

Kumar S (2008). *Biomechanics in Ergonomics*, 2nd ed. Boca Raton, FL, CRC Press. https://doi.org/10.1201/9780849379093

Kurowski A, Buchholz B, Punnett L, ProCare Research Team (2014). A physical workload index to evaluate a safe resident handling program for nursing home personnel. *Human Factors*, 56(4):669–683. https://doi.org/10.1177/0018720813509268

Lavender SA (2006). Training lifting techniques. Chapter 23 in Marras WS, Karwowski W (eds), *The Occupational Ergonomics Handbook: Interventions, Controls, and Applications in Occupational Ergonomics*, 2nd ed. Boca Raton, FL, CRC Press, pp. 848. https://doi.org/10.1201/9781420003642

Liberty Mutual. *Manual Materials Handling Guidelines*. Regularly updated on www.libertymutual.com

Lu ML (2012). *A Practical Guide to the Revised NIOSH Lifting Equation (NLE)*. Cincinnati, OH, National Institute for Occupational Safety and Health.

Marras WS (2008). *The Working Back: A Systems View*. New York, Wiley & Sons. https://doi.org/10.1002/9780470258576

Marras WS, Karwowski W (2006a). *The Occupational Ergonomics Handbook: Fundamentals and Assessment Tools for Occupational Ergonomics*, 2nd ed. Boca Raton, FL, CRC Press. https://doi.org/10.1201/9781420003635

Marras WS, Karwowski W (2006b). *The Occupational Ergonomics Handbook: Interventions, Controls, and Applications in Occupational Ergonomics*, 2nd ed. Boca Raton, FL, CRC Press. https://doi.org/10.1201/9781420003642

Marras WS, Radwin RG (2006). Biomechanical modelling. Chapter 1 in Nickerson RS (ed), *Reviews of Human Factors and Ergonomics*. Santa Monica, CA, Human Factors and Ergonomics Society, pp. 588–589.

Martimo KP, Verbeek J, Karppinen J, Furlan AD, Takala EP, Kuijer PP, Jauhiainen M, Viikari-Juntura E (2008). Effect of training and lifting equipment for preventing back pain in lifting and handling: systematic review. *British Medical Journal*, 336(7641):429–31. https://doi.org/10.1136/bmj.39463.418380.BE

McGill SM (2006). Back belts. Chapter 30 in Marras WS, Karwowski W (eds), *The Occupational Ergonomics Handbook: Interventions, Controls, and Applications in Occupational Ergonomics*, 2nd ed. Boca Raton, FL, CRC Press, pp. 421–442. https://doi.org/10.1201/9781420003642

Nachemson A, Elfström G (1970). Intravital dynamic pressure measurements in lumbar discs. A study of common movements, manoeuvres and exercises. *Scandinavian Journal of Rehabilitation Medicine Supplement*, 1:1–40.

Waters TR, Putz-Anderson V, Garg A, NIOSH (1994, revised 2021). *Applications Manual for the Revised NIOSH Lifting Equation*. Cincinnati, OH, U.S. Department of Health and Human Services, Centers for Disease Control and Prevention, National Institute for Occupational Safety and Health, DHHS (NIOSH) Publication No. 94-110. https://doi.org/10.26616/NIOSHPUB94110revised092021

NIOSH (National Institute for Occupational Safety and Health) (1981). *Work Practices Guide for Manual Lifting*. DHHS (NIOSH) Publication # 81-122, Washington, DC, US Government Printing Office. https://www.cdc.gov/niosh/docs/81-122/default.html

Oezkaya N, Leger D, Goldsheyder D, Nordin M (2017). *Fundamentals of Biomechanics: Equilibrium, Motion, and Deformation*, 4th ed. Switzerland, Springer. https://doi.org/10.1007/978-3-319-44738-4

Ramsey T, Davis KG, Kotowski SE, Anderson VP, Waters T (2014). Reduction of spinal loads through adjustable interventions at the origin and destination of palletizing tasks. *Human Factors*, 56(7):1222–1234. https://doi.org/10.1177/0018720814528356

Snook SH (2000). Back risk factors: an overview. Chapter 11 in Violante F, Armstrong T, Kilbom A (eds), *Occupational Ergonomics. Work Related Musculoskeletal Disorders of the Upper Limb and Back*. London, UK, Taylor & Francis, pp. 120–128. https://doi.org/10.1201/9781482268287

Snook SH (2000). Leading Causes of Workplace Injuries, Liberty Mutual Insurance Company Newsletter Winter 2013/14. www.libertymutual.com

Snook SH (2004). Psychophysical tables: lifting, lowering, pushing, pulling, and carrying. Chapter 13 in Stanton N, Hedge A, Brookhuis K, Salas E, Hendrick HW (eds), *Handbook of Human Factors and Ergonomics Methods*. Boca Raton, FL, CRC Press, pp. 768. https://doi.org/10.1201/9780203489925

Snook SH, Ciriello VM (1991). The design of manual handling tasks: revised tables of maximum acceptable weights and forces, *Ergonomics*,34(9):1197–1213.https://doi.org/10.1080/00140139108964855

Stanton N, Hedge A, Brookhuis K, Salas E, Hendrick HW (2004). *Handbook of Human Factors and Ergonomics Methods*. Boca Raton, FL, CRC Press. https://doi.org/10.1201/9780203489925

Stefana E, Marciano F, Rossi D, Cocca P, Tomasoni G (2021). Wearable devices for ergonomics: a systematic literature review. *Sensors (Basel)*, 21(3):777. https://doi.org/10.3390/s21030777

Violante F, Armstrong T, Kilbom A (2000). *Occupational Ergonomics. Work Related Musculoskeletal Disorders of the Upper Limb and Back*. London, UK, Taylor & Francis. https://doi.org/10.1201/9781482268287

Waters TR (2006). Revised NIOSH lifting equation. Chapter 46 Marras WS, Karwowski W (eds), *The Occupational Ergonomics Handbook: Fundamentals and Assessment Tools for Occupational Ergonomics*, 2nd ed. Boca Raton, FL, CRC Press, pp. 421–442. https://doi.org/10.1201/9781420003635

Waters TR, Putz-Anderson V (1999). Revised NIOSH lifting equation. Chapter 57 in Marras WS, Karwowski W (eds), *The Occupational Ergonomics Handbook*, Boca Raton, FL, CRC Press, pp. 421–442.

Winter DA (2009). *Biomechanics and Motor Control of Human Movement*, 4th ed. New York, Wiley. https://doi.org/10.1002/9780470549148

Zhang X, Chaffin DB (2006). Digital human modelling for computer-aided design. Chapter 10 in Marras WS, Karwowski W (eds), *The Occupational Ergonomics Handbook: Interventions, Controls, and Applications in Occupational Ergonomics*, 2nd ed. Boca Raton, FL, CRC Press, pp. 848. https://doi.org/10.1201/9781420003642

CHAPTER TWENTY-TWO

Healthcare for patients and providers

Healthcare

In recent years, significant advances have been made in understanding, diagnosing, and treating health disorders and the consequences of accidents and aging. This progress goes hand in hand with advancements in medical technology and the management of emergency and routine healthcare procedures.

In the United States, over 6100 hospitals were registered in 2022, collectively housing nearly 1 million patient care beds. These hospitals recorded more than 30 million patient admissions per year (along with many more healthcare visits) and employed about 7 million people. Additionally, around 20,000 nursing homes and residential care facilities in the United States employed nearly 3 million workers. Globally, there are more than 150,000 hospitals, supported by over 100 million health workers, including 30 million nurses and 13 million doctors.*

Healthcare has traditionally adopted roles and rules* that differ from those of other businesses: above all, there is the overriding mission to help people in need. These ethical concerns foster a conviction of knowing what is best for others, regardless of costs (financial and otherwise), which may lead to an attitude that achieving these goals justifies work practices for care providers that exceed or disregard accepted norms. However, in recent decades, it has become increasingly recognized that the much-respected nurses and doctors should limit their working hours and avoid overexertion for their own sake and the well-being of their patients. Additional safeguards, such

DOI: 10.1201/9781032674575-28

as checklists for medical procedures, systems for counting work tools and instruments at the end of operations, and basic patient care guidelines, are being implemented. Healthcare providers are human and fallible and need (and deserve) ergonomic assistance.

Patients also need and deserve ergonomic assistance. We are at our most vulnerable when we require medical care; we may be unable to fully communicate or participate in our care, and we may be surrounded by highly stressed family members and caregivers, each with different (and sometimes conflicting) wishes and needs.

Consequently, ergonomics and human factors engineering have much to contribute to improving the quality of care and well-being for patients, as well as enhancing care providers' working conditions and safety. The field of medical applications is broad and ever-changing, so the following text discusses several topics but is certainly still incomplete. Although details and specific applications will evolve over time, the fundamental principles of human factors will continue to apply.

22.1 Basics of patient care and safety

Patients

Patients are treated in many different settings and interactions. Hence, considerations of ergonomics and human factors engineering are needed in many healthcare locations and in the transitions between them, including:

- Ambulatory care (doctor's offices, day-surgery centers, dialysis clinics, etc.),
- Emergency medical care (at the incident site, in transit, and in treatment areas),
- Intensive or critical care,
- Operating rooms and surgical settings,
- Recovery and ongoing in-hospital care,
- Rehabilitation and nursing home care, and
- At-home care.

Patient transitions include both the physical aspects (such as beds, gurneys, and wheelchairs) as well as the informational aspects (such as care records and medication schedules)—both of which are further discussed below.

22.2 Basics of healthcare provider performance and safety

Care providers

In each aspect of patient care, someone is providing this care. The working conditions and safety needs of these care providers must be considered. In the pursuit of safe, efficient, effective, and compassionate care, healthcare staff are impacted by the following overlapping topics (and more) in public health:

- Medical decision-making,
- Medical information technology,
- Electronic prescription technology and medication alerts,
- Electronic medical records, checklists, and questionnaires,
- Training by simulation and other means,
- Workplace design and workflow analysis,
- Design of operating theaters and surgical sites,
- Design of medical devices, instrumentation, equipment,
- Technology of medical and drug delivery devices and equipment,
- Usability of medical and drug delivery devices and equipment,
- Healthcare supply logistics,
- Drug handling and exposure,
- Hygiene, cleanliness, and sanitation,
- Disinfection and sterilization,
- Emergency procedures,
- Organizational and management issues,
- Handling of non-cooperative patients,
- Boredom and fatigue during long shifts,
- Work schedules and shifts,
- Stress management,
- Error recovery, and
- Safety guidelines, standards, and laws.

Several of these topics are addressed in other chapters in this book: making decisions in Chapter 9; boredom and fatigue in Chapter 11; workload and stress in Chapter 12; organizational and management issues in Chapters 13 and 14; work schedules in Chapters 15 and 16; computer use in Chapter 19; office and workplace design in Chapters 18 and 20; load handling in

Chapter 21. The specific conditions associated with healthcare require special considerations and applications that go beyond the discussions in this book so far. In the following text, several of these topics are discussed in more detail to serve as examples of ergonomic approaches.

22.3 Addressing emergency medical services

Emergency care

Emergency medical service (EMS) is often a life-saving component of public safety, especially during accidents in transport and industry or catastrophic events such as earthquakes or wild fires. The victims' chances of injury recovery and survival depend on the speed at which emergency responders arrive, their ability to make critical decisions, and their skill in administering on-site medical treatment. The working environment for emergency personnel is unusually complex, fast-paced, high-stress, and often physically dangerous for everyone involved. It requires first-aid providers to make decisions quickly and take actions efficiently at a high level of competence. Interventions performed by EMS paramedics and first-aid physicians often involve procedures that, if done incorrectly or at the wrong time, can cause serious harm to both patients and caregivers.

Ergonomic assistance to emergency services can be provided at several levels. The proper size, design, and contents of the paramedics' response bag*, which includes medications, bandages, instruments, and tools, is essential. Training for first-aid providers in the correct and efficient treatment of victims as well as ensuring their own safety is equally important, as they are at high risk for both acute and chronic injuries.

Human factors engineering can also assist in selecting and equipping ambulances and other vehicles needed to transport EMS personnel to the accident site and to the hospital. One major concern is the intensity and quality of emergency treatment: Is an experienced physician necessary to perform life-saving interventions on-site? Does the specialized physician or other care provider need to be present in person, or can telemedicine and other advanced remote capabilities be utilized? What implications does this have for ambulance equipment? What types of life-support apparatus are required on-site and during transfer to the nearest hospital? How will the patients, along with all relevant information related to their treatment or tests already performed, be transferred to the hospital staff?

22.4 Designing wheelchairs and hospital beds

Wheelchairs

Wheelchair design has come a long way from the uncomfortable folding chairs that did not provide adequate support to users. In older designs, the seat and backrest were either overly rigid or sagged under the weight of the body, adding painful discomfort to the distress already experienced by the patient.

Solid construction can stabilize patients of various body sizes in different postures by allowing adjustments in the height and angles of the seat, backrest, and foot supports. For long-term sitting, these settings should be adjustable by the user to provide the necessary pressure relief to avoid pressure sores (ulcerations of skin and subcutaneous tissue caused by poor circulation due to prolonged pressure on body parts, especially bony protuberances). Users of motorized wheelchairs* appreciate features such as mechanisms for lowering the seat to fit their legs and the chair under a table, raising the seat to reach a high shelf or converse eye-to-eye with a standing person, and tilting the seat and back support to shift the user's body to help avoid pressure sores. Some motorized wheelchairs can accommodate rough terrain (even stairs) to allow for independent movement, especially when facilities are not designed to be friendly to wheelchair users.

The selection and arrangement of the controllers for adjustments can be a challenging human engineering task, particularly for wheelchairs, because many patients have limited movement capabilities. On motorized chairs, the type of controls and their locations are often customized to suit individual users' abilities and needs.

Beds

Hospital beds have a number of adjustments, made by hand or motorized: these typically include height adjustments of the whole bed or of the sections that support the head or feet. The adjustment features are meant to allow patients to assume different body postures, relieve pressure points, and make it easier for caregivers to turn patients and help them get on and off the bed. These adjustment features may be accessible to both the care provider and the patient, requiring intuitive and easy-to-understand control designs that require minimal force to operate.

22.5 Moving patients

Moving patients

Many tasks involved in caring for patients at home, in hospitals, and in nursing homes require considerable strenuous physical labor from caregivers, especially when transferring, lifting, sliding, and otherwise moving individuals who cannot actively cooperate. Healthcare providers who handle patients experience a high number of work-related back pain cases; overall, the injury rate* for caregivers is about double the rate for all full-time workers in other occupations. Beyond the pain to the worker, the cost of lost workdays is a concern for management because it impacts the quality, availability, and expense of healthcare.

Ergonomic interventions help to reduce the risk of work-related strain injuries by designing patient handling devices and work procedures that can be performed safely. Certainly, patients are not "loads," but most of the biomechanical discussions in Chapter 21 regarding proper "load handling" help to avoid overexertion and injuries to the caregiver. Back injuries and shoulder strains often occur while lifting, transferring, and repositioning patients.

Risks in patient handling

Risk factors for different aspects of patient handling* are related to the task itself, the patient, and the environment. Risks related to the physical *task* of moving patients are listed below:

- *Force*: the amount of physical effort required to perform the task (such as heavy lifting, pulling, and pushing) or to maintain control of equipment and tools.
- *Repetition*: performing the same motion or series of motions continually or frequently during the work shift, and over multiple work shifts.
- *Awkward positions*: assuming positions that place stress on the body, such as leaning over a bed, kneeling, reaching, or twisting the trunk while lifting.

There are additional risks related to patient factors. Patients cannot be lifted like loads, so although the biomechanical discussions in Chapter 21 help to avoid overexertion, other aspects of the ergonomic advice for materials handling do not apply. Caregivers cannot reduce the body weight of the patient, they often cannot hold the patient close to their body, and they cannot always lift with their upper body upright. Patients present the following special circumstances:

- Patients may be medically fragile, have wounds, and/or be attached to tubing and equipment.
- Patients do not have handles.
- Patients may be unable or unwilling to cooperate.
- The patient's response to being moved may be hard to predict, especially if the patient is on medication, in pain, or sedated.

There are additional risks related to the *environment*. Caregivers may not be able to adjust workflow or workplace conditions (see Chapters 18 and 20) to accommodate immediate or urgent patient needs. Risks from the surroundings may include the following:

- Slip, trip, and fall hazards,
- Uneven and soft support surfaces,
- Space limitations (small rooms, bulky equipment, other patients, or caregivers),
- Inadequate footwear, clothing, protective gear, and lighting,
- Inadequate equipment, and
- Lack of knowledge or training.

Transfer of patients

It is useful to distinguish three ways of patient handling:

1. Manual transfer is carried out by one or more caregivers using their own muscular force and, wherever possible, any residual movement capacity of the patient involved.
2. Transfer using small patient handling aids is carried out with specific equipment such as low-friction fabric sheets, ergonomic belts, rotatable footboards, and trapeze bars attached above the bed. This is also augmented by any residual movement capacity of the patient involved.
3. Transfer using sophisticated patient handling aids is carried out with electro-mechanical lifting equipment.

Selection of the proper handling technique involves an assessment of the needs and abilities of both the patient and the caregiver(s):

- A patient who is able and willing to partially support his or her own weight may be able to transfer from bed to a chair using a standing assist device (such as a grab bar or other support), while a patient who is non-cooperative may need a mechanical lift.
- A patient who is too large or too heavy for caregiver(s) to lift may need mechanical assistance.
- A patient with medical conditions such as pregnancy, wounds, tubes, bandages, or attached equipment may need specialized assistance for transfer or repositioning.
- The physical environment (such as space limitations) may necessitate awkward positions and postures, thus creating unsafe conditions for both the patient and caregiver(s).

Manual lifting of patients should be eliminated when feasible and minimized in all other cases. There are many available options, from simple aid devices to advanced (and often expensive) technical patient handling aids, that should be carefully considered.

Sole caregiver

For a sole caregiver treating a patient, the greatest risks for overexerting the caregiver's back occur when pulling a patient who is reclining in bed into a sitting position, transferring a patient from a bed to a chair, and leaning over a patient for extended periods. These actions involve awkward postures, difficult reaches, and potentially heavy and sudden exertions. When lifting or moving a person, the sole caregivers can reduce their risk of back injury by

- Maintaining proper alignment of their head and neck with the spine,
- Maintaining the natural curvature of their spine,
- Avoiding bending at the waist,
- Avoiding twisting their body when carrying or repositioning a person,
- Keeping the person who is being moved close to their body,
- Having their feet shoulder-width apart to maintain their balance, and
- Using the muscles in their legs to lift and/or pull.

If the task is too straining, the sole caregiver should get help.

22.6 Designing electronic health records

EHR

Electronic health records (EHRs) are widely used in medical offices and healthcare facilities. EHRs are valuable tools when they are properly set up and maintained, and when the information is complete and accurate. They allow for the documentation of a patient's medical history, test results, medications, and other relevant information in an accessible and searchable format. Hence, incomplete or incorrect records can be dangerously misleading. Since new patients (and often their accompanying companions) are usually asked to fill out multiple EHR forms as a prerequisite for medical treatment or admission, the design of these forms requires careful consideration of the users' ability to understand the questions they are asked to answer. People in acute distress, with low language fluency or literacy, limited knowledge of health-related medical terms, or limited technical skills may find it difficult to provide the requested information*. If they cannot (or do not) obtain the necessary assistance, they may unintentionally provide false or misleadingly incomplete data, which can have severe consequences.

One somewhat unexpected side effect of the widespread adoption of electronic medical records by healthcare organizations is an increase in musculoskeletal disorders among healthcare staff who use the computerized systems. Pain and discomfort in the neck, shoulder, upper and lower back, and wrist are the most frequently reported repetitive strain injuries among medical computer users*. This is similar to what occurred several decades ago when a large number of computers were introduced into regular offices (as discussed in Chapters 18 and 19).

22.7 Addressing medication alerts

Alerts

With EHRs in increasing use, pharmacists and others who prescribe and dispense medication have better visibility into a patient's medical history. Accurate and complete medication records can be set up to produce alerts* when prescribing a particular medication that may pose a health risk to a patient, such as

- *Drug–drug alerts*: when the effects of one drug alter or interact with those of another drug when both are administered together.
- *Drug allergy alerts*: when the patient has or may have an allergy to a drug.
- *Drug–disease alerts*: when a drug may negatively impact a pre-existing medical condition.
- *Drug–food alerts*: when the effects of a drug are altered by food intake.
- *Drug indication alerts*: when indications and/or contraindications of a drug impact the patient.
- *Overdose alerts*: when the intended dose of the drug exceeds standard or recommended dosing values for that patient.

Alert fatigue

It is a promising concept to use EHRs to automatically generate a risk alert when prescribing a particular medication to a specific patient. However, to be usable and to be used, the software systems must be reliable, with up-to-date information and data privacy safeguards. Furthermore, these systems should not generate an overwhelming number of alerts, which may appear complex, unexplained, confusing, or even cryptic to the prescriber. Generating too many alerts can lead to "alert fatigue," causing caregivers to overlook or ignore clinically useful warnings. This can result in serious consequences for patients.

22.8 Designing medical devices

Medical diagnoses and treatments have improved our lives immeasurably. Global life expectancy is now over 70 years of age; 200 years ago, it was half that. We want to enjoy healthier, longer lives; advances in medical technology help us do that. Ideally, all medical treatments would always be safe and effective. However, no diagnosis or treatment is ever perfect. Even the most ideal patient or the best surgeon can have a bad outcome from any medical intervention. As humans, we misunderstand, we overlook something important, we become tired and distracted, we make errors. Since humans cannot be made fail-proof, we expect the medical industry to reduce risks and accommodate human factors in the design of medical systems. Not only is this the right thing to do for patients and care providers, but also there are regulations and guidelines in place that govern the medical industry.

Safe, effective

Agencies like the Food and Drug Administration in the United States, the European Medicines Agency and Medical Device Regulations, the National Medical Products Administration in China, international standards organizations like ISO, and other agencies around the world regulate and standardize medicines and medical devices for their local regions. For example, medical devices must generally be shown to be safe and effective before they can be made available to surgeons and other medical professionals. *Safe* means the absence of unreasonable risk of illness or injury when the device is used as intended. *Effective* means that the device, when used as intended and when accompanied by adequate directions for use and warnings against unsafe use, will provide clinically significant results.

Usability testing

How human factors are addressed in the design of medical devices is outlined in, among others, FDA, ISO, and ANSI documents*: user-centered design of devices must consider the user, the use environment, and the device-user interface. Usability engineering is conducted in parallel to the actual development of the device and the formal process for managing risks. The usability testing assesses not only the physical attributes of the device and use environment but also the cognitive and physiological human factors of the device and its labeling (such as the instructions for use that accompany devices and equipment).

Use errors

Usability testing is intended to uncover any *use errors*, which are defined as situations where an unexpected user action (or lack of action) caused a result that was unexpected, was not caused solely by device failure, and did or could result in harm to the user(s) or patient. These use errors can be intended or unintended actions, including a *slip* (failure in attention), *lapse* (failure in memory), *mistake* (misapplication of a good rule or application of a bad rule), or *ignorance*. Note that "use" errors are not called "user" errors; the focus (or blame) is placed on the task (or the device and its instructions) and not the operator.

Usability testing often starts with mapping the medical procedure and how it is experienced by both the patient and the care provider. Additional testing may include task and use error analysis, contextual inquiry and ethnography, moderated simulated use studies, ergonomic assessments, mental workload assessments, task load index, and others.

Risk based approach

Medical device designers take a risk-based approach, seeking to balance benefits against any residual risks that remain after the device has been designed and tested. Taking a risk-based approach means the designers understand the user, use environment, and the device-user interface well enough to reduce the probability of use errors. No device or medicine is ever completely risk-free for the patient or caregiver; even a simple tongue depressor can cause injury if inserted too deeply into the mouth. As another example, for riskier devices such as those that deliver drugs, the manufacturer would pay special attention to use errors that might:

- Impact dosing of a medicine (overdose, underdose, missed dose),
- Impact administration of the product (improper preparation or delivery of the drug), and/or
- Have the potential to result in harm (physical injury, adverse events).

Given the criticality of safe and effective medicines and devices, it is important (as well as required by regulations) to consider human factors in their design. When we recognize that a user may skip steps or perform them out of order, plug in the wrong components, or incorrectly read or input data, then we can create interfaces that address the risk of use error. As with any application of ergonomics, we should not blame the user for being human.

22.9 Addressing stress in the workplace

Stress

Patient care routinely requires healthcare staff to cope with some of the most stressful situations found in any workplace. The staff must care for dependent and often demanding patients; they must deal with life-threatening injuries and illnesses, and often, with patient deaths. At the same time, their work is complicated by intricate (and possibly malfunctioning) equipment, tight schedules, understaffing and overwork, and complex hierarchies of authority and skills. All of these factors contribute to stress. It is no surprise that stress can affect caregivers' attitudes and behavior. Frequently reported consequences of stress include difficulties in judging the seriousness of a potential emergency, communicating with very ill patients, and maintaining pleasant relationships with patients and coworkers. Stressed individuals report loss of appetite, ulcers, mental disorders, migraines, difficulty sleeping, emotional instability, disruption of social and family life, and increased use of tobacco, alcohol, and drugs.

Ergonomic interventions for workplace stress are also discussed in Chapter 12 and are acutely needed in a healthcare setting. The consequences of not addressing healthcare work stress can be severe. When incidents and injuries occur in the workplace, they need to be properly documented and investigated. Analyses of the workplace and reviews of accident and incident reports by management and employees can provide insight and direction for ergonomic efforts to reduce stressors for both patients and care staff, as well as guide the proper selection and use of facilities and equipment.

Training, assistance

Training programs, carefully designed and implemented, should be in place to provide continual education and training about workplace stresses, hazards, and ergonomic controls to managers, supervisors, and all healthcare providers, including orientation for new employees. Well-designed and confidential employee assistance programs can provide psychosocial support. Training and assistance programs should be reviewed and updated as changes occur in the workplace.

22.10 Addressing safety guidelines, standards, and laws

Regulations

A plethora of safety-related guidelines, requirements, and regulations exists in many countries and specific localities. These documents vary in scope and depth, but they all aim to achieve a high level of safety and health for patients and caregivers. Applicable standards are continually reviewed and updated: international standards include ISO 45001 to govern occupational health and safety management systems, ISO 7101 to govern healthcare management, and ISO 9001 to govern quality management systems. Regional laws may also apply; for example, the US states of California (in 2012) and New York (in 2014) enacted specific laws to promote safe patient handling and reduce musculoskeletal injuries in healthcare workers. Such regulations may require specific training for care staff in safe patient handling (lifting, mobilizing, repositioning, transferring) and the use of a lift team. Thus, everyone responsible for setting up and managing care facilities must understand and comply with the relevant guidelines and laws and apply ergonomic principles for patient and caregiver safety and well-being.

Summary

Patients in ambulatory care and acute care in hospitals and nursing homes partially, and often completely, depend on healthcare staff to provide medications and general nursing assistance. Patients may need help with daily living tasks such as dressing, bathing, feeding, and toileting. Each of these activities involves multiple person-to-person interactions, often requiring handling or transferring of patients, which can result in injuries to either or both parties. Ergonomic solutions to reduce risks include the appropriate design of facilities, installations, and devices according to human factors principles and practices, as well as training staff in the proper use of these implements.

When there is a mismatch between the physical requirements of the job and the physical capacities of the worker, work-related disorders, overexertions, and accidents (including trips, falls, and back injuries) can result. Ergonomic interventions can adjust work demands and practices to reduce such incidents.

The shift from paper to electronic records in the healthcare profession has brought similar human factors problems as seen earlier when other businesses transitioned to computerized records. Poorly designed questionnaires and insufficient support lead to incomplete and incorrect input of information about the patient's medical history. The next step, data management, requires appropriate software algorithms to detect, for example, medication interactions to which the prescribing physician should be alerted. If that purpose is not met, vital information may be overlooked, which can have dire consequences.

Medical devices, equipment, and drug delivery systems need to be designed with consideration for human factors and a user-centered approach. Ergonomic contributions to the design of interfaces with medical devices and their instructions for use can significantly enhance the safe and effective use of medical instruments and medication delivery.

Healthcare settings are inherently stressful, and workplace stress must be considered and addressed. Providers, patients, and their families all require additional support. Training and assistance programs need to be implemented and reviewed regularly to reflect changes in the workplace. Workplaces must comply with applicable regulations and guidelines.

Steps to fit the human

Step 1: Determine what tasks will be performed, with what equipment, by which persons, and what information must be conveyed and acted upon via displays and controls.

Step 2: Consider the specific needs and preferences of the patient and healthcare provider, along with any other tools or aids that will be used. Involve patients and providers in the design of workflows, workplaces, equipment, tools, and training. Where possible, add flexibility to the design of work equipment, furnishings, displays, and controls to allow for user adjustments.

Step 3: If the users, their needs, or the tasks change, consider how this can be accommodated. Ask workers if the workplace set-up, furnishings, and training meet their needs and those of the patients, and adjust as necessary.

Step 4: Ensure compliance with applicable regulations and guidelines. Provide appropriate assistance and training.

Step 5: Seek to continuously improve. Monitor conditions, modify, and iterate as necessary.

Notes

The text contains markers, *, to indicate specific references and comments, which follow.

Hospitals, care facilities, and workers: Data retrieved in July 2024 from the websites of the American Hospital Association, the US Bureau of Labor Statistics, the US Occupational Safety and Health Administration, and the World Health Organization.

Rules in healthcare: Carayon (2011), Charney (2010), Dillon (2008), Waterson (2014). See also regulatory requirements and guidelines such as from the US Food and Drug Administration, European Medicines Agency and Medical Device Regulations, and other national regulatory and healthcare agencies around the world.

22.3 Basics of emergency medical services

Paramedics' "response bag": Bitan, Ramey, Philp et al. (2015).

22.4 Designing wheelchairs and hospital beds

Motorized wheelchairs: Robert C Williges, personal communication with KHE Kroemer July 8, 2015.

22.5 Moving patients

Injury rate for caregivers: Data retrieved in July 2024 from the website of the US Bureau of Labor Statistics. See also Kurowski, Buchholz, and Punnett (2014), Waters (2010).

Handling patients: Guidelines for patient handling are available from the European Agency for Safety and Health at Work (http://osha.europa.eu), the US Occupational and Safety Health Agency OSHA (https://www.osha.gov/healthcare/safe-patient-handling), and many nursing and medical associations.

22.6 Designing electronic health records

Furnish the asked-for information: Czaja, Zarcadoolas, Vaughon et al. (2015).

Repetitive strain injuries of medical computer users: Hedge and James (2012) and the white paper by CIEHF (2021).

22.7 Addressing medication alerts

Electronic health records and alerts: Mauney, Furlough, and Barnes (2015), Patterson, Latkany, Brick et al. (2015).

22.8 Designing medical devices

Usability in medical devices: Weinger, Wiklund, and Gardner-Bonneau (2010), Wiklund, Kendler, and Strochlic (2016), FDA Human Factors Information Guidance (2016, 2022), IEC 62366-1, ISO 14971, ISO 13485, ANSI/AAMI HE75, AAMI TIR59, Kroemer Elbert, Kroemer, and Kroemer Hoffman (2018).

References and further reading

AAMI TIR59 (2017). Integrating Human Factors into Design Controls in accordance with FDA 21 CFR 820.30.

ANSI/AAMI HE75 (2009). R2018, Human Factors Engineering - Design of Medical Devices. ANSI/ASHRAE Standard 55, Thermal Environmental Conditions for Human Occupancy, latest edition 2023.

Bitan Y, Ramey S, Philp G, Uukkivi T (2015) Working with paramedics on implementing human factors improvements to their response bags. *Proceedings of the International Symposium on Human Factors and Ergonomics in Health Care*, 4:179–181. https://doi.org/10.1177/2327857915041016

Carayon P (2011). *Handbook of Human Factors and Ergonomics in Healthcare and Patient Safety*, 2nd ed. Boca Raton, FL, CRC Press. https://doi.org/10.1201/b11219

Charney W (2010). *Handbook of Modern Hospital Safety*, 2nd ed. Boca Raton, FL, CRC Press. https://doi.org/10.1201/9781420047868

CIEHF (2021). *Human Factors and Ergonomics in Healthcare AI*, white paper from Chartered Institute of Ergonomics & Human Factors. https://ergonomics.org.uk/resource/human-factors-in-healthcare-ai.html

Czaja SJ, Zarcadoolas C, Vaughon WL, Lee CC, Rockoff ML, Levy J (2015). The usability of electronic personal health record systems for an underserved adult population. *Human Factors*, 57(3):491–506. https://doi.org/10.1177/0018720814549238

Dhillon BS (2008). *Reliability Technology, Human Error, and Quality in Health Care*. Boca Raton, FL, CRC Press. https://doi.org/10.1201/9781420065596

Gawande A (2015). Overkill. *The New Yorker*, May 11.42–53.

Hedge A, James T (2012). Ergonomic issues of computer use in a major healthcare system. In Duffy VG (ed) *Advances in Human Aspects of Healthcare*. Boca Raton, FL, CRC Press, pp. 760. https://doi.org/10.1201/b12318

IEC 62366-1 (2015). *Application of Usability Engineering to Medical Devices*. Geneva, CH, International Labour Office.

ISO 13485 (2016). *Medical Devices - Quality Management Systems - Requirements for Regulatory Purposes*. Geneva, CH, International Organization for Standardization.

ISO 14971 (2019). *Application of Risk Management to Medical Devices*. Geneva, CH, International Organization for Standardization.

ISO 45001 (2018). *Occupational Health and Safety Management Systems — Requirements with Guidance for Use*. Geneva, CH, International Organization for Standardization.

ISO 7101 (2023). *Healthcare Organization Management — Management Systems for Quality in Healthcare Organizations — Requirements*. Geneva, CH, International Organization for Standardization.

Kroemer Elbert K, Kroemer HB, Kroemer Hoffman AD (2018). *Ergonomics: How to Design for Ease and Efficiency*, 3rd ed. London, UK, Elsevier. https://doi.org/10.1016/C2016-0-02516-0

Kurowski A, Buchholz B, Punnett L, ProCare Research Team (2014). A physical workload index to evaluate a safe resident handling program for nursing home personnel. *Human Factors*, 56(4):669–683. https://doi.org/10.1177/0018720813509268

Mauney J, Furlough C, Barnes J (2015). Developing a better clinical alert system in EHRs. *Proceedings of the International Symposium on Human Factors and Ergonomics in Health Care*, 4(1):29–36. https://doi.org/10.1177/2327857915041023

Patterson ES, Gibbons MC, Latkany P, Ramaiah M, Brick D, Lowry SZ (2015). Integrating electronic health records into clinical workflow: an application of human factors modelling methods to two specialty care areas. *Proceedings of the International Symposium on Human Factors and Ergonomics in Health Care*, 4(1):42–49. https://doi.org/10.1177/2327857915041018

Robinson PA (2009). *Writing and Designing Manuals and Warnings*, 4th ed. Boca Raton, FL, CRC Press. https://doi.org/10.1201/9781420069860

US Food and Drug Administration (FDA) Center for Devices and Radiological Health Office of Device Evaluation (2016, updated draft 2022). Applying Human Factors and Usability Engineering to Medical Devices - Guidance for Industry and Food and Drug Administration Staff. https://www.fda.gov/media/80481/download and https://www.fda.gov/media/163694/download

Waters TR (2010). Introduction to ergonomics for healthcare workers. *Rehabilitation Nurse*, 35(5):185–191. https://doi.org/10.1002/j.2048-7940.2010.tb00046.x

Waterson P (2014). *Patient Safety Culture: Theory, Methods and Application.* Boca Raton, FL, CRC Press. https://doi.org/10.1201/9781315599717

Weinger MB, Wiklund ME, Gardner-Bonneau DJ (2010). *Handbook of Human Factors in Medical Device Design*. Boca Raton, FL, CRC Press. https://doi.org/10.1201/b10439

Wiklund ME, Kendler J, Strochlic AY (2016). *Usability Testing of Medical Devices*, 2nd ed. Boca Raton, FL, CRC Press. https://doi.org/10.1201/b19082

CHAPTER TWENTY-THREE

Emerging ergonomic issues with artificial intelligence

In recent years, significant advances have been made in software and hardware that extend human abilities, specifically what is often termed *artificial intelligence* (AI). As various aspects of AI become more enmeshed in our daily lives, integrating human factors becomes increasingly important. The question is not whether AI will affect our work but how it will be used, and how the principles of fairness and transparency, privacy and security, and responsibility will be addressed. This chapter differs from most others in this book because it highlights the ergonomic challenges that come with emerging technologies. Therefore, it does not present solutions but rather discusses concerns; how people might cope with changes in technology and organization, and suitable ways to accommodate individuals under these new conditions. This chapter attempts to preview and plan for ergonomic progress in a changing world.

Since ergonomists are not able to predict the future, we should instead focus on what is unchanging: the need for human factors to be part of the design. After all, the human is the critical component in any complex, interactive system. AI can excel at automating routine knowledge work and generating insights from existing data sets. However, it cannot imagine that for which it has no data, it does not know what it has not been taught, and it lacks common sense.

DOI: 10.1201/9781032674575-29

New technologies, new challenges

This chapter introduces the basic concepts of ergonomic applications in AI, both generally and through specific examples. One example is the progression of technology in motor vehicles, evolving from driver assistance to the fully autonomous automobile (AA) and beyond. Such systems have the potential to increase transit efficiency and safety but also raise issues regarding cost, fairness, privacy, autonomy, and control. Other examples include the design of user interfaces, such as augmented reality (AR) and digital assistants. All of these raise numerous human factors issues: psychological, sociological, physiological, and anthromechanical. Ergonomists working on projects within this developing technology must be aware of the multi-faceted context.

23.1 Basics of artificial intelligence

What is AI

What we call artificial intelligence (AI) is not well defined: in its broadest sense, it refers to the intelligence of machines (specifically, computer systems). AI is an umbrella term that encompasses the programs or systems allowing machines to perform tasks that typically require human intelligence, such as understanding language or solving complex problems. The term was first used at a 1956 conference at Dartmouth College in New Hampshire, where researchers from universities and industry convened "to proceed on the basis of the conjecture that every aspect of learning or any other feature of intelligence can, in principle, be so precisely described that a machine can be made to simulate it."* Philosophers and writers have long imagined intelligent non-humans: biochemist and science fiction writer Isaac Asimov (1920–1992) famously proposed his "Laws of Robotics," which served as the mathematical foundation for a robot's positronic brain. He later added the zeroth law to supersede the first, second, and third laws, considering humanity rather than a specific individual.

- *Zeroth Law*: A robot may not injure humanity or, through inaction, allow humanity to come to harm.
- *First Law*: A robot may not injure a human being or, through inaction, allow a human being to come to harm.
- *Second Law*: A robot must obey the orders given it by human beings except where such orders would conflict with the First Law.
- *Third Law*: A robot must protect its own existence as long as such protection does not conflict with the First or Second Law.

These are fictional laws applied to robots in a fictional universe, but they parallel how we design AI today as well as some of the challenges we face—more on this below.

AI effect AI can be thought of as a system's ability to respond appropriately in an uncertain environment in support of larger goals. It can also be considered an extension of *autonomy*, which refers to a system's ability to choose among courses of action to achieve a goal. Autonomy itself can be considered an extension of *automation*, which is a system that replaces human work in specific tasks without direct human control.* As we have progressed from automation to autonomy to AI, our expectations have continued to increase. This phenomenon is known as the *AI effect*: as soon as we create a machine or computer that accomplishes a new task previously deemed impossible, we will now tend to downplay the achievement of the AI by arguing that it is not "real" intelligence.

Full or narrow AI Since AI is evolving rapidly, the terminology associated with it will evolve as well. As of this writing (in 2024), the general term AI can be divided into categories of full (or strong) AI and narrow (or weak) AI. Full AI, also known as artificial general intelligence (AGI), is an AI that can understand, learn, and perform any intellectual task a human can do, akin to humanoid robots in Isaac Asimov's science fiction or what we might consider "real" intelligence. AGI is not (yet) a reality. Narrow AI is an AI that can perform a specific task, often outperforming humans in that task, such as a computer programmed to play chess, recognize abnormal heart rhythms, or drive a vehicle through complex traffic patterns.

Machine learning Machine learning (ML) is a type of AI where the system learns from examples using specific ML algorithms. The system improves its performance ("learns") over time by analyzing training data. This technique assumes there are sufficient and relevant data from which to train the model and that the data are free from biases and errors. ML contrasts with conventional programming, in which the system is designed to follow explicit rules; thus, conventional programming requires that the rules are known. Since ML can identify trends and correlations in data, it has applications in data classification and labeling (such as sorting wanted from unwanted email), regression analyses to predict quantities (such as forecasting demand for a new product), and document generation (such as, in 2024, OpenAI's ChatGPT, Microsoft's Copilot, and Google's Gemini).

A subset of ML is deep learning (DL), in which artificial neural networks (ANNs) use unsupervised learning to generate accurate outputs from raw, unstructured data. Deep neural networks have multiple (thousands of) layers of interconnected nodes, attempting to mimic the interconnected neurons of the human brain. The non-linear nature of DL can make it challenging to understand or explain the model's output; see below for more discussion on explainability and trust.

Natural language processing

Natural language processing (NLP) is a type of AI focused on understanding, interpreting, and generating human language. With advances in ML, NLP has become prominent, exemplified by voice-activated digital assistants such as Apple's Siri and Google-Translate. NLP can address questions that seem easy for humans but are actually complex for machines. For instance, answering the question "Will I need a jacket today?" requires the AI to know your current time and location, where you are likely to go during the time period you define as "today," the relevant local weather forecast, the chance and type of precipitation, and even an assessment of how chilly or how significant that risk of snow or rain is meaningful to you.

23.2 Addressing challenges of AI

AI is a powerful tool that, like all tools, requires proper design and use. Physicist Stephen Hawking (1942–2018) stated, "Success in creating AI could be the biggest event in the history of our civilization. But it could also be the last—unless we learn how to avoid the risks." These risks include issues of fairness, transparency, privacy, security, and responsibility.

Fairness

The risks for AI related to *fairness* involve the data sources on which the AI is trained. If the training data do not fully and robustly represent the intended use and user or if the training algorithms are biased, then inherent biases and gaps may be unknowingly propagated by the AI itself. Therefore, humans must monitor and evaluate the input data and the output results to ensure fairness, not only during the AI's initial development but also over time to understand and adjust for drift and changes in the data set.

Transparency

The risks for AI related to *transparency* include the need for human users to understand the data set and algorithms that train the AI. Only then can the results produced by the AI be explained. AI is not a "black box" that merely generates answers; instead, the human users must grasp the underlying assumptions of the decisions made. Transparency is how we build trust in the outputs of the AI, as discussed below.

Privacy

The risks for AI related to *privacy* include the need for careful assessment of privacy and data protection regulations and laws. The training data should be free from copyright or licensing concerns, and should be derived from ethically and legally obtained sources. Data containing personal information (such as an individual's driving records or medical history) must be collected, used, and stored with appropriate consent and safeguards, in compliance with applicable laws and regulations.

Security

The risks for AI related to *security* include how the AI and its training data and algorithms are created, stored, and accessed. Cybersecurity is crucial for any computer-based system, including AI. The AI, along with its data and programming, must be securely stored with appropriately controlled access, in compliance with applicable laws and regulations.

Responsibility

The risks for AI related to *responsibility* include how the AI is used to generate results. It is humans who are responsible for understanding the data set used to train the AI, interpreting the output, and making and implementing (or overriding, if necessary) decisions. We cannot blame the AI for negative outcomes. We, as human users, are both responsible for the work that goes into AI and accountable for the results produced*.

Building trust

23.2.1 Trusting AI As AI becomes a larger part of our lives, we have come to trust it in smaller matters that we can easily observe (e.g., weather forecasting or book recommendations) and also in larger matters that most of us do not directly control (e.g., global economic policy). If we believe that the AI being used has been designed to address the risks outlined above, particularly fairness and transparency, we are more likely to trust its outputs. Trust is reinforced when the AI reliably demonstrates valid outputs over time.

Explaining outputs

Generative AI can produce results that seem to be convincing but are not actually based in reality. When the AI is used for entertainment, these "hallucinations" can be amusing. However, when the AI is used for decisions with consequences, these hallucinations can be dangerous. We must have confidence that AI outputs are accurate and based on fair and ethical data.

Crucially, since the human user is responsible for how the AI outcomes are used, the user must be able to explain how those outcomes were generated. Explaining the output means that the user understands and can justify the assumptions and decision-making processes used by the AI. This further emphasizes the importance of fact-checking and confirming that the data sources and training algorithms are fair and appropriate.

Trusting too much?

There is a limit to how much we should trust AI. If we, the human users, rely too heavily on it, we may reduce our awareness and alertness to the point of lacking sufficient understanding of the current situation. If the AI recommends an inappropriate action or fails to make a recommendation, the user must have enough situational awareness and information to take control. The user must know when to reject the AI's output and must be able to act appropriately and in time to prevent accidents or injuries. For example, if an industrial materials handling system misidentifies a heavy, dangerous load as being light and easy to move, the human operator must intervene in time to prevent an accident. If an autonomous flight system misidentifies the angle of inclination of the airplane, the human pilot must intervene in time to prevent a stall or even a crash.

23.3 Implementing autonomy and AI in automobiles

One example of how autonomy and AI are being implemented is on our roads. Automotive technology has advanced from assisting drivers to taking over for them, with the possibility of fully autonomous vehicles on automated roadways ahead. Future roads could link vehicles and instruments to provide hands-off and feet-off navigation, where road infrastructure and traffic control systems, instead of the driver, guide the vehicle automatically. This could virtually connect many vehicles to accelerate or brake simultaneously and to follow each other at small but safe distances, akin to a train of cars. Such a system has the potential to increase transit efficiency and safety, but it also raises human factors implications.

Early automobiles

In 1769, Nicolas-Joseph Cugnot built the first steam-powered automobile. In 1807, François Isaac de Rivaz designed the first car propelled by an internal combustion engine running on hydrogen. In 1835, Sibrandus Stratingh and Christopher Becker built an all-electric model car powered by batteries. In 1886, Karl Benz designed a reliable gasoline-powered automobile and produced several copies of his "Patent Motorwagen." Motorized wagons soon replaced animal-pulled carriages as automobiles became affordable for many; mass production of Henry Ford's Model T began around 1908. Today, it is estimated that there are roughly 1.4 billion cars on the world's roads, one-third of which are in Asia alone.

Traffic speed

In 1901, Connecticut may have been the first US state to regulate motor vehicle speed, limiting it to 19 km/h (12 mph) in cities and 24 km/h (15 mph) in rural areas. This was, in fact, approximately the maximum speed of horseless carriages then, powered by steam, gas, gasoline, or electric engines. Such vehicles were still rare, so there was little need for new roads, and air pollution was not yet a concern. Furthermore, the likelihood of personal injury in a crash was relatively low because the kinetic energy of a moving body depends on the square of its velocity: at 24 km/h, the kinetic energy is about 1/18 of that at 100 km/h.

Cost in human life

Today's heavier and faster road traffic is reflected in grim statistics regarding the carnage inflicted on people (both inside and outside vehicles) by current driving conditions. It is estimated that each year, 1.35 million people are killed or disabled in traffic accidents, half of whom are cyclists, motorcyclists, or pedestrians*. Low- and middle-income countries are disproportionately affected, having only 60% of the world's vehicles but 90% of road traffic deaths. Definitions of human error* differ, but a significant majority of crashes appear to result from drivers' actions.

Other costs

Further reasons for changing the way we use automobiles include reducing the time wasted on excessively long travel (often prolonged by clogged traffic conditions) and decreasing combustion byproducts that harm the environment.

Car technology

As cars and other motor vehicles have become faster, they have also become safer due to technological improvements. Vehicles now generally come equipped with three-point seat belts (since the 1960s), airbags (since the 1980s), and (in the 2000s) electronic stability and traction control (including computer-controlled and anti-lock braking), back-up cameras, and blind spot warnings. Driver-assistance features are widely used in newer cars, such as collision warnings, automatic emergency braking, road lane departure warnings, and adaptive cruise control to maintain safe distances between vehicles.

AAs are considered the next frontier: in 2014, Google prototyped a driverless car with no steering wheel. In 2020, Waymo (formerly known as the Google self-driving car project) first offered driverless taxis in the US; the following year, DeepRoute.ai did so in China. Tragically, in 2018, Elaine Herzberg became the first recorded pedestrian fatality after being hit by a self-driving test car when the vehicle (in autonomous driving mode) did not recognize the pedestrian, and the human back-up driver did not intervene in time to avoid the collision.

Paradox of automation

This sad incident of a self-driving car killing a pedestrian highlights that automation can be wonderful when it works, however its failures can lead to unintended and tragic outcomes. The *paradox of automation* is this: the more efficient the automated system, the more crucial the human contribution of its operators. As automation (or AI or a digital worker) becomes more efficient, human involvement becomes less necessary, but the remaining human involvement that is required becomes even more critical. In cases of AI failure, there must be sufficient time for control to transition to a capable, engaged, trained, situationally aware, and vigilant human operator, as well as sufficient time for this operator to take effective action. Ergonomic principles are essential to ensure that factors related to real-world humans are considered.

Outdated vehicle design

23.3.1 Addressing automobile ergonomics Over the past few decades, little has changed regarding the driver's manual tasks (except that the clutch pedal has mostly been obsoleted by automatic transmission). In conventional road vehicles, the driver is still constrained in the seat by the resolute requirements to (1) keep a foot on a pedal, (2) keep hands on the steering wheel, and (3) keep eyes on the road. The driver's body position cannot change significantly, and any sudden movement (such as sneezing or reaching for a distant control) can dramatically affect navigation in terms of direction and velocity.

Foot pedals

The traditional pedal arrangement* is an astonishing example of poor ergonomic design: the driver's foot must move quickly and accurately between pedals that are largely invisible. Pushing one pedal accelerates the vehicle, while a similar push on the other pedal brakes it. These actions require delicate control and are often critical to avoid damage and injury, even death. Yet, the same foot action on adjacent pedals yields opposite results.

No foot pedals

No significant advances in technology are needed to free up the driver's feet by eliminating all traditional pedals. Riders of motorcycles and all-terrain vehicles use handlebars to control both direction and speed. In airplanes, pilots commonly manipulate attitude and speed using either a control stick (joystick) or a control yoke (a partial wheel). Automobile design can draw from these experiences and simplify vehicle control by integrating the manipulation of both direction and speed. For example, a center column (such as a stick or yoke) or controls located at the driver's side(s) could replace the steering wheel: such actuators could control left/right turns as usual and also regulate acceleration/braking through push/pull or tilt actions. Technology to unfreeze the driver's position and make driving much easier is available and proven, but not widely used.

These basic issues of driver position and foot pedals, along with various other technical and operational flaws, can be explained (but not justified) as the result of many technological steps* that followed each other in a rather haphazard manner. Over time, newer vehicle designs continued to incorporate basic traditional elements (such as steering wheels and foot controls) so that drivers could transfer their acquired skills to operate new vehicle models with minimal additional training. Significant changes in how drivers control vehicles would entail a learning curve and training for the operator but could be combined with other advances in AA.

Driving challenges

Vehicle drivers, a highly diverse population, may be the largest group of individuals ever tasked with performing an exceedingly difficult job. Our current automotive technology and traffic conditions require drivers to navigate hazardous vehicles under often dangerous circumstances. Drivers must make swift decisions based on uncertain information: will the other driver change lanes, and is the road surface slippery? Drivers must perform difficult steering, braking, and acceleration quickly and accurately, often with limited vision to the rear and sides, at night, and in fog. It takes many hours of driver education to learn the basics of these actions, and it takes years of (often hair-raising) experiences to improve them.

Human factors challenges

Driving is a complex task, usually divided into hierarchical levels of control: operational control (such as following another car) is the simplest task, while tactical control maneuvers (such as passing a car) and strategic control (such as trip planning) require higher levels of mastery. Distractions* from the driving task can be visual, manual, or cognitive: visual distractions cause drivers to look away; manual distractions force drivers to take their hands off the steering wheel; and cognitive distractions divert drivers' mental attention from driving. Driving requires that vigilant alertness and high-level control skills be maintained, often for hours and even when fatigued*, inexperienced, or emotional. The driver must focus nearly uninterrupted attention to traffic, road, and vehicle, even though relevant information can be ambiguous and difficult to interpret. The driver's performance must be nearly flawless, as even seemingly slight mistakes can have serious, even fatal, consequences. When things do go wrong, is it really always just human error, or are the circumstances humanly unmanageable? If traffic conditions are unwieldy, they can be mitigated; if the vehicle is poorly designed, it can be improved.

The emergency driver

23.3.2 Addressing AI automobile ergonomics If full automation of vehicles can be achieved, the former driver now becomes a passenger who expects to enjoy effortless, convenient, and safe travel. However, this ideal scenario collapses in the event of a failure of the guidance system. The once-passive passenger must instantly become an emergency manager and take control of the vehicle: steering, braking, avoiding obstacles, and predicting outcomes—just as in the current mode of non-automated driving. The task of quickly grasping the vehicle controls and effectively maneuvering is likely to be much more challenging than in pre-automation times due to the unexpected nature of the event, and the surprised driver may have lost essential skills during their time as a passenger*.

Driving an advanced vehicle

It will take some time to reach a level of automated driving where no human intervention is required*. Until then, human vehicle control—including the emergency driving mentioned above—must adhere to the same human factors design requirements, regardless of the automation level:

- Propose or select appropriate routing.
- Provide intuitive, unambiguous controls for vehicle direction and speed.
- Allow comfortable and adjustable driving postures.
- Detect and react to driver impairment (including drowsiness and distraction).
- Offer means to preview road conditions and to improve visibility in rain, fog, and at night.

- Supply assistance to avoid collisions and stay in the driving lane.
- Eliminate blind spots or offer blind spot monitoring and detection.
- Protect humans in case of a crash.

Combine changes The emergence of increasingly sophisticated automated or AI transportation presents a timely psychological opportunity to also introduce innovative vehicle control designs. These new designs could replace foot pedals with alternative controls and implement other ergonomic enhancements. This progress can be paired with other improvements in traffic controls that incorporate automation and AI. While drivers will require some retraining, we should all be open to changes that enhance the safety and efficiency of driving.

23.4 Implementing autonomy and AI in other applications

AI everywhere Autonomy and AI are being implemented in various applications, not just in cars and other road vehicles. Airplanes have long used autopilot systems of ever-increasing sophistication. Healthcare providers use AI to assist with diagnoses, treatment plans, and patient monitoring. Power systems utilize AI to help manage the long-term stability of the electrical supply grid. Military operations incorporate AI in applications including simulations, monitoring, and cybersecurity. Businesses leverage AI to understand and predict commercial trends and supply chain logistics. Even materials handling and workplace safety assessments use AI to monitor and evaluate the physical ergonomics of individuals who lift loads and perform other physical work: wearable physical sensors and visual systems can measure positions and motions, while biophysical sensors can assess brain activity (electroencephalogram), muscular responses (electromyogram), and cardiac function (electrocardiogram) to help prevent and analyze risks of musculoskeletal and cumulative trauma disorders.

AR and virtual reality We use AI in augmented reality (AR) systems to overlay digital content in real-life environments and in virtual reality (VR) systems for full immersion in a simulated environment: for education and training, simulations, medical procedures, work, design, and entertainment. Digital assistants powered by automation and AI answer our questions, manage our appointment calendars, guide us in filling out forms, play our favorite songs, and remind us to order needed supplies. These systems can be tailored to ergonomically accommodate varying levels of physical capabilities and special needs (including those of the elderly) and to allow user-specific interfaces with computers and other devices.

How much AI

All of the many uses of automation and AI raise a range of human factors issues: psychological, sociological, physiological, and anthromechanical. One way to determine what kind and how much autonomy and AI to incorporate is based on a four-stage model of human task performance: we (and/or the AI) acquire information, analyze it to extract meaning, make decisions, and then take action*. How to distribute the stages of work between AI and human operators should consider human performance criteria, including mental workload, situation awareness, complacency and vigilance, and skills and capabilities. The human operator should also understand how and how much the AI is performing at each of the four stages to ensure appropriate human interaction and oversight.

Another approach to the design of human–AI interactions is based on the skill-, rule-, and knowledge-based (SRK) classification model of human error. Per the SRK framework, skill-based tasks, such as flying an airplane at cruising altitude, are well-defined and thus often suitable for automation. Rule-based tasks, such as deciding when to abort the airplane landing sequence, can likely be automated if the rules are clearly defined. Knowledge-based tasks arise from unfamiliar or unanticipated situations and require deeper experience and insight; thus, they necessitate closer cooperation between humans and AI assistance.

Responsibility

Regardless of the AI's sophistication, human operators will still need to monitor and supervise the system's performance. Those who develop and operate the system bear responsibility for the outcomes.

Summary

Below are some human factors recommendations and considerations* for designing automation and AI systems to assist human work. Taking a human-centered design approach (more on this is in Chapter 24) and understanding the needs, behaviors, and capabilities of humans within the overall system is important. These considerations support effective human–AI teams, ensuring that AI applications work in concert with humans to promote effective and safe teamwork. In this way, ergonomic considerations can help ensure that the benefits of automation and AI outweigh the risks.

- *Situation awareness*: Ensure the human–AI system supports, rather than undermines, the human operator's awareness of the current situation.
- *Workload*: Assess how the AI will alter the workload of human operators, relieving some duties while imposing new ones.
- *Automation bias*: Ensure transparency of assumptions to guard against the human operator relying uncritically on the AI.
- *Explanation and trust*: Provide explanations of AI applications to allow users to understand and trust (or override) AI output.
- *Training*: Ensure human operators are trained on what the AI can and cannot do, how to monitor it, and when and how to intervene.
- *Ethical issues*: AI models must be designed, maintained, tested, and operated with consideration of fairness and bias, transparency, privacy, security, and responsibility to provide benefits and minimize harm.

Steps to fit the human

Step 1: Determine what tasks and work systems will be addressed with automation and AI. Understand the context of the system and how AI will change, but not eliminate, the role of the human operator.

Step 2: Design the AI with a clear understanding of which of the four core functions (acquiring information, analyzing information, making decisions, taking action) it will undertake and how much autonomy is appropriate. Be transparent about assumptions and decision-making. Explicitly define the balance between authority, responsibility, and control for the human and the AI as a system.

Step 3: Provide the human operator with training, tools, and insights to effectively monitor and supervise the AI, along with ample time to assess, respond, and override when necessary. Ensure that there is sufficient oversight and trust in the human within the AI system.

Step 4: Seek to continuously improve the human–AI team. Monitor conditions, and calibrate, maintain, update, train, and test the AI system components and the human–AI team.

Notes

The text contains markers, *, to indicate specific references and comments, which follow.

23.1. Basics of AI

First use of the term AI: https://home.dartmouth.edu/about/artificial-intelligence-ai-coined-dartmouth, retrieved July 2024.

Automation, autonomy, and artificial intelligence: For an extended discussion, see Sawyer, Miller, Canham et al. (2021).

23.2. Addressing challenges of artificial intelligence

Responsibilities of AI: See society guidelines and standards such as the IEEE Global Initiative on Ethics of Autonomous and Intelligent Systems and Human Factors and Ergonomics Society AI guardrails and statements, ISO/TR 9241

23.3. Implementing autonomy and AI in automobiles

Road traffic deaths: Data retrieved in June 2024 from the website of the World Health Organization.

Definitions of human error: Dekker (2014).

23.3.1. Addressing Automobile Ergonomics

Pedal arrangement: Wu, Yang, and Yoshitake (2014).

Technological steps: Bhise (2011), Nagamachi (2011), Pew and Mayor (2007), Walker, Stanton, and Salmon (2015).

Driver distraction: Regan, Lee, and Victor (2013).

Driver fatigue: Popkin (2015).

23.3.2 Addressing AI automobile ergonomics

The surprised emergency driver: Skottke, Debus, Wang et al. (2014).

Automation levels: see SAE J3016_202104 for levels 0 to 5 of automotive automation, where level 0 offers no driving automation, level 1 has driver assistance, and levels 2, 3, 4, and 5 have partial, conditional, high, and full driving automation, respectively.

23.4. Implementing autonomy and AI in other applications

Model for human processing and AI: See Parasuraman, Sheridan, and Wickens (2000) and the white papers by CIEHF (2021, 2022).

Human factors recommendations and considerations: Adapted and expanded from healthcare AI recommendations in CIEHF (2021).

References and further reading

Bhise VD (2011). *Ergonomics in the Automotive Design Process.* Boca Raton, FL, CRC Press. https://doi.org/10.1201/b11237

CIEHF (2021). *Human Factors and Ergonomics in Healthcare AI*, white paper from Chartered Institute of Ergonomics & Human Factors. https://ergonomics.org.uk/resource/human-factors-in-healthcare-ai.html

CIEHF (2022). *Human Factors in Highly Automated Systems*, white paper from Chartered Institute of Ergonomics & Human Factors. https://ergonomics.org.uk/resource/human-factors-in-highly-automated-systems-white-paper.html

Dekker S (2014). *The Field Guide to Understanding "Human Error"*. Williston, VT, Ashgate.

ISO/TR 9241-810 (2020). *Ergonomics of Human-System Interaction, Part 810: Robotic, Intelligent and Autonomous Systems.* Geneva, CH, International Organization for Standardization.

Nagamachi M (2011). *Kansei Engineering*, 2 volumes. Boca Raton, FL, CRC Press. https://doi.org/10.1201/b16799

Parasuraman R, Sheridan TB, Wickens CD (2000). A model for types and levels of human interaction with automation. *IEEE Transactions on Systems, Man, and Cybernetics - Part A: Systems and Humans*, 30(3):286–297. https://doi.org/10.1109/3468.844354

Pew PW, Mavor AS (2007). *Human-System Integration in the System Development Process*. Washington, DC, The National Academies.

Popkin SM (2015). *Worker Fatigue and Transportation Safety. Volume 10 of Reviews of Human Factors and Ergonomics,* Santa Monica, CA, Human Factors and Ergonomics Society.

Regan MA, Lee J, Victor TW (2013). *Driver Distraction and Inattention*. Abingdon, UK, Ashgate. https://doi.org/10.1201/9781315578156.

SAE J3016_202104 (2021). Taxonomy and Definitions for Terms Related to Driving Automation Systems for On-Road Motor Vehicles.

Sawyer BD, Miller DB, Canham M, Karwowski W (2021). Human factors and ergonomics in design of A^3: automation, autonomy, and artificial intelligence. Chapter 52 in Salvendy G, Karwowski W (eds), *Handbook of Human Factors and Ergonomics*, 5th ed. Hoboken, NJ, John Wiley & Sons, pp. 1385–1416. https://doi.org/10.1002/9781119636113.ch52

Skottke EM, Debus G, Wang L, Huestegge L (2014). Carryover effects of highly automated convoy driving on subsequent manual driving performance. *Human Factors*, 56(7):1272–1283. https://doi.org/10.1177/0018720814524594

Walker GH, Stanton NA, Salmon P (2015). *Human Factors in Automotive Engineering and Technology, Human Factors in Transport Series*. Abingdon, UK, Ashgate Publishing Ltd. https://www.ashgate.com/isbn/9781409447573

Wu J, Yang J, Yoshitake M (2014). Pedal errors among younger and older individuals during different pedal operating conditions. *Human Factors*, 56(4):621–630. https://doi.org/10.1177/0018720813487200

CHAPTER TWENTY-FOUR

Designing work to be efficient and pleasant

Fitting

So far, the book primarily addressed the physical aspects of *fitting the task to the human* through careful human factors engineering, although occasional references were made to *fitting the human to the task*, such as through personnel selection and training. This last chapter of the book consolidates ergonomic information to demonstrate how work can be both pleasant (often a primary concern for the individual) and efficient (often a primary concern for the organization).

As a science, ergonomics is the application of scientific principles, data, and methods to develop engineered systems where people interact, optimizing human well-being and system performance. Physical ergonomics focuses on the body (as discussed in Chapters 1–12, 15–22), cognitive ergonomics focuses on the mind (as discussed in Chapters 7–16, 19–24), and macroergonomics focuses on the system and culture (as discussed in Chapters 13–24). There is tremendous overlap in how these approaches interact, and we need to bring this integrate them to make work efficient and pleasant.

24.1 Addressing how to design

Throughout this book, we discuss the importance of designing tasks, tools, equipment, workplaces, organizations, work shifts, and facilities. We have not yet discussed how to conduct this design as we deal with real people facing real ergonomic problems. And so, when tackling any design task, we start by

DOI: 10.1201/9781032674575-30

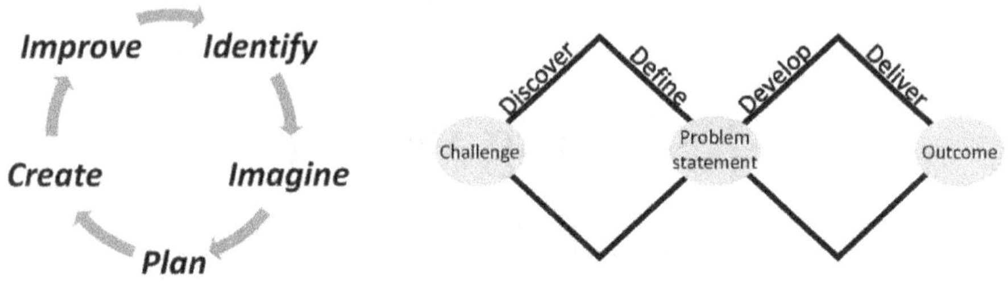

Basic engineering design cycle   Double diamond design process

FIGURE 24.1 Schematic representations of the design process: basic design cycle, double-diamond process.

determining the problem we are trying to solve. Only then can we attempt to devise a design solution*.

Human-centered design

Human-centered design (HCD) first considers the needs, capabilities, and behaviors of humans*. The stakeholders (users and all others affected by the design) are identified and involved in the design of the system to ensure that the appropriate inputs and outputs, as well as constraints, are considered.

Design cycle

The design process is generally understood to be a *design cycle*: the steps occur (mostly) in sequence and iteratively. Steps can be repeated to incorporate what has been learned along the way. By repeatedly testing and improving designs, we view failures (i.e., designs that did not work as intended or expected) as learning opportunities instead. The design cycle can be depicted as a circle (as in Figure 24.1), as a spiral (to indicate that progress is made with each design cycle), or as a series of small design "sprints" that focus on increments of the design as part of the so-called *agile* methodology.

The steps of the design cycle are described in various ways, but generally include the below:

- *Identify*: ask questions to define the problem, criteria, and constraints.
- *Imagine*: explore and generate ideas.
- *Plan*: gather materials and map out the work.
- *Create*: build prototypes and test.
- *Improve*: analyze test results, modify, and repeat.

Double-diamond model

One structured approach to design is the *double-diamond* design process, popularized by the British Design Council. This approach separates the discovery of the problem from the discovery of the solution. Thus, there are two design cycles: first to design the right thing (i.e., to define the problem to be addressed), and then to design the thing right. Since both cycles initially involve expanding activities before refining the concept back down, each cycle can be visualized as a diamond (a square balanced on one corner). This process generally follows four phases, as depicted in Figure 24.1:

- *Discover*: work to deeply understand (discover) the challenge or problem to be solved; spend time with and speak to the people who are affected—this initiates the first diamond where we diverge to expand the problem.
- *Define*: take the insights generated in the discovery phase to distill (define) the problem statement; this will likely not be the same as initially proposed—this completes the first diamond where we converge on a definition of the problem into a problem statement.
- *Develop*: with the now clearly defined problem statement, work to uncover (develop) new answers; spend time and collaborate with a broader range of people—this initiates the second diamond, where we diverge to expand the proposed solutions.
- *Deliver*: take the insights generated in the develop phase to test and refine different proposed solutions, ultimately settling on the final solution—this completes the second diamond, where we converge on a solution (deliver) to the problem as the outcome.

Design thinking

Another approach is the *design thinking* design process, popularized by IDEO and the Institute of Design at Stanford University (also known as d.school). The basic concept is to take a human-centered approach to design, integrating what is *desirable* to humans with what is *feasible* (technologically) and *viable* (economically or practically). This involves a process of empathizing (to understand the user's needs, challenges, and desires), defining (creating a clear problem statement), ideating (generating creative ideas), prototyping (building simple versions of the ideas to refine them), testing (trying out the prototypes and gathering feedback), and refining (iterating to improve the proposed solutions) to arrive at a feasible and viable solution. This approach may follow three phases:

- *Inspiration*: work to better understand the problem and the people involved through observation, interviews, and other research.

- *Ideation*: collect what has been learned about the problem, generate ideas, develop and test prototypes, and refine proposed solutions.
- *Implementation*: put the proposed solution to use, iterating to incorporate feedback along the way.

Assessing designs Once the problem statement is clear, designs can be selected by ranking them based on effort and impact. Designs that require low effort but have low impact should be considered; designs that require low effort but have high impact should be pursued. Designs that require high effort but only have low impact should be set aside; designs that require high effort but have high impact should be investigated.

If we have candidate designs, we can also use a forced-ranking solution, such as the Pugh decision matrix, already discussed in Chapter 18. In this case, the proposed designs are assessed against specific criteria, which are then weighted according to their relative importance. Having an objective way to make decisions compels us to discuss and agree on the criteria and their relative importance, and to follow the same rules. By using a transparent and logical system to judge the candidate designs, we eliminate irrational and highly subjective (and personal) decision-making that otherwise comes down to "I like this better."

Accept humans Throughout any design process, we should remember the words of Don Norman in 2013: "The problem with the designs of most engineers [*or ergonomists or designers*] is that they are too logical. We have to accept human behavior the way it is, not the way we would wish it to be."

24.2 Using our skills and interests, getting along with others at work

Every one of us is a unique individual, different from everybody else due to upbringing, environment, experiences, and personality, all of which make us special. Yet, surveys consistently show—in North America, Europe, Australia/New Zealand, and probably everywhere else—that, in general, better job skills and education lead to higher job satisfaction and income. More importantly, features intrinsic to our tasks provide us with basic satisfaction. Doing a job well is often a reward in and of itself.

Engaging in what we like to do, exercising our talents, learning new skills and techniques, and applying them—along with accomplishing tasks—make us content, even happy.

Work worth doing

Individuals who have a calling* for what they are doing generally feel deeply fulfilled even if their pay is relatively low and their workload rather high. Among the most satisfying jobs are those that help other people get better and feel better; these jobs often involve caring for others, helping and protecting others, and teaching others. If our work leads to improvements in procedures, services, and products, that outcome renders us proud and serves as recompense in itself. Ideally, we will be able to balance our interests and talents with what our world needs for us to find our purpose, as depicted in Figure 24.2.

Satisfied employees

From the organization's perspective, a company's market value is determined by its hard assets, such as property and equipment. However, a large part of its value consists of soft attributes: patents and intellectual property, customer base, reputation, and especially the organization's human resources. Satisfied and engaged employees are among a company's most valuable assets because they do the work and run the business.

Engaged at work

To feel engaged at work requires a good match between a person's interests and aspirations, and the job. The job is defined by:

- Its tasks, which include the equipment and procedures used,

FIGURE 24.2 Venn diagram of overlapping passion, mission, vocation, and profession: ideally, we find ourselves at the center with our purpose.

- Its rewards for good performance, such as recognition, potential for advancement, and pay,
- Its many-faceted working conditions, which include the social relations among people at all levels, often loosely labeled as *organizational climate*.

Job performance is the product of motivation and ability. As discussed in several chapters of this book (in particular, Chapters 13 and 14), situational factors at work can enhance or hinder performance.

Motivation and job performance

Motivation incites, directs, and maintains our behavior towards goals. Motivation and job performance* are intimately related; a person motivated to do well is willing to expend effort to succeed. Understanding what motivates people can help us comprehend why they behave in certain ways on the job and can guide management in helping employees be effective and satisfied at work. Several theories attempt to explain motivation and needs (as outlined in Chapter 13). They essentially fall into two categories: one focuses on the individual and personal inherent traits, while the other places the work environment at the forefront.

Needs-based motivation

Motivation theories that focus on the individual describe what we need and want. Individuals strive to satisfy their basic needs first and then work their way up the hierarchy of needs, which consists of several levels: at its base are physiological requirements for food, water, and shelter, followed by needs for safety and economic and physical security. Higher up are social desires for belonging and esteem, including self-confidence, recognition, appreciation, and respect. The highest-order ambitions concern the fulfillment of one's talents and potentials; at this level, an individual can achieve full personal success and satisfaction.

Performance and expectance

Expectancy theories assume that an individual's motivation (and resulting satisfaction) depends on the difference between what the person's work environment offers and what he or she expects. This can be expressed, in its simplest form, as a model that explains motivation as a function of two coupled expectations:

1. work will lead to performance, *then* (2) performance will lead to reward

Accordingly, the organization should provide rewards for its employees: pay, promotion, and formal recognition. (There can also be a negative outcome, such as getting fired.) The value of the reward differs depending on how the individual perceives the anticipated reward.

Expected outcomes

The expectancy component is crucial. There must first be a connection between how hard one tries and how well one performs. To illustrate, a factory worker on an assembly line may have little incentive to increase her rate of production if the overall speed of the assembly line—and hence her performance—remains unchanged regardless of her efforts. In contrast, a truck driver can deliver more goods the faster she drives, so she may be motivated to take quicker highways and less congested streets. The other expected connection is between performance and reward. If the organization does not recognize an employee's effort to enhance her performance, then there is little incentive for her to work harder; however, if she receives reward, this will most likely motivate her to strive even more.

Proud to work

Surveys of the top global companies provide a summary of the various factors underlying individual job satisfaction and performance. One US-based company that routinely makes the list of the 100 Best Companies to Work For has as its company motto: "Treat people the way they want to be treated. Strive for mutual respect and foster an atmosphere that makes people proud to work here. Provide good tools and equipment. Offer career opportunities. Say thank you for a job well done."

Life outside work

Individually, we need to balance our work and personal lives. Only a few people live for their work; most want to have a life outside of it. The times of early industrialization are gone when employees labored for 10 or more hours a day, 6 or 7 days a week. Nowadays, the 5-day workweek with about 8 hours of work a day (less in some industrialized countries) is the norm. Even companies notorious for extraordinarily demanding work schedules that often overtax their employees are looking for ways to reduce workload and stress so that their employees can enjoy a healthier life. Some companies have eliminated weekend work (including business travel); they encourage or enforce leaving mobile phones and work computers turned off outside business hours; they emphasize taking personal days and uninterrupted vacations without being on-call.

Bad bosses

Insufficient job performance is often a reason why workers lose their jobs, but bad bosses are not infrequent either. A 2007 survey in the United States* showed that two out of five supervisors did not keep their word or failed to give credit when it was due; one in four bad-mouthed employees or blamed them for mistakes made by the boss. Such managers also produce problems for their companies because they generate poor morale, reduce productivity, and increase employee turnover. Bad managers create severe dissatisfaction among employees: many who look for another job do not want to leave their company but rather just want to get away from their boss.

Taking charge

Good working conditions, including pleasant personal relationships at all levels, are important, often more so than small differences in wages. Abusive relationships between employers and employees were not rare in the past, when the "little people" were dependent on the "big bosses." We hope that these medieval conditions have been replaced by the understanding that employees at all levels are responsible for making a business prosper or fail, and that they should all pull in the same direction*. Yet, if for whatever reason a person feels dissatisfied with the conditions and treatment at work, there are constructive ways, such as those listed in Table 24.1, to determine the causes of discontent so that the conditions can be improved. Both managers and employees have similar factors to consider, though viewed from different perspectives, that make their work lives more enjoyable and efficient.

Table 24.1 Taking charge of personal relations at work

These considerations will contribute to making you feel good about your new employment:

- When you consider taking up new employment, strive to understand the new organization itself and the "environment" it offers.
- Examine the new company's structure, strategies, policies, conventions, and culture to make sure they agree with your own personality, personal style, goals, and beliefs.
- Spend some time with your prospective colleagues and supervisors and determine how you feel about them since you will be with them many hours at work.
- Does the new job meet your physical needs, such as sufficient pay and other benefits? What about your emotional needs, like recognition and a sense of achievement? How are the work conditions, the management set-up, the dress code, and so forth—are they appropriate for and pleasing to you?

(Continued)

Table 24.1 (Continued) Taking charge of personal relations at work

These considerations will contribute to making you feel good about your work:

- Talk with your supervisor/colleagues/subordinates about what should be done and about how to do it. Propose solutions and listen to proposals.
- Do as well as you can do. Take advantage of any training and development programs available to you, especially if offered by your company, to increase your skills.
- Technology can make your life easier, but make sure it does not keep you chained to work day and night.

What to do when you are miserable at your job:

- Talk with your supervisor/colleagues/support staff in a non-confrontational tone about existing problems. It is important that you propose solutions and listen for insights.
- Check your employment contract and the company policies and procedures manual to determine your duties and rights at work.
- Take advantage of any stress reduction and employee assistance programs your company offers.
- If there is no solution to a problem that bothers you so much that you do not want to work with the organization any longer, then look for another position.

Source: Adapted from Kroemer and Kroemer (2017).

24.3 Setting up our own work, workplace, and work environment

Larger, richer jobs

In the early 1900s, Tayloristic principles favored breaking jobs into small task elements that were then standardized. This often meant that a worker performed the same task repeatedly, resulting in extreme specialization and, unfortunately, often acute tedium despite high skill. Since the mid-1900s, *job enlargement* (increasing tasks and task variety) and *job enrichment* (increasing workers' participation and control) have grown popular. Jobs can be designed to positively influence behavior by considering skill variety, task details and significance, workplace autonomy, and feedback. This helps the worker understand how their work is meaningful, with responsibility for work outcomes and knowledge of results. These experiences generate high motivation and satisfaction for the employee and, of particular importance to management, lead to better-quality work, lower absenteeism, and less turnover.

Suggestions for improvements

When management recognizes and rewards expressions of personal interest in improving work, follows up on suggestions, and incorporates proposals for novel work and workplace layouts, a spirit of teamwork and cooperation develops that provides motivation and drive to achieve. The design of most workplaces for manual work in production and repair follows well-established traditions resulting from the work procedures employed in the past, the objects to be worked on, the machinery used, and the hand tools and equipment needed. In some cases, there is little leeway for the individual worker to develop new designs for products, workplaces, or work practices. Yet numerous reports and anecdotes show that major enhancements in productivity, safety, and appeal can result from suggestions for improvements made by involved personnel. Thus, smart management encourages workers' suggestions; collecting and using these suggestions early in the planning and concept stages of products and processes is more effective than during production. One approach to continuous improvements in the workplace is *Kaizen* (see Chapter 20), which encourages small but ongoing changes to make broader positive improvements. Another approach is *Kansei (affective)* engineering*, which explicitly considers people's preferences and emotional responses by translating the user's feelings and needs into the design of the product, work process, or workplace. In this way, the product or place is intentionally designed to evoke the intended feeling.

Work environment

In this book, Chapters 17 through 22 describe workplaces, tasks, and work procedures that suit the human body and mind. Special aspects of the work environment can strongly influence well-being and performance, prominently concerning seeing, hearing, and physical climate, as discussed in Chapters 5 through 8.

Vision requires good lighting

The characteristics of human vision discussed in Chapter 5 provide detailed information for designing environments that support proper vision. The most important concept is the provision of suitable workplace lighting that adequately illuminates visual targets. What matters most is the luminance of an object—the light reflected or emitted from it—that reaches the eye. The ability to see an object clearly is greatly influenced by a strong luminance contrast between the object and its background, including shadows.

Avoid glare, be careful with colors

Another important point is the avoidance of unwanted or excessive glare. Direct glare strikes the eye straight from a light source, such as the sun or a lamp shining into your eyes. Indirect glare is reflected off a surface into your eyes, such as when the sun or a lamp is mirrored in your computer screen. The use of colors on visual targets, if selected properly, can be helpful; however, color vision requires sufficient light, and luminance contrast is usually more important than color alone. Table 24.2 lists recommendations for workplace lighting.

Table 24.2 Taking charge of the lighting at work

These visions considerations will contribute to making you feel good at work:

- General illumination is best at about 500 lx. If there are many dark (light-absorbing) surfaces in the room, illumination may be as high as 1000 lx. If computer displays are present, it could be between lower (200 and 500 lx) or higher (up to 750 lx) as preferred by the users including you.
- Consider indirect lighting, where all light is reflected (at the ceiling or walls of a room, or within the luminaire) before it reaches the work area. This helps to avoid direct and indirect glare.
- Make sure that high-intensity light sources (including windows) are outside a cone-shaped range of 60° around your line of sight. This avoids direct glare.
- Modify your work area to have dull, matte, non-reflecting surfaces. This avoids indirect glare.
- Shine a task light on your paperwork or another visual target if it seems too dim to you.
- Usually the best place for a visual target at which you look often is
    - Not more than 1 meter away from your eyes, usually at about arm's length,
    - Straight ahead, neither to the left or right but straight ahead,
    - Below the height of your eyes.

If you are having difficulties seeing objects clearly, or you feel "eye fatigue" or "eye strain" or "dry eyes"

- Have a physician (ophthalmologist or optometrist) check your eyes and advise you about measures that improve and protect your seeing.

What to do when you are not comfortable:

- If you feel eye fatigue or eye strain, check and correct, as needed
    - The distance from your eyes to visual targets,
    - The direction of the line of sight to visual targets,
    - The luminance contrast between the details of the visual target and the background (such as lines on a source document, letters on the screen, objects on the work bench),
    - Any direct glare shining into your eyes, or reflected glare from the display or other mirroring surface.
- If a window is in front of you, with the sun or bright daylight shining into your eyes
    - Draw a curtain or lower blinds,
    - Turn your workstation so that the window is to your side.
- If your light-colored clothing is mirrored in your computer screen, making it hard to decipher things on the display, change into darker clothes.
- If you feel fatigue or strain in the neck or the back, check and correct, as needed, the posture of your neck and back, which may be affected by
    - The distance from your eyes to visual targets,
    - The direction of the line of sight to visual targets.

Source: Adapted in part from Kroemer and Kroemer (2017).

Sounds convey information

Another essential environmental aspect is hearing and sounds in the workplace—see Chapter 6. Sounds convey information about the functioning of machinery and work progress; they provide verbal communication from coworkers and, possibly of vital importance, present warning signals about dangers in the work environment. Furthermore, the sound environment may be hazardous to one's hearing if it produces temporary or permanent damage to our ears. Table 24.3 lists recommendations for the acoustic environment at work.

Suitable climate at work

The third important environmental condition at work concerns the physical climate, determined by temperature, air movement, and humidity, as discussed in detail in Chapter 8. When we work outside, we must accept the climate as it is, and the adjustments we can make may be limited to our working habits

Table 24.3 Taking charge of the sound environment at work

These sound considerations will contribute to making you feel good at work:

- In the office, the overall sound level it should be around 50 dB.
- In all workplaces, it should not exceed 75 dB. If the surrounding sound level exceeds about 75 dB or is otherwise disturbing, you should wear hearing protection devices.
- If your work requires intense concentration, the sound level should not change appreciably or dramatically (except for a warning).
- Many people find background music pleasant, but preferences for the kind of music, its intensity, and its duration are highly individual. Therefore, music is best presented by individual speakers, such as headphones or caps (muffs), that do not disseminate sound to the environment.

If you are having ringing in your ears or if your hearing is reduced after exposure to loud sounds

- Have a physician (otolaryngologist or otologist) check your hearing and advise you about measures that improve and protect your hearing.

What to do when it is too loud for you:

- Eliminate the noise at its source by
  - Replacing noisy equipment or machines or tasks with quieter ones,
  - Placing a noisy piece of equipment outside your room,
  - Turning down the sound level of your music,
  - Asking your co-workers for quieter behavior.
- Reduce your exposure to noise by
  - Encapsulating the source of sound,
  - "Softening" hard surfaces which reflect or transmit the sound with drapes, carpets, acoustic tiles and the like which dampen or absorb sound,
  - Consider using noise-cancelling headphones, if this is suitable for your workplace.
- If all else fails, move to a different workplace.

Source: Partly adapted from Kroemer and Kroemer (2017).

and clothing. When we work inside, however, various low- and high-tech means are available, as listed in Table 24.4, to enhance well-being and work outcomes.

I like my work

One of the best statements a worker can make is, "I like my work." Too often, work is drudgery, an unpleasant way to earn a living. However, work can be efficient, easy, successful, satisfying, and even pleasant if it aligns with our interests and suits our skills, fitting our body and mind—which are the basic aims of ergonomics*.

Table 24.4 Taking charge of the climate at work

These climate considerations will contribute to making you feel good at work:

- With light physical work, such as in an office, and with appropriate clothing, comfortable environment temperature ranges are from about 20°C to 27°C during the summer; between about 18°C to 25°C in the winter.
- The preferred range of relative humidity is from 20% to 60%, best near 40% to 50%.
- If the sun shines on you, particularly on warm days, try to get out of the sun or to find shade behind blinds, curtains, screens, and the like.
- When you have no control over the climate (like outside), adjust your work intensity, timing, and pace (including rest breaks) and your work clothes to achieve sustainable levels of body exertion and temperature.

What to do when it is too warm for you:

- Lower the room temperature.
- Take off a layer of clothing and bare more skin, or wear clothing that helps with sweat evaporation.
- Move away from a heat source such as a radiator, a warm wall or window; get out of the sun.
- Move closer to a cool surface.
- Lower air humidity with a de-humidifier.
- Increase the air movement around you (unless it is very hot air).
- Wet your exposed skin; place a cool/moist cloth on your forehead, neck, or wrists.
- Take a rest break; do not exercise your body excessively.

What to do when it is too cool for you:

- Increase the room temperature.
- Add a layer of closing; cover more skin.
- Move closer to a heat source, such as a radiator, a warm wall or window; get into the sunshine.
- Move away from a cool surface.
- Decrease the air movement around you (unless it is comfortably warm air).
- Keep your body moving.

Source: Partly adapted from Kroemer and Kroemer (2017).

Note: These recommendations apply to most conditions in moderate climates, such as northern America and Europe.

Summary and steps to fit the human

Ten ergonomic principles apply throughout.

1. Plan tasks, equipment, and working conditions for safety, ease, and efficiency. User selection and training cannot substitute for good ergonomic design.
2. Design for real people, not for statistical phantoms.
3. Design for the physical and psychologic capabilities of the users.
4. Design to fit the physical sizes of the users.
5. Strive to accommodate "extraordinary" individuals: pregnant women, the elderly, people with impairments, people with unusual body proportions, children, etc.
6. Design for movements among suitable work postures.
7. Provide clear visual, acoustic, and sensory information.
8. Consider that people must be able to perform their tasks for hours, weeks, and years without overexertion or health hazards.
9. Allow people some control over their own work.
10. Create work environments that promote comfort and satisfaction, encourage job enlargement and enrichment, and ensure efficient functionality.

Notes

The text contains markers, *, to indicate specific references and comments, which follow.

24.1 Designing how to design: There are many sources available, a good place to start is Norman (2013).

HCD: This is also sometimes called "user-centered design" (UCD). Both HCD and UCD are focused on providing the human end-user with an appropriate design. In some cases, user-centered refers more the tangible (physical) aspects of the specific intended user, whereas human-centered may also consider the psychological aspects as well. HCD is the terminology in ISO 9241-210:2019, Ergonomics of human–system interaction: Human-centred design for interactive systems.

24.2 Using our skills and interests, getting along with others at work

People who have a calling: Smith (2007).

Job performance: Greenberg (2010), Kroemer and Kroemer (2017), Muchinsky and Culberson (2016).

Bad bosses: Harvey, Stoner, Hochwarter et al. (2007).

All should pull in the same direction: Useem (2015).

24.3 Setting up our own work workplace and work environment

Kansei engineering: Nagamachi (2011).

The basic aims of ergonomics: Many books on ergonomics or human (factors) engineering provide guidance and practical advice. These are listed under References.

References and further reading

Bhattacharya A, McGlothlin JD (2011). *Occupational Ergonomics: Theory and Applications*, 2nd ed. Boca Raton, FL, CRC Press. https://doi.org/10.1201/b11717

Bridger R (2008). *Introduction to Ergonomics*, 3rd ed. Boca Raton, FL, CRC Press. https://doi.org/10.1201/9781439894927

Chengular SN, Rodgers SH, Bernard TE (2003). *Kodak's Ergonomic Design for People at Work*, 2nd ed. New York, Wiley.

Fisk AD, Rogers WA, Charness N, Czaja SJ, Sharit J (2009). *Designing for Older Adults: Principles and Creative Human Factors Approaches.* Boca Raton, FL, CRC Press. https://doi.org/10.1201/9781420080681

Greenberg J (2010). *Behavior in Organizations: Understanding and Managing the Human Side of Work*, 10th ed. Upper Saddle River, NJ, Prentice-Hall.

Harvey P, Stoner J, Hochwarter W, Kacmar C (2007). Coping with abusive supervision: the neutralizing effects of ingratiation and positive affect on negative employee outcomes. *The Leadership Quarterly*, 18(3):264–280. https://doi.org/10.1016/j.leaqua.2007.03.008

Helander M (2005). *A Guide to Human Factors and Ergonomics*, 2nd ed. Boca Raton, FL, CRC Press. https://doi.org/10.1201/b12385

Human Factors and Ergonomics Society (HFES) 300 Committee (2004). *Guidelines for Using Anthropometric Data in Product Design.* Santa Monica, CA, Human Factors and Ergonomics Society.

Karwowski W (2006b). *International Encyclopedia of Ergonomics and Human Factors*, 2nd ed. Boca Raton, FL, CRC Press. https://doi.org/10.1201/9780849375477

Karwowski W, Soares MM, Stanton NA (2011). *Human Factors and Ergonomics in Consumer Product Design: Uses and Applications.* Boca Raton, FL, CRC Press. https://doi.org/10.1201/b10949

Konz S, Johnson S (2007). *Work Design: Industrial Ergonomics*, 6th ed. Scottsdale, AR, Holcomb Hataway.

Kroemer AD, Kroemer KHE (2017). *Office Ergonomics: Ease and Efficiency at Work*, 2nd ed. Boca Raton, FL, CRC Press. https://doi.org/10.1201/9781315368603

Kroemer Elbert K, Kroemer HB, Kroemer Hoffman AD (2018). *Ergonomics: How to Design for Ease and Efficiency*, 3rd ed. London, UK, Elsevier. https://doi.org/10.1016/C2016-0-02516-0

Kroemer KHE (2006c). *"Extra-Ordinary" Ergonomics: How to Accommodate Small and Big Persons, the Disabled and Elderly, Expectant Mothers and Children.* Boca Raton, FL, CRC Press. https://doi.org/10.1201/9780203025246

Lehto M, Landry SJ (2013). *Introduction to Human Factors and Ergonomics for Engineers*, 2nd ed. Boca Raton, FL, CRC Press.

Lueder R, Rice VJB (2007). *Ergonomics for Children: Designing Products and Places for Toddler to Teens.* Boca Raton, FL, CRC Press. https://doi.org/10.1201/9780203609163

Marras WS, Karwowski W (2006). *The Occupational Ergonomics Handbook (2 volumes)*, 2nd ed. Boca Raton, FL, CRC Press. https://doi.org/10.1201/9781420003635 and https://doi.org/10.1201/9781420003642

Muchinsky PM, Culberson SS (2016). *Psychology Applied to Work*, 11th ed. Summerfield, NC, Hypergraphic Press.

Nagamachi M (2011). *Kansei Engineering*, 2 volumes. Boca Raton, FL, CRC Press. https://doi.org/10.1201/b16799

Norman DA (2013). *The Design of Everyday Things.* Cambridge, MA, The MIT Press.

Pew PW, Mavor AS (2007). *Human-System Integration in the System Development Process.* Washington, DC, The National Academies.

Proctor RW, Van Zandt T (2008). *Human Factors in Simple and Complex Systems,* 2nd ed. Boca Raton, FL, CRC Press. https://doi.org/10.1201/9781420063608

Rubio S, Díaz E, Martín J, Puente JM (2004). Evaluation of subjective mental workload: a comparison of SWAT, NASA-TLX, and workload profile methods. *Applied Psychology*, 53:61–86. https://doi.org/10.1111/j.1464-0597.2004.00161.x

Salvendy G (2012). *Handbook of Human Factors and Ergonomics*, 4th ed. New York, Wiley. https://doi.org/10.1002/9781118131350

Smith TM (2007). *Job Satisfaction in America: Trends and Socio-Demographic Correlates.* Available at https://www-news.uchicago.edu/releases/07/pdf/070827.jobs.pdf

Stanton N, Hedge A, Brookhuis K, Salas E, Hendrick HW (2004). *Handbook of Human Factors and Ergonomics Methods.* Boca Raton, FL, CRC Press. https://doi.org/10.1201/9780203489925

Useem J (2015). Why it pays to be a jerk: new research confirms what they say about nice guys. *The Atlantic*, June 2015:1–25.

Vink P (2004). *Comfort and Design: Principles and Good Practice.* Boca Raton, FL, CRC Press. https://doi.org/10.1201/9781420038132

Wickens CD, Lee J, Liu Y & Gordon-Becker S (2004). *An Introduction to Human Factors Engineering*, 2nd ed. Upper Saddle River, Prentice-Hall/Pearson education.

Wilson JR, Sharples S (2015). *Evaluation of Human Work*, 4th ed. Boca Raton, FL, CRC Press. https://doi.org/10.1201/b18362

The last page

The motto "Fitting the Human", which is also the book title, suggest two approaches in ergonomics: fitting the human to the task and environment or fitting the human by the task and environment.

The first approach primarily falls within the domain of industrial psychology: matching individuals with task requirements by personnel selection and training, and by having humans adapt to their environments. However, compensating for misguided designs by instructing people on proper usage is a difficult task.

The computer keyboard (Chapter 19) and hand/foot controls of road vehicles (Chapter 20) are examples of long-established designs that once seemed workable but are now inadequate. It is nearly impossible to teach people to use equipment efficiently and safely when it has serious human engineering flaws: consider the clumsy operation of the QWERTY keyboard as a computer (or mobile phone) input device and think about the injuries suffered by thousands of people in vehicular accidents despite extensive driver training.

The better ergonomic approach is the second one: fitting the task and environment to the human. This strategy involves carefully planning all tools and equipment, work tasks and procedures, working hours and shift arrangements, as well as the physical and social conditions, and the overall working organization and environment. Even as technology continues to change our work and personal lives, the fundamental principles of human interactions and ergonomics remain. Accommodating the human is the fundamental and most successful approach, which is the focus of this book: designing the overall work system and allits details to fit the human. This makes work safe, efficient, satisfying, and even enjoyable.

After all, although we humans are similar, we do not look, think, respond, or work exactly alike, and that is what keeps ergonomics interesting.

Similar, but not all the same!

(courtesy of Kristianna Elbert 2024)

References

The literature contains recent anthropometric information, published usually in journal articles or as research reports. Handbooks, encyclopedias and standards are often a few years behind.

AAMI TIR59 (2017). Integrating Human Factors into Design Controls in accordance with FDA 21 CFR 820.30.

Aksoy CG, Barrero JM, Bloom N, Davis SJ, Dolls M, Zarate P (2023). *Working from Home Around the Globe: 2023 Report*. https://wfhresearch.com/wp-content/uploads/2023/06/GSWA-2023.pdf (retrieved June 11 2024).

Alderfer CP (1972). *Existence, Relatedness and Growth: Human Needs in Organizational Settings*. New York, Free Press.

Anderson AM, Mirka GA, Joines SM, Kaber DB (2009). Analysis of alternative keyboards using learning curves. *Human Factors*, 51(1):35–45. https://doi.org/10.1177/0018720808329844

Annett J (2002). Subjective rating scales: science or art? *Ergonomics*, 45(14):966–987. https://doi.org/10.1080/00140130210166951

ANSI/AAMI HE75 (2009). R2018, Human Factors Engineering - Design of Medical Devices. ANSI/ASHRAE Standard 55, Thermal Environmental Conditions for Human Occupancy, latest edition 2023.

ANSI/HFES 100-1998 (2007). Human Factors Engineering of Computer Workstations (2024 revision expected).

Armstrong TJ (2006). The ACGIH TLV for hand activity work. Chapter 41 in Marras WS, Karwowski W (eds), *The Occupational Ergonomics Handbook: Fundamentals and Assessment Tools for Occupational Ergonomics*, 2nd ed. Boca Raton, FL, CRC Press, pp. 421–442. https://doi.org/10.1201/9781420003635

Arndt R, Putz-Anderson V (2014). *Cumulative Trauma Disorders*. Boca Raton, FL, CRC Press.

Arswell CM, Stephens EC (2001). Information processing. In Karwowski W (ed), *International Encyclopaedia of Ergonomics and Human Factors*. London, UK, Taylor & Francis, pp. 256–259.

Astrand PO, Rodahl K, Dahl HA, Stromme SB (2004). *Textbook of Work Physiology: Physiological Bases of Exercise*, 4th ed. Champaign, IL, Human Kinetics.

Bailey RW (1996). *Human Performance Engineering: Designing High Quality, Professional User Interfaces for Computer Products, Applications, and Systems*, 3rd ed. Upper Saddle River, NJ, Prentice Hall.

Baker NA, Cham R, Cidboy EH, Cook J, Redfern MS (2007). Kinematics of the fingers and hands during computer keyboard use. *Clinical Biomechanics*, 22(1):34–43. https://doi.org/10.1016/j.clinbiomech.2006.08.008

Basbaum AI, Julius D (2006). Toward better pain control. *Scientific American*, 294(6):60–67. https://doi.org/10.1038/scientificamerican0606-60

Beltran Martinez K, Nazarahari M, Rouhani H (2023). Breaking the fatigue cycle: investigating the effect of work-rest schedules on muscle fatigue in material handling jobs. *Sensors (Basel)*, 23(24):9670. https://doi.org/10.3390/s23249670

Berger EH, Royster LH, Royster JD, Driscoll DP, Layne M (2003). *The Noise Manual*, 5th ed. Fairfax, VA, American Industrial Hygiene Association.

Bernard TE (2002). Thermal stress. Chapter 12 in Plog BA (ed), *Fundamentals of Industrial Hygiene*, 5th ed. Itasca IL, National Safety Council, pp. 32.

Bernard TE, Dukes-Dobos FN (2002). Heat stress. In Anna D (ed), *Chemical Protective Clothing*, 2nd ed. Fairfax, VA, AIHA Press, pp. 475–492.

Bhattacharya A, McGlothlin JD (2011). *Occupational Ergonomics: Theory and Applications*, 2nd ed. Boca Raton, FL, CRC Press. https://doi.org/10.1201/b11717

Bhise VD (2011). *Ergonomics in the Automotive Design Process.* Boca Raton, FL, CRC Press. https://doi.org/10.1201/b11237

Bitan Y, Ramey S, Philp G, Uukkivi T (2015) Working with paramedics on implementing human factors improvements to their response bags. *Proceedings of the International Symposium on Human Factors and Ergonomics in Health Care*, 4:179–181. https://doi.org/10.1177/2327857915041016

Bjoerkman T (1996). The rationalisation movement in perspective and some ergonomic implications. *Applied Ergonomics*, 27(2):111–117. https://doi.org/10.1016/0003-6870(95)00065-8

Black TR, Shah SM, Busch AJ, Metcalfe J, Lim HJ (2011). Effect of transfer, lifting, and repositioning (TLR) injury prevention program on musculoskeletal injury among direct care workers. *Journal of Occupational and Environmental Hygiene*, 8(4):226–235. https://doi.org/10.1080/15459624.2011.564110

Boff KR, Kaufman L, Thomas JP (1986). *Handbook of Perception and Human Performance.* New York, Wiley.

Boff KR, Lincoln JE (1988). *Engineering Data Compendium: Human Perception and Performance.* Wright-Patterson AFB, OH, Armstrong Aerospace Medical Research Laboratory. https://ntrs.nasa.gov/api/citations/19880019247/downloads/19880019247.pdf

Booher HR (2003). *Handbook of Human Systems Integration.* New York, Wiley.

Borg G (2001). Rating scales for perceived physical effort and exertion. In Karwowski W (ed), *International Encyclopaedia of Ergonomics and Human Factors.* London, UK, Taylor & Francis, pp. 358–541.

Borg G (2004). Scaling experiences during work: perceived exertion and difficulty. Chapter 11 in Stanton N, Hedge A, Brookhuis K, Salas E, Hendrick HW (eds), *Handbook of Human Factors and Ergonomics Methods*. Boca Raton, FL, CRC Press, pp. 768. https://doi.org/10.1201/9780203489925

Boyce PR (2014). *Human Factors in Lighting*. Boca Raton, FL, CRC Press. https://doi.org/10.1201/b16707

Bradtmiller B (2015). Anthropometry in human systems integration. Chapter 8 in Boehm-Davis DA, Durso FT, Lee JD(eds), *APA Handbook of Human Systems Integration*. Washington, DC, American Psychological Association, pp. 117–132. https://doi.org/10.1037/14528-000

Bridger R (2008). *Introduction to Ergonomics*, 3rd ed. Boca Raton, FL, CRC Press. https://doi.org/10.1201/9781439894927

Carayon P (2011). *Handbook of Human Factors and Ergonomics in Healthcare and Patient Safety*, 2nd ed. Boca Raton, FL, CRC Press. https://doi.org/10.1201/b11219

Carayon P, Lim SY (2006). Psychosocial work factors. Chapter 5 in Marras WS, Karwowski W (eds), *The Occupational Ergonomics Handbook: Interventions, Controls, and Applications in Occupational Ergonomics*, 2nd ed. Boca Raton, FL, CRC Press, pp. 848. https://doi.org/10.1201/9781420003642

Casali JG, Gerges SNY (2006). Protection and enhancement of hearing in noise. Chapter 7 in Williges RC (ed), *Reviews of Human Factors and Ergonomics*, Volume 2. Santa Monica, CA, Human Factors and Ergonomics Society, pp. 195–240. https://doi.org/10.1177/1557234X0600200108

Casali JG, Robinson GS (2006). Noise in industry. Chapter 31 in Marras WS, Karwowski W (eds), *The Occupational Ergonomics Handbook: Fundamentals and Assessment Tools for Occupational Ergonomics*, 2nd ed. Boca Raton, FL, CRC Press, pp. 421–442. https://doi.org/10.1201/9781420003635

Centers for Disease Control and Prevention (2015). NIOSH, National Health Interview Survey (NHIS) Report dated 5 Aug 2015. https://wwwn.cdc.gov/NIOSH-WHC/chart/ohs-workorg?OU=*&T=OU&V=R

Chaffin DB, Andersson GBJ, Martin BJ (2006). *Occupational Biomechanics*, 4th ed. New York, Wiley.

Charney W (2010). *Handbook of Modern Hospital Safety*, 2nd ed. Boca Raton, FL, CRC Press. https://doi.org/10.1201/9781420047868

Chengular SN, Rodgers SH, Bernard TE (2003). *Kodak's Ergonomic Design for People at Work*, 2nd ed. New York, Wiley.

CIEHF (2021). *Human Factors and Ergonomics in Healthcare AI*, white paper from Chartered Institute of Ergonomics & Human Factors. https://ergonomics.org.uk/resource/human-factors-in-healthcare-ai.html

CIEHF (2022). *Human Factors in Highly Automated Systems*, white paper from Chartered Institute of Ergonomics & Human Factors. https://ergonomics.org.uk/resource/human-factors-in-highly-automated-systems-white-paper.html

Ciriello VM (2001). The effects of box size, vertical distance, and height on lowering tasks. *International Journal of Industrial Ergonomics*, 28(2):61–67. https://doi.org/10.1016/S0169-8141(01)00012-9

Ciriello VM (2007). The effects of container size, frequency and extended horizontal reach on maximum acceptable weights of lifting for female industrial workers. *Applied Ergonomics*, 38(1):1–5. https://doi.org/10.1016/j.apergo.2006.02.001

Ciriello VM, McGorry RW, Martin SE, Bezverkhny IB (1999). Maximum acceptable forces of dynamic pushing: comparison of two techniques. *Ergonomics*, 42(1):32–29. https://doi.org/10.1080/001401399185784

Colapinto J (2015). Lighting the brain. *The New Yorker*, 18 May issue, 74–83.

Colligan MJ, Tepas DI (1986). The stress of hours of work. *American Industrial Hygiene Association Journal*, 47(11):686–695. https://doi.org/10.1080/15298668691390485

Colombini D, Occhipinti E, Alvarez-Casado E, Waters T (2012). *Manual Lifting: A Guide to the Study of Simple and Complex Lifting Tasks*. Boca Raton, FL, CRC Press.

Corlett EN (2005). The evaluation of industrial seating. Chapter 27 in Wilson JR, Corlett N (eds), *Evaluation of Human Work*, 3rd ed. London, UK, Taylor & Francis.

Costa G (2010). Shift work and health: current problems and preventive actions. *Safety and Health at Work*, 1(2):112–123. https://doi.org/10.5491/SHAW.2010.1.2.112

Cox T, Griffiths A (2005). The nature and measurement of work-related stress: theory and practice. Chapter 19 in Wilson JR, Sharples S (eds), *Evaluation of Human Work*, 4th ed. Boca Raton, FL, CRC Press, pp. 1031. https://doi.org/10.1201/b18362

Czaja SJ, Lee CC (2001). The Internet and older adults: design challenges and opportunities. In Charness N, Parks DC, Sabel BA (eds), *Communication, Technology, and Aging: Opportunities and Challenges for the Future*. New York, Springer, pp. 60–80.

Czaja SJ, Lee CC (2002). Designing computer systems for older adults. Chapter 21 in Jacko J, Sears A (eds), *The Human-Computer Interaction Handbook: Fundamentals, Evolving Technologies and Emerging Applications*. New York, Lawrence Erlbaum and Associates (IEA), pp. 1384.

Czaja SJ, Zarcadoolas C, Vaughon WL, Lee CC, Rockoff ML, Levy J (2015). The usability of electronic personal health record systems for an underserved adult population. *Human Factors*, 57(3):491–506. https://doi.org/10.1177/0018720814549238

Czeisler CA, Gooley JJ (2007). Sleep and circadian rhythms in humans. *Cold Spring Harbour Symposia on Quantitative Biology*, 72:579–597.

Daams BJ (1993). Static force exertion in postures with different degrees of freedom. *Ergonomics*, 36:397–406.

Daams BJ (1994). *Human Force Exertion in User-Product Interaction: Background for Design.* Delft, NL, Delft University Press – IOS Press.

Daams BJ (2001). Push and pull data and torque data. In Karwowski W (ed), *International Encyclopaedia of Ergonomics and Human Factors.* London, UK, Taylor & Francis, pp. 299–316, pp. 334–342.

Dekker S (2014). *The Field Guide to Understanding "Human Error".* Williston, VT, Ashgate.

Delleman NJ, Haslegrave CM, Chaffin DB (2004). *Working Postures and Movements.* Boca Raton, FL, CRC Press. https://doi.org/10.1201/9781482265095

Dempsey PG (1998). A critical review of biomechanical, epidemiological, physiological and psychophysical criteria for designing manual materials handling tasks. *Ergonomics*, 41(1):73–88. https://doi.org/10.1080/001401398187332

Dempsey PG (2006). Psychophysical approach to task analysis. Chapter 47 in Marras WS, Karwowski W (eds), *The Occupational Ergonomics Handbook: Fundamentals and Assessment Tools for Occupational Ergonomics*, 2nd ed. Boca Raton, FL, CRC Press, pp. 421–442. https://doi.org/10.1201/9781420003635

Dennerlein J (2006). The computer keyboard system design. Chapter 39 in Marras WS, Karwowski W (eds), *The Occupational Ergonomics Handbook: Interventions, Controls, and Applications in Occupational Ergonomics*, 2nd ed. Boca Raton, FL, CRC Press, pp. 848. https://doi.org/10.1201/9781420003642

Dennerlein JT (2005). Finger flexor tendon forces are a complex function of finger joint motions and fingertip forces. *Journal of Hand Therapy*, 18(2):120–127. https://doi.org/10.1197/j.jht.2005.01.011

Deyo RA, Weinstein JN (2001). Low back pain. *New England Journal of Medicine*, 344:363–370. https://doi.org/10.1056/NEJM200102013440508

Dhillon BS (2008). *Reliability Technology, Human Error, and Quality in Health Care.* Boca Raton, FL, CRC Press. https://doi.org/10.1201/9781420065596

Doidge N (2007). *The Brain that Changes Itself: Stories of Personal Triumph from the Frontiers of Brain Science.* New York, Penguin.

Doutres O, Terroir J, Jolly C, Gauvin C, Martin L, Negrini A (2022). Towards a holistic model explaining hearing protection device use among workers. *International Journal of Environmental Research and Public Health*, 19(9):5578. https://doi.org/10.3390/ijerph19095578

Driskell JE, Mullen B (2005). The efficacy of naps as a fatigue countermeasure: a meta-analytic integration. *Human Factors*, 47:360–377.

Dvorak A (1943). There is a better typewriter keyboard. *National Business Education Quarterly*, 12:51–58.

Eastman Kodak Company (1983). *Ergonomic Design for People at Work.* New York, Van Nostrand Reinhold.

Ettema AM, Zhao C, Amadio PC, O'Byrne MM, An KN (2007). Gliding characteristics of flexor tendon and tenosynovium in carpal tunnel syndrome: a pilot study. *Clinical Anatomy*, 20(3):292–299. https://doi.org/10.1002/ca.20379

European Commission (2021). Directive 2003/10/EC of the European Parliament and of the Council of 6 February 2003 on the Minimum Health and Safety Requirements Regarding the Exposure of Workers to the Risks Arising from Physical Agents (Noise) European Commission; Brussels, Belgium. https://eur-lex.europa.eu/legal-content/EN/TXT/PDF/?uri=CELEX:02003L0010-20190726

Fagarasanu M, Kumar S (2004). Hand strength. Chapter 10 in Kumar S (ed), *Muscle Strength*. Boca Raton, FL, CRC Press, pp. 538. https://doi.org/10.1201/9780203503591

Feathers D, D'Souza C, Paquet V (2015). Anthropometry in ergonomic design. Chapter 27 in Wilson JR, Sharples S (eds), *Evaluation of Human Work*, 4th ed. Boca Raton, FL: CRC Press, pp. 1031. https://doi.org/10.1201/b18362

Figueira V, Silva S, Costa I, Campos B, Salgado J, Pinho L, Freitas M, Carvalho P, Marques J, Pinho F (2024). Wearable's for monitoring and postural feedback in the work context: a scoping review. *Sensors (Basel)*, 24(4):1341. https://doi.org/10.3390/s24041341

Finomore VS, Shaw TH, Warm JS, Matthews G, Boles DB (2013). Viewing the workload of vigilance through the lenses of the NASA-TLX and the MRQ. *Human Factors*, 55(6):1044–1063. https://doi.org/10.1177/0018720813484498

Fisk AD, Rogers WA, Charness N, Czaja SJ, Sharit J (2009). *Designing for Older Adults: Principles and Creative Human Factors Approaches*. Boca Raton, FL, CRC Press. https://doi.org/10.1201/9781420080681

Fletcher H, Munson WA (1933). Loudness, its definition, measurement and calculation. *Journal of the Acoustic Society of America*, 5:82–108.

Flier JF, Maratos-Flier E (2007). What fuels fat. *Scientific American*, 297(3):72–81. https://doi.org/10.1038/scientificamerican0907-72

Folkard S, Lombardi DA, Tucker PT (2005). Shift work: safety, sleepiness and sleep. *Industrial Health*, 43(1):20–23. https://doi.org/10.2486/indhealth.43.20

Folkard S, Monk TH (1985). *Hours of Work*. Chichester, UK, Wiley.

Folkard S, Tucker P (2003). Shift work, safety and productivity. *Occupational Medicine (London)*, 53(2):95–101. https://doi.org/10.1093/occmed/kqg047

Foster R, Kreitzman L (2004). *Rhythms of Life: The Biological Clocks that Control the Daily Lives of Every Living Thing*. London, UK, Profile.

Fox JG (1983). Industrial music. In Oborne DJ, Gruneberg MM (eds), *The Physical Environment at Work*. New York, Wiley, pp. 221–226.

Freivalds A (1999). Ergonomics of hand controls. Chapter 27 in Marras WS, Karwowski W (eds), *The Occupational Ergonomics Handbook: Fundamentals and Assessment Tools for Occupational Ergonomics*. Boca Raton, FL, CRC Press, pp. 319–338.

Freivalds A (2006). Upper extremity analysis of the wrist. Chapter 45 in Marras WS, Karwowski W (eds), *The Occupational Ergonomics Handbook: Fundamentals and Assessment Tools for Occupational Ergonomics*, 2nd ed. Boca Raton, FL, CRC Press, pp. 361–404. https://doi.org/10.1201/9781420003635

Freivalds A (2011). *Biomechanics of the Upper Limbs: Mechanics, Modelling and Musculoskeletal Injuries*, 2nd ed. Boca Raton, FL, CRC Press. https://doi.org/10.1201/b11547

Garg A, Marras WS (2014). Epidemiological studies of workplace musculoskeletal disorders. *Special Issue of Human Factors*, 56(1):5.

Gawande A (2008). The itch. *The New Yorker*, June 30:58–65.

Gawande A (2015). Overkill. *The New Yorker*, May 11:42–53.

Gibbons JD (1997). *Nonparametric Methods for Quantitative Analysis*, 3rd ed. Columbus, OH, American Sciences Press.

Gokmen Y, Turen U, Erdem H, Tokmak I (2021). National preferred interpersonal distance curbs the spread of COVID-19: a cross-country analysis. *Disaster Medicine and Public Health Preparedness*, 15(6):e20–e26. https://doi.org/10.1017/dmp.2020.295

Goodman HJ, Choueka J (2005). Biomechanics of the flexor tendons. *Hand Clinics*, 21(2):129–149. https://doi.org/10.1016/j.hcl.2004.11.002

Gordon CC, Blackwell CL, Bradtmiller B, Parham JL, Barrientos P, Paquette SP, Corner BD, Carson JM, Venezia JC, Rockwell BM, Mucher M, Kristensen S (2014). 2012 Anthropometric Survey of U.S. Army Personnel: Methods and Summary Statistics. Technical Report NATICK/TR-15/007. U.S. Army Natick Soldier Research, Development and Engineering Center Natick, MA. https://apps.dtic.mil/sti/citations/ADA611869

Grandjean E (1987). *Ergonomics in Computerized Offices*. London, UK, Taylor & Francis. https://doi.org/10.1201/b12564

Greenberg J (2010). *Behaviour in Organizations: Understanding and Managing the Human Side of Work*, 10th ed. Upper Saddle River, NJ, Prentice-Hall.

Griffin MJ (1990). *Handbook of Human Vibration*. San Diego, CA, Academic Press. https://doi.org/10.1016/C2009-0-02730-5

Hackman JR, Oldham GR (1980). *Work Redesign*. Reading, MA, Addison-Wesle.

Hall, JE (2015). *Guyton and Hall Textbook of Medical Physiology*, 13th ed. Amsterdam, NL, Elsevier.

Harvey P, Stoner J, Hochwarter W, Kacmar C (2007). Coping with abusive supervision: The neutralizing effects of ingratiation and positive affect on negative employee outcomes. *The Leadership Quarterly*, 18(3):264–280. https://doi.org/10.1016/j.leaqua.2007.03.008

Havenith G (2005). Thermal conditions measurement. Chapter 60 in Stanton N, Hedge A, Brookhuis K, Salas E, Hendrick HW (eds), *Handbook of Human Factors and Ergonomics Methods*. Boca Raton, FL, CRC Press, pp. 768. https://doi.org/10.1201/9780203489925

Hedge A, James T (2012). Ergonomic issues of computer use in a major healthcare system. In Duffy VG (ed) *Advances in Human Aspects of Healthcare.* Boca Raton, FL, CRC Press, pp. 760. https://doi.org/10.1201/b12318

Helander M (2003). Forget about ergonomics in chair design? Focus on aesthetics and comfort! *Ergonomics*, 46(13–14):1306–1319. https://doi.org/10.1080/00140130310001610847

Helander M (2005). *A Guide to Human Factors and Ergonomics,* 2nd ed. Boca Raton, FL, CRC Press. https://doi.org/10.1201/b12385

Helander MG, Zhang L (1997). Field studies of comfort and discomfort in sitting. *Ergonomics*, 40(9):895–915. https://doi.org/10.1080/001401397187739

Hendrick HW, Kleiner BM (2005). *Macroergonomics: Theory, Methods, and Applications*, 2nd ed. Boca Raton, FL, CRC Press.

Herzberg F (1966). *Work and the Nature of Man.* New York, Thomas.

Herzberg F (1968). One more time: how do you motivate employees? *Harvard Business Review*, 81:87–96; reprint 87507 HBR 1987 and 2003.

Herzog W (2007). Determinants of muscle strength. Chapter 7 in Kumar S (ed), *Biomechanics in Ergonomics*, 2nd ed. Boca Raton, FL, CRC Press, pp. 34.

Hill SG, Kroemer KHE (1986). Preferred declination of the line of sight. *Human Factors*, 28(2):127–134. https://doi.org/10.1177/001872088602800201

Hinkelmann K, Kempthorne O (1994). *Design and Analysis of Experiments, Vol. 1: Introduction to Experimental Design.* New York, Wiley.

Hinkelmann K, Kempthorne O (2005). *Design and Analysis of Experiments, Vol. 2: Advanced Experimental Design.* New York, Wiley-Interscience. https://doi.org/10.1002/0471709948

Hornberger S, Knauth P, Costa G, Folkard S (2000). *Shiftwork in the 21st Century.* Frankfurt, DE, Lang.

Horne J (1988). *Why We Sleep – The Functions of Sleep in Humans and Other Mammals.* Oxford, UK, Oxford University.

Horne JA (2006). *Sleepfaring: A Journey Through the Science of Sleep.* Oxford, UK, Oxford University Press. https://doi.org/10.1016/j.eeh.2007.03.002

Houy DA (1983). Range of joint motion in college males. *Proceedings of the 27th annual meeting Human Factors Society*, pp. 374–378. Santa Monica, CA, Human Factors Society.

Howarth PA (2015). Assessment of the visual environment. Chapter 25 in Wilson JR, Sharples S (eds), *Evaluation of Human Work*, 4th ed. Boca Raton, FL, CRC Press, pp. 1031. https://doi.org/10.1201/b18362

Hsiao H, Long D, Snyder K (2002). Anthropometric differences among occupational groups. *Ergonomics* 45, 136–152. https://doi.org/10.1080/00140130110115372

Huberman M, Minns, C (2007). The times they are not changin': days and hours of work in old and new worlds, 1870–2000. *Explorations in Economic History*, 44(4):538–567.

Hughes RE, An KN (2007). Biomechanical models of the hand, wrist, and elbow in ergonomics. Chapter 14 in Kumar S (ed), *Biomechanics in Ergonomics*, 2nd ed. Boca Raton, FL, CRC Press, pp. 20.

Human Factors and Ergonomics Society (HFES) 300 Committee (2004). *Guidelines for Using Anthropometric Data in Product Design.* Santa Monica, CA, Human Factors and Ergonomics Society.

IEC 62366-1 (2015). *Application of Usability Engineering to Medical Devices.* Geneva, CH, International Labour Office.

ILO (1979, 1992). *Introduction to Work Study*, 3rd and 4th ed. Geneva, CH, International Labour Office.

ILO (1988). *Maximum Weights in Load Lifting and Carrying, Occupational Safety and Health Series #59.* Geneva, CH, International Labour Office.

Imrhan SN, Nguyen MT, Nguyen NN (1993). Hand anthropometry of Americans of Vietnamese origin. *International Journal of Industrial Ergonomics*, 12:281–287. https://doi.org/10.1016/0169-8141(93)90098-X

ISO 10075-1 (2017). *Ergonomic Principles Related to Mental Workload: General Issues and Concepts, Terms and Definitions.* Geneva, CH, International Organization for Standardization.

ISO 13485 (2016). *Medical Devices - Quality Management Systems - Requirements for Regulatory Purposes.* Geneva, CH, International Organization for Standardization.

ISO 14971 (2019). *Application of Risk Management to Medical Devices.* Geneva, CH, International Organization for Standardization.

ISO 15535 (2023). *General Requirements for Establishing Anthropometric Databases.* Geneva, CH, International Organization for Standardization.

ISO 20685-1 (2018). *3-D Scanning Methodologies for Internationally Compatible Anthropometric Databases – Part 1: Evaluation Protocol for Body Dimensions Extracted from 3-D Body Scans.* Geneva, CH, International Organization for Standardization.

ISO 226 (2023). *Acoustics — Normal Equal-Loudness-Level Contours.* Geneva, CH, International Organization for Standardization.

ISO 45001 (2018). *Occupational Health and Safety Management Systems — Requirements with Guidance for Use.* Geneva, CH, International Organization for Standardization.

ISO 7101 (2023). *Healthcare Organization Management — Management Systems for Quality in Healthcare Organizations — Requirements.* Geneva, CH, International Organization for Standardization.

ISO 7243 (2017). *Ergonomics of the Thermal Environment — Assessment of Heat Stress Using the WBGT (Wet Bulb Globe Temperature) Index.* Geneva, CH, International Organization for Standardization.

ISO 7250-1 (2017). *Basic Human Body Measurements for Technological Design – Part 1: Body Measurement Definitions and Landmarks.* Geneva, CH, International Organization for Standardization.

ISO 7933 (2023). *Ergonomics of the Thermal Environment — Analytical Determination and Interpretation of Heat Stress Using Calculation of the Predicted Heat Strain*. Geneva, CH, International Organization for Standardization.

ISO 8996 (2021). *Ergonomics of the Thermal Environment — Determination of Metabolic Rate*. Geneva, CH, International Organization for Standardization.

ISO 9920 (2007). *Ergonomics of the Thermal Environment — Estimation of Thermal Insulation and Water Vapour Resistance of a Clothing Ensemble*. Geneva, CH, International Organization for Standardization.

ISO/CIE 8995-3 (2018). *Lighting of Work Places, Part 3: Lighting Requirements for Safety and Security of Outdoor Work Places*. Geneva, CH, International Organization for Standardization.

ISO/CIE DIS 8995-1 (2024). *Lighting of Work Places, Part 1: Indoor (Expected 2024)*. Geneva, CH, International Organization for Standardization.

ISO/DIS 7730 (2005). *Ergonomics of the Thermal Environment — Analytical Determination and Interpretation of Thermal Comfort Using Calculation of the PMV and PPD Indices and Local Thermal Comfort Criteria*. Geneva, CH, International Organization for Standardization.

ISO/TR 9241-810 (2020). *Ergonomics of Human-System Interaction, Part 810: Robotic, Intelligent and Autonomous Systems*. Geneva, CH, International Organization for Standardization.

Jarrett A (1973). *The Physiology and Pathology of the Skin*. London, UK, Academic Press.

Jenkins WL (1953). *Design Factors in Knobs and Levers for Making Settings on Scales and Scopes, WADC-TR-53-2*. Wright-Patterson AFB, OH, Aero Medical Laboratory.

Kapandji IA (2007). *The Physiology of the Joints*, 6th ed. Edinburgh, UK, Churchill Livingstone.

Karwowski W (2006a). From past to future: building a collective vision for HFES 2020. *HFES Bulletin*, 49(11):1–3.

Karwowski W (2006b). *International Encyclopaedia of Ergonomics and Human Factors,* 2nd ed. Boca Raton, FL, CRC Press. https://doi.org/10.1201/9780849375477

Karwowski W, Soares MM, Stanton NA (2011). *Human Factors and Ergonomics in Consumer Product Design: Uses and Applications*. Boca Raton, FL, CRC Press. https://doi.org/10.1201/b10949

Kenny WL, Wilmore J, Costill D (2015). *Physiology of Sport and Exercise*, 6th ed. Champaign, IL, Human Kinetics.

Kinkead RD, Gonzalez BK (1969). *Human Factors Design Recommendations for Touch-Operated Keyboards – Final Report* (Document No. 12091-FR). Minneapolis, MN, Honeywell, Inc.

Klemmer ET (1958). *A Ten-Key Typewriter* (Research Memorandum RC-65). Yorktown Heights, NY, IBM Research Center.

Klockenberg EA (1926). *Rationalisierung der Schreibmaschine und ihrer Bedienung* [Rationalization of the typewriter and its operation]. Berlin, Springer.

Knauth P (2006). Workday length and shift work issues. Chapter 29 in Marras WS, Karwowski W (eds), *The Occupational Ergonomics Handbook: Interventions, Controls, and Applications in Occupational Ergonomics*, 2nd ed. Boca Raton, FL, CRC Press, pp. 848. https://doi.org/10.1201/9781420003642

Knauth P (2007a). Extended work periods. *Industrial Health*, 45(1): 125–136. https://doi.org/10.2486/indhealth.45.125. Erratum in: *Industrial Health*. 2007 Apr;45(2):375.

Knauth P (2007b). Schicht- und Nachtarbeit (shift work and night work, in German). In Landau K (ed), *Lexikon Arbeitsgestaltung*, Stuttgart, DE, Gentner.

Koepke N, Zwahlen M, Wells JC, Bender N, Henneberg M, Rühli FJ, Staub K (2017). Comparison of 3D laser-based photonic scans and manual anthropometric measurements of body size and shape in a validation study of 123 young Swiss men. *Peer Journal*, 5:e2980. https://doi.org/10.7717/peerj.2980

Konz S, Johnson S (2007). *Work Design: Industrial Ergonomics*, 6th ed. Scottsdale, AR, Holcomb Hataway.

Kroemer AD, Kroemer KHE (2017). *Office Ergonomics: Ease and Efficiency at Work*, 2nd ed. Boca Raton, FL, CRC Press. https://doi.org/10.1201/9781315368603

Kroemer Elbert K, Kroemer HB, Kroemer Hoffman AD (2018). *Ergonomics: How to Design for Ease and Efficiency*, 3rd ed. London, UK, Elsevier. https://doi.org/10.1016/C2016-0-02516-0

Kroemer KHE (1965). *Ergonomic Aspects of Control Operation* (in German). PhD dissertation, Technical University Hannover.

Kroemer KHE (1974). *Designing for Muscular Strength of Various Populations*. AMRL-Technical Report 72-46, Wright-Patterson AFB, OH, Aerospace Medical Research Laboratory.

Kroemer KHE (1985). Testing individual capability to lift material: repeatability of a dynamic test compared with static testing. *Journal of Safety Research*, 16(1):1–7. https://doi.org/10.1016/0022-4375(85)90032-5

Kroemer KHE (1986). Coupling the Hand with the handle: an improved notation of touch, grip, and grasp. *Human Factors*, 28(3):337–339. https://doi.org/10.1177/001872088602800309

Kroemer KHE (1997). *Ergonomic Design for Material Handling Systems*. Boca Raton, FL, CRC Press. https://doi.org/10.1201/9780203756683

Kroemer KHE (1998). Relating muscle strength and its internal transmission to design data. In Kumar S (ed), *Advances in Occupational Ergonomics and Safety*. Amsterdam, NL, IOS, pp. 349–352.

Kroemer KHE (1999). Assessment of human muscle strength for engineering purposes: a review of the basics. *Ergonomics*, 42(1):74–93. https://doi.org/10.1080/001401399185810

Kroemer KHE (2001). Keyboards and keying: an annotated bibliography of the literature from 1878 to 1999. *Universal Access in the Information Society UAIS*, 1(2):99–160. https://doi.org/10.1007/s102090100012

Kroemer KHE (2006a). Designing children's furniture and computers for school and home. *Ergonomics in Design*, 14(3):8–16. https://doi.org/10.1177/106480460601400304

Kroemer KHE (2006b). Designing for older people. *Ergonomics in Design*, 14(4):25–31. https://doi.org/10.1177/106480460601400407

Kroemer KHE (2006). *"Extra-Ordinary" Ergonomics: How to Accommodate Small and Big Persons, the Disabled and Elderly, Expectant Mothers and Children.* Boca Raton, FL, CRC Press. https://doi.org/10.1201/9780203025246

Kroemer KHE (2008). Anthropometry and biomechanics: anthromechanics. Chapter 2 in Kumar S (ed), *Biomechanics in Ergonomics*, 2nd ed. Boca Raton, FL, CRC Press, pp. 48. https://doi.org/10.1201/9780849379093

Kroemer KHE (2010). 40 Years of human engineering the keyboard. *Proceedings of the 54th annual meeting of the Human Factors and Ergonomics Society*, 54(15):1134–1138. https://doi.org/10.1177/154193121005401509

Kroemer KHE, Kroemer HJ, Kroemer-Elbert KE (2020). *Engineering Physiology: Bases of Human Factors Engineering/Ergonomics*, 5th ed. Heidelberg, Springer. https://doi.org/10.1007/978-3-030-40627-1

Kroemer KHE, Robinette JC (1968). *Ergonomics in the design of office furniture: a review of European literature* (AMRL-TR 68-90). Wright-Patterson Air Force Base, OH: Aerospace Medical Research Laboratories. (Also published with shortlist of references in 1969). *International Journal of Industrial Medicine and Surgery*, 38(4):115–125.

Kroemer KHE, Robinson DE (1971). *Horizontal static forces exerted by men standing in common working positions on surfaces of various tractions.* Technical Report AMRL-TR-70–114, Wright-Patterson Air Force Base, OH, Aerospace Medical Research Laboratory.

Kumar S (2001). Theories of musculoskeletal injury causation. *Ergonomics*, 44(1):17–47. https://doi.org/10.1080/00140130120716

Kumar S (2004). *Muscle Strength.* Boca Raton, FL, CRC Press. https://doi.org/10.1201/9780203503591

Kumar S (2008). *Biomechanics in Ergonomics,* 2nd ed. Boca Raton, FL, CRC Press. https://doi.org/10.1201/9780849379093

Kumar S, Mital A (1996). *Electromyography in Ergonomics.* London, UK, Francis & Taylor. https://doi.org/10.1201/9780203758670

Kurowski A, Buchholz B, Punnett L, ProCare Research Team (2014). A physical workload index to evaluate a safe resident handling program for nursing home personnel. *Human Factors*, 56(4):669–683. https://doi.org/10.1177/0018720813509268

Landy FJ, Conte JM (2006). *Work in the 21st Century: An Introduction to Industrial and Organizational Psychology.* Malden, MA, Blackwell (7th edition in 2024 from Wiley)

Lavender SA (2006). Training lifting techniques. Chapter 23 in Marras WS, Karwowski W (eds), *The Occupational Ergonomics Handbook: Interventions, Controls, and Applications in Occupational Ergonomics*, 2nd ed. Boca Raton, FL, CRC Press, pp. 848. https://doi.org/10.1201/9781420003642

Lee JC, Healy JC (2005). Normal sonographic anatomy of the wrist and hand. *Radiographics*, 25(6):1577–1590. https://doi.org/10.1148/rg.256055028

Lehmann G (1953). *Praktische Arbeitsphyiologie* (2nd ed. 1962), Stuttgart, DE, Thieme.

Lehto M, Landry SJ (2013). *Introduction to Human Factors and Ergonomics for Engineers*, 2nd ed. Boca Raton, FL, CRC Press.

Lepore J (2009). Not so fast. *The New Yorker*, 85(32):114–122.

Levitt SD, List JA (2009). Was there really a Hawthorne Effect at the Hawthorne Plant? An analysis of the original illumination experiments. National Bureau of Economic Research, Cambridge, MA, Working Paper 15016. https://doi.org/10.3386/w15016

Leyk D, Küchmeister G, Jürgens HW (2006). Combined physiological and anthropometrical databases as ergonomic tools. *Journal of Physiological Anthropology*, 25(6):363–369. https://doi.org/10.2114/jpa2.25.363

Liberty Mutual. *Manual Materials Handling Guidelines*. Regularly updated on www.libertymutual.com

Lockhead GR, Klemmer ET (1959). *An Evaluation of an Eight-Key Word-Writing Typewriter* (Research Report RC-180). Yorktown Heights, NY, IBM Research Center.

Lohse C (1968). Proposed USA standard: general-purpose alphanumeric keyboard arrangement for information interchange. *Communications of the ACM*, 11:126–129.

Longo L, Wickens CD, Hancock G, Hancock PA (2022). Human mental workload: a survey and a novel inclusive definition. *Frontiers in Psychology*, 13:883321 with erratum in *Frontiers in Psychology*, 13:969140. https://doi.org/10.3389/fpsyg.2022.883321

Lu ML (2012). *A Practical Guide to the Revised NIOSH Lifting Equation (NLE)*. Cincinnati, OH, National Institute for Occupational Safety and Health.

Lueder R, Rice VJB (2007). *Ergonomics for Children: Designing Products and Places for Toddler to Teens*. Boca Raton, FL, CRC Press. https://doi.org/10.1201/9780203609163

Lundervold AJ (1951). Electromyographic investigations of position and manner of working in typewriting. *Acta physiologica Scandinavica Supplementum*, 24(84):1–171.

Marklin RW, Simoneau GG (2004). Design features of alternative computer keyboards: a review of experimental data. *Journal of Orthopaedic & Sports Physical Therapy*, 34(10):638–649. https://doi.org/10.2519/jospt.2004.34.10.638

Marquié JC, Tucker P, Folkard S, Gentil C, Ansiau D (2015). Chronic effects of shift work on cognition: findings from the VISAT longitudinal study. *Occupational and Environmental Medicine*, 72(4):258–264. https://doi.org/10.1136/oemed-2013-101993

Marras WS (2008). *The Working Back: A Systems View*. New York, Wiley & Sons. https://doi.org/10.1002/9780470258576

Marras WS, Karwowski W (2006a). *The Occupational Ergonomics Handbook: Fundamentals and Assessment Tools for Occupational Ergonomics*, 2nd ed. Boca Raton, FL, CRC Press. https://doi.org/10.1201/9781420003635

Marras WS, Karwowski W (2006b). *The Occupational Ergonomics Handbook: Interventions, Controls, and Applications in Occupational Ergonomics*, 2nd ed. Boca Raton, FL, CRC Press. https://doi.org/10.1201/9781420003642

Marras WS, McGlothlin JD, McIntyre DR, Nordin M, Kroemer KHE (1993). *Dynamic Measures of Low Back Performance*, Fairfax, VA, American Industrial Hygiene Association.

Marras WS, Radwin RG (2006). Biomechanical modelling. Chapter 1 in Nickerson RS (ed), *Reviews of Human Factors and Ergonomics.* Santa Monica, CA, Human Factors and Ergonomics Society, pp. 588–589.

Martimo KP, Verbeek J, Karppinen J, Furlan AD, Takala EP, Kuijer PP, Jauhiainen M, Viikari-Juntura E (2008). Effect of training and lifting equipment for preventing back pain in lifting and handling: systematic review. *British Medical Journal*, 336(7641):429–31. https://doi.org/10.1136/bmj.39463.418380.BE

Maslow AH (1943). A theory of human motivation. *Psychological Review*, 50(4):370–396. https://doi.org/10.1037/h0054346

Maslow AH (1954). *Motivation and Personality.* New York, Harper & Row.

Matthews G, Reinerman-Jones LE, Barber DJ, Abich J (2015). The psychometrics of mental workload: multiple measures are sensitive but divergent. *Human Factors*, 57(1):125–143. https://doi.org/10.1177/0018720814539505

Mauney J, Furlough C, Barnes J (2015). Developing a better clinical alert system in EHRs. *Proceedings of the International Symposium on Human Factors and Ergonomics in Health Care*, 4(1):29–36. https://doi.org/10.1177/2327857915041023

McCambridge J, Witton J, Elbourne DR (2014). Systematic review of the Hawthorne effect: new concepts are needed to study research participation effects. *Journal of Clinical Epidemiology*, 67:e267–e277. https://doi.org/10.1016/j.jclinepi.2013.08.015

McGill SM (2006). Back belts. Chapter 30 in Marras WS, Karwowski W (eds), *The Occupational Ergonomics Handbook: Interventions, Controls, and Applications in Occupational Ergonomics*, 2nd ed. Boca Raton, FL, CRC Press, pp. 421–442. https://doi.org/10.1201/9781420003642

McMillan GR (2001). Brain and muscle signal-based control. In Karwowski W (ed), *International Encyclopaedia of Ergonomics and Human Factors.* London, UK, Taylor & Francis, pp. 379–381.

McMillan GR, Calhoun GL (2001). Gesture-based control. In Karwowski W (ed), *International Encyclopaedia of Ergonomics and Human Factors.* London, UK, Taylor & Francis, pp. 237–239.

McMin T (2013). "A-weighting": is it the metric you think it is? In *Proceedings of Acoustics 2013* – Victor Harbor, 165–168, Australian Acoustical Society, Victor Harbor, Australia.

McMulkin ML, Kroemer KHE (1994). Usability of a one-hand ternary chord keyboard. *Applied Ergonomics*, 25(3):177–181. https://doi.org/10.1016/0003-6870(94)90016-7

Megaw T (2005). The definition and measurement of mental workload. Chapter 18 in Wilson JR, Sharples S (eds), *Evaluation of Human Work*, 3rd ed. Boca Raton, FL, CRC Press, pp. 1031. https://doi.org/10.1201/b18362 (4th edition in 2015)

MIL- HDBK 1472 (2020). *Human Engineering*. Philadelphia, PA, DLA Document Services.

MIL-HDBK 759 (2018). *Human Factors Engineering Design for Army Material (Metric)*. Philadelphia, PA, DLA Document Services.

Mishan EJ, Quah E (2007). *Cost-Benefit Analysis*. London, UK, Routledge.

Molday RS, Moritz OL (2015). Photoreceptors at a glance. *Journal of Cell Science*, 128(22):4039–4045. https://doi.org/10.1242/jcs.175687

Monk TH (2006). Shiftwork. Chapter 32 in Marras WS, Karwowski W (eds), *The Occupational Ergonomics Handbook: Fundamentals and Assessment Tools for Occupational Ergonomics*, 2nd ed. Boca Raton, FL, CRC Press, pp. 421–442. https://doi.org/10.1201/9781420003635

Muchinsky PM, Culberson SS (2016). *Psychology Applied to Work*, 11th ed. Summerfield, NC, Hypergraphic Press.

Nachemson A, Elfström G (1970). Intravital dynamic pressure measurements in lumbar discs. A study of common move ments, manoeuvres and exercises. *Scandinavian Journal of Rehabilitation Medicine Supplement,* 1:1–40.

Nagamachi M (2002). Relationships among job design, macroergonomics, and productivity. Chapter 6 in Hendrick HW, Kleiner B (eds), *Macroergonomics. Theory, Methods, and Applications.* Bora Raton, FL, CRC Press, pp. 428. https://doi.org/10.1201/b12477

Nagamachi M (2011). *Kansei Engineering*, 2 volumes. Boca Raton, FL, CRC Press. https://doi.org/10.1201/b16799

NASA/SP-2010-3407/REV1 (2014). Human Integration Design Handbook (HIDH). https://www.nasa.gov/wp-content/uploads/2023/03/human-integration-design-handbook-revision-1.pdf?emrc=664e5339d33ee

NASA-STD-3000 (2023). The Man-System Integration Standards. Archived at https://msis.jsc.nasa.gov/

NASA-STD-3001 (2023). Space Flight Human-System Standard Volume 1 (Crew Health); Volume 2 (Human Factors, Habitability and Environmental Health) https://standards.nasa.gov/standard/NASA/NASA-STD-3001-VOL-1 and https://standards.nasa.gov/standard/NASA/NASA-STD-3001-VOL-2

National Research Council (US) Steering Committee for the Workshop on Work-Related Musculoskeletal Injuries: The Research Base (1999). Work-Related Musculoskeletal Disorders: Report, Workshop Summary, and Workshop Papers. Washington, DC, US, National Academies Press. https://doi.org/10.17226/6431

Waters TR, Putz-Anderson V, Garg A, NIOSH (1994, revised 2021). *Applications Manual for the Revised NIOSH Lifting Equation*. Cincinnati, OH, U.S. Department of Health and

Human Services, Centers for Disease Control and Prevention, National Institute for Occupational Safety and Health, DHHS (NIOSH) Publication No. 94-110. https://doi.org/10.26616/NIOSHPUB94110revised092021

NIOSH (National Institute for Occupational Safety and Health) (1981). *Work Practices Guide for Manual Lifting*. DHHS (NIOSH) Publication # 81-122, Washington, DC, US Government Printing Office. https://www.cdc.gov/niosh/docs/81-122/default.html

NIOSH (National Institute for Occupational Safety and Health) *Occupational Noise Exposure publication 98-126 (1998, last reviewed 2023)*. https://stacks.cdc.gov/view/cdc/6376/cdc_6376_DS1.pdf

Nordin M, Frankel VH (1989). *Basic Biomechanics of the Musculoskeletal System*, 2nd ed. Philadelphia, PA, Lea & Febiger. https://doi.org/10.1002/art.1790030209

Norman DA (1991). Cognitive science in the cockpit. *CSERIAC Gateway*, 2:1–6.

Norman DA (2013). *The Design of Everyday Things.* Cambridge, MA, The MIT Press.

Noyes J (1983a). Chord keyboards. *Applied Ergonomics*, 14(1): 55–59. https://doi.org/10.1016/0003-6870(83)90221-1

Noyes J (1983b). The QWERTY keyboard: a review. *International Journal of Man-Machine Studies*, 18(3): 265–281. https://doi.org/10.1016/S0020-7373(83)80010-8

Oezkaya N, Leger D, Goldsheyder D, Nordin M (2017). *Fundamentals of Biomechanics: Equilibrium, Motion, and Deformation*, 4th ed. Switzerland, Springer. https://doi.org/10.1007/978-3-319-44738-4

Osczevski R, Bluestein M (2005). The new wind chill equivalent temperature chart. *Bulletin American Meteorological Society*, 86:1453–1458. https://doi.org/10.1175/BAMS-86-10-1453

Osler W (1892). *The Principles and Practice of Medicine, IX. Professional Spasms: Occupation Neuroses.* New York, Appleton.

Owen D (2006). The soundtrack of your life: Muzak in the realm of retail theatre. *The New Yorker*, 10:66–71.

Paquette S, Gordon C, Bradtmiller B (2009). Anthropometric Survey (ANSUR) II Pilot Study: Methods and Summary Statistics. Technical Report Natick/TR-09/014. Anthrotch, US Army Natick Soldier Research, Development and Engineering Center, Natick, MA. https://apps.dtic.mil/sti/pdfs/ADA498172.pdf

Parasuraman R, Sheridan TB, Wickens CD (2000). A model for types and levels of human interaction with automation. *IEEE Transactions on Systems, Man, and Cybernetics - Part A: Systems and Humans*, 30(3):286–297. https://doi.org/10.1109/3468.844354

Parmanen J (2012). Some reasons to revise the international standard ISO 226 2003: acoustics—normal equal-loudness-level contours. *Open Journal of Acoustics*, 2:143–149. https://doi.org/10.4236/oja.2012.24016

Parsons HM (1974). What happened at Hawthorne? *Science*, 18:922–932. https://doi.org./10.1126/science.183.4128.92

Parsons KC (2003). *Human Thermal Environments: The Effects of Hot, Moderate, and Cold Environments on Human Health, Comfort and Performance.* London, UK, Taylor & Francis. https://doi.org/10.1201/9781420025248

Parsons KC (2005). Ergonomic assessments of thermal environments. Chapter 23 in Wilson JR, Corlett N (eds), *Evaluation of Human Work,* 3rd ed. London, UK, Taylor & Francis, pp. 387.

Parsons KC (2014). *Human Thermal Environments*, 3rd ed. Boca Raton, FL, CRC Press. https://doi.org/10.1201/b16750

Patterson ES, Gibbons MC, Latkany P, Ramaiah M, Brick D, Lowry SZ (2015). Integrating electronic health records into clinical workflow: an application of human factors modelling methods to two specialty care areas. *Proceedings of the International Symposium on Human Factors and Ergonomics in Health Care*, 4(1):42–49. https://doi.org/10.1177/2327857915041018

Peebles L, Norris B (1998). *Adultdata. The Handbook of Adult Anthropometric and Strength Measurements – Data for Design Safety.* DTI/Pub 2917/3k/6/98/NP, London, UK, Department of Trade and Industry.

Peebles L, Norris B (2000). *Strength Data.* DTI/URN 00/1070. London, UK, Department of Trade and Industry.

Peebles L, Norris B (2003). Filling "gaps" in strength data for design. *Applied Ergonomics*, 34(1):73–88. https://doi.org/10.1016/s0003-6870(02)00073-x

Pereira A, Hsieh CM, Laroche C, Rempel D (2014). The effect of keyboard key spacing on typing speed, error, usability, and biomechanics, Part 2: vertical spacing. *Human Factors,* 56(4):752–759. https://doi.org/10.1177/0018720813502524

Pereira A, Wachs JP, Park K, Rempel D (2015). A user-developed 3-D hand gesture set for human-computer interaction. *Human Factors*, 57(4):607–621. https://doi.org/10.1177/0018720814559307

Pew PW, Mavor AS (2007). *Human-System Integration in the System Development Process.* Washington, DC, The National Academies.

Pheasant S, Haslegrave CM (2006). *Anthropometry, Ergonomics and the Design of Work*, 3rd ed. London, UK, Taylor & Francis. https://doi.org/10.1201/9781315375212

Poore GV (1872). "Writer's cramp": its pathology and treatment. *Practitioner*, August issue:341–350.

Poore GV (1887). Clinical lecture on certain conditions of the hand and arm which interfere with the performance of professional acts, especially piano-playing. *The British Medical Journal*, February 26 issue, 441–444

Popkin SM (2015). *Worker Fatigue and Transportation Safety. Volume 10 of Reviews of Human Factors and Ergonomics,* Santa Monica, CA, Human Factors and Ergonomics Society.

Popkin SM, Howarth HD, Tepas DI (2006). Ergonomics of work systems. In Salvendy G (ed), *Handbook of Human Factors and Ergonomics,* 3rd ed. Hoboken, NY, Wiley, pp. 761–800.

Potvin JR, Ciriello VM, Snook SH, Maynard WS, Brogmus GE (2021). The liberty mutual manual materials handling (LM-MMH) equations. *Ergonomics*, 64(8):955–970. https://doi.org/10.1080/00140139.2021.1891297

Proctor RW, Van Zandt T (2008). *Human Factors in Simple and Complex Systems*, 2nd ed. Boca Raton, FL, CRC Press. https://doi.org/10.1201/9781420063608

Putz-Anderson V (1988). *Cumulative Trauma Disorders*. Boca Raton, FL, CRC Press. https://doi.org/10.1201/9781315140704

Ramazzini B (1713). *De morbis artificum diatriba* [diseases of workers]. Translated in *American Journal of Public Health*, 2001, 91(9):1380–1382. https://doi.org/10.2105/AJPH.91.9.1380

Ramsey T, Davis KG, Kotowski SE, Anderson VP, Waters T (2014). Reduction of spinal loads through adjustable interventions at the origin and destination of palletizing tasks. *Human Factors*, 56(7):1222–1234. https://doi.org/10.1177/0018720814528356

Rea MS (2005). Photometric characterization of the luminous environment. Chapter 68 in Stanton N, Hedge A, Brookhuis K, Salas E, Hendrick HW (eds), *Handbook of Human Factors and Ergonomics Methods*. Boca Raton, FL, CRC Press, pp. 768. https://doi.org/10.1201/9780203489925

Reddy NP, Gupta V (2007). Toward direct biocontrol using surface EMG signals: control of finger and wrist joint models. *Medical Engineering & Physics*, 29(3):398–403. https://doi.org/10.1016/j.medengphy.2005.10.016

Refinetti R (2016). *Circadian Physiology*, 3rd ed. Boca Raton, FL, CRC Press.

Regan MA, Lee J, Victor TW (2013). *Driver Distraction and Inattention*. Abingdon, UK, Ashgate. https://doi.org/10.1201/9781315578156.

Remington RJ, Rogers M (1969) Keyboard Literature Survey, Phase 1: Bibliography, TR 29.0042. Research Triangle Park, NC, IBM Systems Development Division.

Rempel D (2008). The split keyboard: an ergonomics success story. *Human Factors*, 50(3):385–392. https://doi.org/10.1518/001872008X312215

Rempel D, Willms K, Anshel J, Jaschinski W, Sheedy J (2007). The effects of visual display distance on eye accommodation, head posture, and vision and neck symptoms. *Human Factors*, 49(5):830–838. https://doi.org/10.1518/001872007X230208

Robinson PA (2009). *Writing and Designing Manuals and Warnings*, 4th ed. Boca Raton, FL, CRC Press. https://doi.org/10.1201/9781420069860

Rodahl K (1989). *The Physiology of Work*. London, UK, CRC Press. https://doi.org/10.1201/b12599

Roebuck JA (1995). *Anthropometric Methods – Designing to Fit the Human Body*. Santa Monica, CA, Human Factors and Ergonomics Society.

Roethlisberger FJ, Dickson WJ (1939). *Management and the Worker*. Cambridge, MA, Harvard University Press. https://doi.org/10.4324/9780203503010

Roser M, Appel C, Ritchie H (revised 2024). Human Height: The average height of a population can inform us about the nutrition and living conditions of populations in the past for which we have little other data. https://ourworldindata.org/human-height#how-does-human-height-vary-across-the-world (retrieved May 3 2024)

Rubio S, Díaz E, Martín J, Puente JM (2004). Evaluation of subjective mental workload: a comparison of SWAT, NASA-TLX, and workload profile methods. *Applied Psychology*, 53:61–86. https://doi.org/10.1111/j.1464-0597.2004.00161.x

SAE J3016_202104 (2021). Taxonomy and Definitions for Terms Related to Driving Automation Systems for On-Road Motor Vehicles.

Salvendy G (2012). *Handbook of Human Factors and Ergonomics, 4th ed.* New York, Wiley. https://doi.org/10.1002/9781118131350

Sanseverino D, Caputo A, Cortese CG, Ghislieri C (2023). "Don't stop the music," please: the relationship between music use at work, satisfaction, and performance. *Behavioural Sciences,* 13(1):15. https://doi.org/10.3390/bs13010015

Saval N (2014). *Cubed: A Secret History of the Workplace.* New York, Doubleday.

Sawyer BD, Miller DB, Canham M, Karwowski W (2021). Human factors and ergonomics in design of A^3: automation, autonomy, and artificial intelligence. Chapter 52 in Salvendy G, Karwowski W (eds), *Handbook of Human Factors and Ergonomics*, 5th ed. Hoboken, NJ, John Wiley & Sons, pp. 1385–1416. https://doi.org/10.1002/9781119636113.ch52

Selye H (1956, revised 1978). *The Stress of Life.* New York, McGraw-Hill.

Selye H (1974). *Stress without Distress.* Philadelphia, PA, Lippincott.

Sheedy J (2006). Vision and work. Chapter 18 in Marras WS, Karwowski W (eds), *The Occupational Ergonomics Handbook: Fundamentals and Assessment Tools for Occupational Ergonomics*, 2nd ed. Boca Raton, FL, CRC Press, pp. 421–442. https://doi.org/10.1201/9781420003635

Siegel JM (2003). Why we sleep. *Scientific American*, November issue, 92–97.

Silva I, Costa D (2023). Consequences of shift work and night work: a literature review. *Healthcare*, 11(10):1410. https://doi.org/10.3390/healthcare11101410

Sinclair DC (1973). Mapping of spinal nerve roots. In Jarrett A (ed), *The Physiology and Pathophysiology of the Skin*, Volume 2. London, UK, Academic Press, pp. 445.

Skottke EM, Debus G, Wang L, Huestegge L (2014). Carryover effects of highly automated convoy driving on subsequent manual driving performance. *Human Factors*, 56(7):1272–1283. https://doi.org/10.1177/0018720814524594

Smith TM (2007). *Job Satisfaction in America: Trends and Socio-Demographic Correlates.* Available at https://www-news.uchicago.edu/releases/07/pdf/070827.jobs.pdf

Snook SH (2000). Back risk factors: an overview. Chapter 11 in Violante F, Armstrong T, Kilbom A (eds), *Occupational Ergonomics. Work Related Musculoskeletal Disorders of the Upper Limb and Back.* London, UK, Taylor & Francis, pp. 120–128. https://doi.org/10.1201/9781482268287

Snook SH (2000). Leading Causes of Workplace Injuries, Liberty Mutual Insurance Company Newsletter Winter 2013/14. www.libertymutual.com

Snook SH (2004). Psychophysical tables: lifting, lowering, pushing, pulling, and carrying. Chapter 13 in Stanton N, Hedge A, Brookhuis K, Salas E, Hendrick HW (eds), *Handbook of Human Factors and Ergonomics Methods*. Boca Raton, FL, CRC Press, pp. 768. https://doi.org/10.1201/9780203489925

Snook SH, Ciriello VM (1991). The design of manual handling tasks: revised tables of maximum acceptable weights and forces, *Ergonomics*, 34(9):1197–1213. https://doi.org/10.1080/00140139108964855

Sprent P, Smeeton NC (2007). *Applied Nonparametric Statistical Methods*. New York, Chapman & Hall / CRC. https://doi.org/10.1201/b15842

Staff KR (1983). *A comparison of range of joint mobility in college females and males*. Master's Thesis, Industrial Engineering, A&M University, College Station, TX

Stanton N, Hedge A, Brookhuis K, Salas E, Hendrick HW (2004). *Handbook of Human Factors and Ergonomics Methods*. Boca Raton, FL, CRC Press. https://doi.org/10.1201/9780203489925

Stefana E, Marciano F, Rossi D, Cocca P, Tomasoni G (2021). Wearable devices for ergonomics: a systematic literature review. *Sensors (Basel)*, 21(3):777. https://doi.org/10.3390/s21030777

Strokina AN, Pakhomova BA (1999). *Anthropo-ergonomic Atlas*. Moscow, Russia, Moscow State University Publishing House.

Strong EP (1956). *A Comparative Experiment and Simplified Keyboard Retraining and Standard Keyboard Supplementary Training*. Washington, DC, (US) General Services Administration.

Sweileh WM (2022). Analysis and mapping of global research publications on shift work (2012–2021). *Journal of Occupational Medicine and Toxicology*, 17:22. https://doi.org/10.1186/s12995-022-00364-0

Swink JR (1966). Intersensory comparisons of reaction time using an electro-pulse tactile stimulus. *Human Factors*, 8(2):143–145. https://doi.org/10.1177/001872086600800205

Tamhane AC (2009). *Statistical Analysis of Designed Experiments: Theory and Applications*. Hoboken, NJ, Wiley. https://doi.org/10.1111/j.1751-5823.2010.00109_3.x

Taylor F (1911). *The Principles of Scientific Management*. New York, Norton.

Toomingas A, Mathijssen SE, Tornquist EW (2011). *Occupational Physiology*. Boca Raton, FL, CRC Press. https://doi.org/10.1201/b11474

Ugbolue UC, Hsu WH, Goitz RJ, Li ZM (2005). Tendon and nerve displacement at the wrist during finger movements. *Clinical Biomechanics*, 20(1):50–56. https://doi.org/10.1016/j.clinbiomech.2004.08.006

University of Nottingham (2002). *Strength Data for Design Safety, Phase 2*. DTI URN 01/1433, London, UK, Department of Trade and Industry.

US Food and Drug Administration (FDA) Center for Devices and Radiological Health Office of Device Evaluation (2016, updated draft 2022). Applying Human Factors and Usability

Engineering to Medical Devices - Guidance for Industry and Food and Drug Administration Staff. https://www.fda.gov/media/80481/download and https://www.fda.gov/media/163694/download

Use J (2015). Are bosses necessary? *The Atlantic*, October issue:28–32.

Useem J (2015). Why it pays to be a jerk: new research confirms what they say about nice guys. *The Atlantic,* June 2015:1–25.

Vercruyssen M, Hendrick HL (2011). *Behavioural Research and Analysis,* 4th ed. Boca Raton, FL, CRC Press.

Vicente KJ (2002). Ecological interface design: progress and challenges. *Human Factors*, 44(1):62–78. https://doi.org/10.1518/0018720024494829

Vink P (2004). *Comfort and Design: Principles and Good Practice.* Boca Raton, FL, CRC Press. https://doi.org/10.1201/9781420038132

Violante F, Armstrong T, Kilbom A (2000). *Occupational Ergonomics. Work Related Musculoskeletal Disorders of the Upper Limb and Back.* London, UK, Taylor & Francis. https://doi.org/10.1201/9781482268287

Walji AH (2007). Functional anatomy of the upper limb (extremity). Chapter 8 in Kumar S (ed), *Biomechanics in Ergonomics*, 2nd ed. Boca Raton, FL, CRC Press, pp. 744. https://doi.org/10.1201/9780849379093

Walker GH, Stanton NA, Salmon P (2015). *Human Factors in Automotive Engineering and Technology, Human Factors in Transport Series.* Abingdon, UK, Ashgate Publishing Ltd. https://www.ashgate.com/isbn/9781409447573

Wang MJJ, Wang EMY, Lin YC (2002). *Anthropometric Data Book of the Chinese People in Taiwan.* Hsinchu, Taiwan, The Ergonomics Society of Taiwan.

Wargo MJ (1967). Human operator response speed, frequency and flexibility: a review and analysis. *Human Factors*, 9(3):221–238. https://doi.org/10.1177/001872086700900303

Waters TR (2006). Revised NIOSH lifting equation. Chapter 46 Marras WS, Karwowski W (eds), *The Occupational Ergonomics Handbook: Fundamentals and Assessment Tools for Occupational Ergonomics*, 2nd ed. Boca Raton, FL, CRC Press, pp. 421–442. https://doi.org/10.1201/9781420003635

Waters TR (2010). Introduction to ergonomics for healthcare workers. *Rehabilitation Nurse*, 35(5):185–191. https://doi.org/10.1002/j.2048-7940.2010.tb00046.x

Waters TR, Putz-Anderson V (1999). Revised NIOSH lifting equation. Chapter 57 in Marras WS, Karwowski W (eds), *The Occupational Ergonomics Handbook*, Boca Raton, FL, CRC Press, pp. 421–442.

Waterson P (2014). *Patient Safety Culture: Theory, Methods and Application.* Boca Raton, FL, CRC Press. https://doi.org/10.1201/9781315599717

Weinger MB, Wiklund ME, Gardner-Bonneau DJ (2010). *Handbook of Human Factors in Medical Device Design.* Boca Raton, FL, CRC Press. https://doi.org/10.1201/b10439

Whitcome KK, Shapiro LJ, Lieberman DE (2007). Fetal load and the evolution of lumbar lordosis in bipedal hominins. *Nature*, 450(7172):1075–1078. https://doi.org/10.1038/nature06342 Corrigendum in *Nature* 487, 128 (2012). https://doi.org/10.1038/nature11168

Wickens CD, Lee J, Liu Y & Gordon-Becker S (2004). *An Introduction to Human Factors Engineering*, 2nd ed. Upper Saddle River, Prentice-Hall/Pearson education.

Wiklund ME, Kendler J, Strochlic AY (2016). *Usability Testing of Medical Devices,* 2nd ed. Boca Raton, FL, CRC Press. https://doi.org/10.1201/b19082

Williges RC (2007). *CADRE: Computer-Aided Design Reference for Experiments.* Electronic Book CD-ROM-07-01, Blacksburg, VA, Virginia Tech.

Wilmore JH, Costill D, Kenney WL (2008). *Physiology of Sport and Exercise*, 4th ed. Champaign, IL, Human Kinetics.

Wilson JR, Corlett N (2005). *Evaluation of Human Work,* 3rd ed. London, UK, Taylor & Francis.

Wilson JR, Sharples S (2015). *Evaluation of Human Work,* 4th ed. Boca Raton, FL, CRC Press. https://doi.org/10.1201/b18362

Winter DA (2009). *Biomechanics and Motor Control of Human Movement,* 4th ed. New York, Wiley. https://doi.org/10.1002/9780470549148

Wu G, Siegler S, Allard P, Kirtley C, Leardini A, Rosenbaum D, Whittle M, D'Lima DD, Cristofolini L, Witte H, Schmid O, Stokes I (2002). Standardization and terminology committee of the international society of biomechanics. ISB recommendation on definitions of joint coordinate system of various joints for the reporting of human joint motion--part I: ankle, hip, and spine. *Journal of Biomechanics,* 35(4):543–548. https://doi.org/10.1016/s0021-9290(01)00222-6

Wu J, Yang J, Yoshitake M (2014). Pedal errors among younger and older individuals during different pedal operating conditions. *Human Factors,* 56(4):621–630. https://doi.org/10.1177/0018720813487200

Youle A (2005). The thermal environment. Chapter 20 in Gardiner K, Harrington M (eds), *Occupational Hygiene,* 3rd ed. Oxford, UK, Wiley-Blackwell, pp. 286–306. https://doi.org/10.1002/9780470755075

Zhang X, Chaffin DB (2006). Digital human modelling for computer-aided design. Chapter 10 in Marras WS, Karwowski W (eds), *The Occupational Ergonomics Handbook: Interventions, Controls, and Applications in Occupational Ergonomics,* 2nd ed. Boca Raton, FL, CRC Press, pp. 848. https://doi.org/10.1201/9781420003642

Zhao C, Ettema AM, Osamura N, Berglund LJ, An KN, Amadio PC (2007). Gliding characteristics between flexor tendons and surrounding tissues in the carpal tunnel: a biomechanical cadaver study. *Journal of Orthopaedic Research*, 25(2):185–190. https://doi.org/10.1002/jor.20321

Index

Note: **Bold** page numbers refer to tables and *italic* page numbers refer to figures.